Mathematical Modelling of Energy Systems and Fluid Machinery 2022

Mathematical Modelling of Energy Systems and Fluid Machinery 2022

Editors

Michele Pinelli
Alessio Suman
Nicola Casari

MDPI • Basel • Beijing • Wuhan • Barcelona • Belgrade • Manchester • Tokyo • Cluj • Tianjin

Editors

Michele Pinelli
Department of Engineering
University of Ferrara
Ferrara
Italy

Alessio Suman
Department of Engineering
University of Ferrara
Ferrara
Italy

Nicola Casari
Department of Engineering
University of Ferrara
Ferrara
Italy

Editorial Office
MDPI
St. Alban-Anlage 66
4052 Basel, Switzerland

This is a reprint of articles from the Special Issue published online in the open access journal *Energies* (ISSN 1996-1073) (available at: www.mdpi.com/journal/energies/special_issues/Mathematical_Modelling_and_Fluid_Machinery_II).

For citation purposes, cite each article independently as indicated on the article page online and as indicated below:

LastName, A.A.; LastName, B.B.; LastName, C.C. Article Title. *Journal Name* **Year**, *Volume Number*, Page Range.

ISBN 978-3-0365-8359-4 (Hbk)
ISBN 978-3-0365-8358-7 (PDF)

Contents

About the Editors

Michele Pinelli

Michele Pinelli received his MSc Degree in Mechanical Engineering at the University of Bologna in 1997 and his Ph.D. in 2001. Since 2021, he has been Full Professor of Fluid Machinery at the Engineering Department in Ferrara. His research activity has been carried out in the field of fluid machinery and energy systems and recently it has mainly dealt with the development of techniques for the numerical estimation of turbine and compressor behavior in deteriorated conditions (fouling and erosion), the development of a small-scale compressor test rig for the study of deposition, water ingestion, stall and surge, and the study of innovative energy systems in micro-cogeneration applications, with particular attention to the Organic Rankine Cycle systems. His research activity is documented by more than 340 scientific articles. He is the coordinator of a research group comprising PhD students and research fellows. He is the scientific director of several international research collaborations, including Imperial College of London and St. John's College of the University of Oxford.

Alessio Suman

Dr. Alessio Suman holds a B.S. and M.S. in Mechanical Engineering from the University of Ferrara. In addition, Dr. Suman attended the Engineering Science Ph.D. course at the University of Ferrara, where he worked on the performance degradation of gas turbines. Since 2019, he has been a researcher at the University of Ferrara as a member of the Fluid Machinery Research Group. His main research interests include the analysis of particle–surface interaction by experimental and numerical approaches. Original measurement techniques and data interpretations have been used to develop models for predicting the adhesion of micro-sized solid particles and substrates. His research activity is documented by more than 150 scientific articles on turbomachinery and energy systems.

Nicola Casari

Nicola Casari works as an application engineer at ToffeeAM ltd., where he is involved in the design of components exploiting and developing Topology Optimization applications. His area of interest covers heat exchangers and heat sinks, cold plates, manifolds and mixers for electronic, oil and gas, aviation and automotive industry. Nicola Casari holds a Ph.D. in Mechanical Engineering from the University of Ferrara where he worked for 4 years as an Assistant Professor. He was a visiting scholar at Oxford University and Imperial College of London. He holds one patent and authored more than 60 papers and won 2 individual ASME Turbo Expo Best Paper awards.

Preface to "Mathematical Modelling of Energy Systems and Fluid Machinery 2022"

The digitalization of the energy sector is of paramount importance and, at the same time, an ever-increasing phenomenon. For example, it is worth noting how IoT technologies and the widespread utilization of digital twins are changing the way in which energy systems and fluid machinery are conceived and designed. However, this can only be achieved if these new ICT technologies are posed on solid bases for the representation of energy systems and fluid machinery. Therefore, mathematical modelling is still relevant, and its importance cannot be underestimated. This Special Issue, which follows a previous one on the same topic, is intended for a collection of contributions about the mathematical modelling of energy systems and fluid machinery in order to build and consolidate this base of knowledge.

Michele Pinelli, Alessio Suman, and Nicola Casari
Editors

energies

Article

Performance Analysis of WHR Systems for Marine Applications Based on sCO_2 Gas Turbine and ORC

Fabrizio Reale *, **Raffaela Calabria** and **Patrizio Massoli**

Institute of Sciences and Technologies for Sustainable Energy and Mobility, STEMS-CNR, 80125 Naples, Italy
* Correspondence: fabrizio.reale@cnr.it

Abstract: Waste heat recovery (WHR) can represent a solution to improve the efficiency of ships' propulsion, helping to exceed stringent greenhouse gas emission limits. This is particularly suitable in the case of propulsion based on gas turbines due to their medium-high temperature level of the exhaust gases. This study analyzes the performance of a hybrid energy grid, in which the heat is recovered by the exhaust gases of an aeroderivative gas turbine, a GE LM2500+, when the bottoming system is a supercritical CO_2 gas turbine. Given the issues and peculiarities related to the onboard installation, where size and weight are fundamental concerns, six WHR schemes have been analyzed. They span from the simple cycle to partial preheated and regenerative, to a cascade layout in which an ORC system receives thermal power by the sCO_2 GT. The influence of the seawater temperature on the performance of the hybrid energy system has been also considered. The energetic and exergetic performance comparison of the different schemes has been carried out by using the commercial software Thermoflex. The results showed that an increase in overall performance by up to 29% can be obtained and that the increase in seawater temperature can lead to a decrease in the overall performance.

Keywords: waste heat recovery; WHR for marine applications; sCO_2 gas turbine; ORC; hybrid energy system; performance analysis of WHR system

Citation: Reale, F.; Calabria, R.; Massoli, P. Performance Analysis of WHR Systems for Marine Applications Based on sCO_2 Gas Turbine and ORC. *Energies* **2023**, *16*, 4320. https://doi.org/10.3390/en16114320

Academic Editor: George Kosmadakis

Received: 28 April 2023
Revised: 21 May 2023
Accepted: 23 May 2023
Published: 25 May 2023

1. Introduction

The International Maritime Organization (IMO) established the first set of mandatory international measures in 2011 to improve the energy efficiency of ships and to reduce greenhouse gas (GHG) emissions in accordance with the International Convention for the Prevention of Pollution from Ships (MARPOL). In 2018, the IMO adopted the Initial Strategy on the reduction of GHG emissions from shipping, which aims to achieve a 40% reduction of CO_2 emissions per transport work compared to 2008 levels by 2030. In addition, it sets the targets to decrease the total annual GHG emissions by 50% and the CO_2 emissions per transport work by 70% by 2050, also compared to 2008 levels [1]. Carbon dioxide emissions from shipping are primarily generated by container ships, bulk carriers, and oil tankers, accounting their emissions for approximately 60% in 2012 [2].

Waste heat recovery (WHR) may be a viable solution to enhance overall energy efficiency since it allows to increase the available power with the same amount of fuel and the same quantity of greenhouse gas emissions. The importance of WHR in the shipping context becomes evident considering the recent efficiency index imposed by IMO as mandatory since 2023, the Energy Efficiency Existing Ship Index (EEXI) [3]. This index allows us to measure the ship's energy efficiency. EEXI is indeed a measure of the energy efficiency of a vessel and depends on its design, which is fixed, while the carbon intensity indicator (CII) is used to consider operational factors. EEXI represents the CO_2 emissions divided by the transport work and is expressed in grams-CO_2/tonne-mile. WHR, saving power and reducing both engine power and specific fuel consumption, is a measure that can be applied to reduce EEXI, avoiding more complex and costly actions such as vessel designs changes.

Alongside these beneficial impacts, the implementation of WHR systems in marine applications should be evaluated considering the constraints of high efficiency, high power density, and small footprint in terms of volume and weight. WHRs should be capable of functioning effectively even in the case of transient heat source and sink properties. Indeed, the ambient conditions can vary with respect to the inlet conditions of the main prime mover and the temperature of the seawater, which serves as the cold sink in of the cycle [4].

Diesel engines are currently the most widely used for power generation in several types of vessels. In fact, for all civilian ships exceeding 100 gross tons, approximately 96% of the installed output is generated by diesel plants, which typically operate at about 50% of efficiency. Implementing WHR systems can further enhance the overall efficiency, considering that most of the wasted heat is between low and medium quality. In fact, in diesel engines, the waste heat can be recovered from different sources, such as exhaust gas, air cooler, lubricating oil cooler, or jacket water cooler. All sources are at low temperatures, with the exception of exhaust gases [4]. Gas turbines (GTs) have a more limited use in shipping due to their typical lower efficiency compared to diesel engines. However, GTs have a significantly higher power density, requiring less space and weighing less than diesel engines [5]. The medium-high temperature waste heat, which is recoverable from the exhausts of gas turbines, allows us to consider combined cycle power plants overcoming this constraint. In marine applications, a conventional combined cycle, in which the bottoming cycle is a steam Rankine cycle, can be used when the GTs directly drive the propeller shafts, in the case of combined gas turbine and steam configuration (COGAS) or in the case of turboelectric transmission for the combined gas turbine electric and steam (COGES) configuration [6].

The electric configuration offers the advantage of a single engine able to simultaneously supply power to multiple shafts and to the onboard services' load demand. However, the main drawback is related to the reduced efficiency due to the conversion of mechanical power first into electricity and then into mechanical power again. The electrical transmission optimizes the positioning of the engines on board [7]. In this context, waste heat recovery from the exhausts opens up possibilities for more interesting and more innovative solutions such as Organic Rankine Cycles (ORC) and, more recently, closed Brayton cycles, in which carbon dioxide in supercritical conditions is the working fluid. In this way, the COGAS or COGES configuration gives way to hybrid energy systems. The exhaust temperature at the exit of aeroderivative gas turbines is generally greater than 450 °C [8], and this may be a constraint on the individuation of the most appropriate waste heat recovery system. Indeed, ORC systems are a valid technical solution for waste heat recovery, primarily in the case of a heat source temperature range between 100 °C and 400 °C, because the upper limit is imposed by the flammability, low chemical stability and the risk of decomposition of the organic fluids at high temperature [9,10].

However, supercritical carbon dioxide gas turbines (sCO_2 GT) have a high efficiency, specifically within the medium-high temperature range. Initially, the supercritical CO_2 power cycle was recognized as a promising technology to recover heat in high-efficiency IV-generation nuclear reactors because the operating temperatures of these reactors are 500–900 °C and, in this range, this choice is more efficient than others [11]. In particular, for temperatures exceeding 700 °C, sCO_2 power cycles may be considered the only available option for waste heat recovery [10]. Furthermore, the turbomachinery is compact and approximately 10 times smaller than that of steam Rankine cycle turbomachinery because the fluid remains dense throughout the entire system and the volumetric flow rate decreases [12].

A Brayton cycle that utilizes sCO_2 as a working fluid combines, in fact, the advantages of both the Rankine cycle and gas turbines. This is due to the specific properties of a fluid under supercritical conditions [11]. Moreover, the pressure ratio is considerably lower compared to other Brayton or Rankine cycles. For these reasons, in recent years, the researchers' interest in this type of power cycle significantly grew up, extending the range of utilization to include geothermal applications, renewable plants, or waste heat recovery. The sCO_2 Brayton cycle enables the possibility of further cascading waste heat recovery, and an impressive number of layout configurations have been explored for both stand-alone and combined cycles [13].

Since the initial studies conducted by Feher [14] and Angelino [15] on sCO_2 power cycles over fifty years ago, several layouts have been defined and analyzed to determine the best performance conditions [12]. The simplest architectures are both simple and regenerative cycles, which are particularly suitable for WHR from gas turbine exhausts [16], preheating can also be considered as an effective solution [17]. The simple Brayton cycle is understandably the most straightforward and lightweight option, while the recuperative cycle incorporates only one additional heat exchanger with respect to the first one.

The introduction of other additional components (e.g., preheater and other heat exchangers or further compressors or turbines) can result in more complex layouts and higher performance levels. Several studies have shown that the recompression cycle is the most efficient architecture, mainly for nuclear applications, but the slight temperature difference in the heater, coupled with the lower heat recovered by the exhaust gases, makes it less suitable for waste heat recovery [16]. Indeed, the volume and weight footprints have to be considered in marine applications, but also the characteristics of the working fluid, in terms of toxicity and flammability, and the simplicity of the layout, which may mean less maintenance, acquire great importance in the layout choice. It is clear that the high working pressure and the consequent compact layout allow us to consider this solution suitable in the case of space-limited applications [8]. Moreover, carbon dioxide is non-flammable and non-toxic, and pure CO_2 has reduced corrosion issues [13]. Of course, one of the major issues in systems placed on board ships is the sealing of the system due to the impulses and vibrations to which the system is subject due to the motion and vibrations typical of the vessel. Particular attention is paid to the control of sCO_2 leaks in the system, which can generate unwanted CO_2 emissions.

The opportunities related to the utilization of sCO_2 GTs in marine applications have been discussed in the recent literature, exploring various layouts, and analyzing potentialities and the economic aspects. Wang et al. [9] focused their attention on a thermodynamic configuration method to design recuperative sCO_2 gas turbines for WHR of marine engines. Results of simulations of their proposed combined system proved that a thermal efficiency of up to 33.17% can be obtained. Sakalis [18] proposed an integrated energy system in which the sCO_2 gas turbine recovers waste heat from the exhausts of a turbocharged marine diesel engine, focusing on the techno-economic performance. Results highlighted that the introduction of the sCO_2 power cycle can be economically justified, leading to a reduction of operational fuel costs. Hou et al. [19] proposed a combined cooling, heating, and power system using the waste heat of a marine gas turbine. Their layout was based on a recompression cycle, two trans-critical CO_2 refrigeration cycles, and a steam generator. Hou et al. [20] defined a combined sCO_2 recompression and regenerative cycle for WHR. Results of the multi-objective optimization highlighted that their proposed layout could improve the part-load performance of the ship. Du et al. [21] defined a thermodynamic model of a marine sCO_2 recompression cycle, focusing on the size optimization of the sCO_2 cycle in a limited space. Pan et al. [22] proposed a modified sCO_2 recompression Brayton cycle as WHR in a ship-defined dual turbine-alternator-compressor recompression sCO_2 system. Results of their simulations showed that their scheme can lead to an increase in energetic and exergetic efficiency with a more compact layout and an increase of the Energy Efficiency Design Index of about 1%. Hu et al. [23] analyzed the effects of the rolling motion on the heat transfer and, consequently, on the efficiency of the sCO_2 Brayton cycle used as waste heat recovery. Their study has been carried out considering extreme ocean conditions, to highlight the Instabilities related to the rolling motion. Guo et al. [24] proposed a WHR system based on the sCO_2 Brayton cycle, transcritical CO_2 cycle, compressed CO_2 energy storage, and thermal storage system. Results of their thermodynamic model highlighted that the thermal efficiency reaches 40%. Sao et al. [25] proposed a recompression-regeneration sCO_2 combined cycle as a WHR system of a marine gas turbine in substitution of diesel-based power packs. The thermal efficiency of the hybrid energy system of 52.5% can be achieved as a result of numerical simulations. The availability of an additional amount of waste heat to be recovered from the sCO_2 Brayton cycle makes it possible the introduction of further bottoming cycles. In particular, an ORC system can be located between the first bottoming cycle and the cold heat sink [12,26–28].

The purpose of this article is to examine the opportunities related to the waste heat recovery for marine applications, with the main focus on improving performance levels when the bottoming cycle is a closed Brayton cycle, in which supercritical carbon dioxide is the working fluid. The main energy system subject of the study is the GE LM2500+ aeroderivative gas turbine of 30 MW. Given the issues and peculiarities related to the onboard installation, six different WHR layouts have been considered and examined, with an increasing complexity in terms of the number of components but also with an increasing efficiency. Five of these are based on sCO_2 gas turbines, while the sixth is a more complex sCO_2 GT–ORC cascade layout.

An energetic and exergetic performance comparison has been carried out for the different WHR schemes. The numerical approach allows a thermodynamic analysis, considering the peculiarities of carbon dioxide in supercritical conditions and also the pressure and thermal losses and efficiencies of each component.

The last section of the analysis looks at the influence of seawater temperature variations on the hybrid energy system behavior. In this way, it is analyzed the performance of WHR systems based on sCO_2 gas turbines when the layouts are simpler with respect to the recompression one, highlighting the improvement related to each layout modification and considering that the effects of seawater temperature variations on the WHR system efficiency and overall load cannot be neglected.

2. Hybrid Energy System Layout

As stated above, the supercritical CO_2 gas turbine is suitable to be a good technological solution for waste heat recovery, especially in the case of medium-high temperature levels of the exhaust gases. The following six bottoming gas turbine layouts are modeled and analyzed, considering both the need to simplify the scheme and to keep the weight within certain limits:

1. Simple sCO_2 Brayton cycle;
2. Recuperated sCO_2 Brayton cycle;
3. Preheated and Recuperated sCO_2 Brayton cycle;
4. Partially Preheated and Recuperated sCO_2 Brayton cycle;
5. Dual heated Cascade cycle;
6. Organic Rankine cycle coupled to the preheated and recuperated sCO_2 Brayton cycle.

For the sake of simplicity, the LM2500+ gas turbine has been depicted in the figures as a turbine coupled to a compressor.

The first scheme provides the simple Brayton cycle, consisting of a compressor, turbine, and two heat exchangers for the hot and cold source, respectively (Figure 1).

Figure 2 reports a recuperated Brayton cycle (layout 2), in which part of the heat available at the turbine outlet is used to heat the gas leaving the compressor, to increase the global efficiency of the entire system.

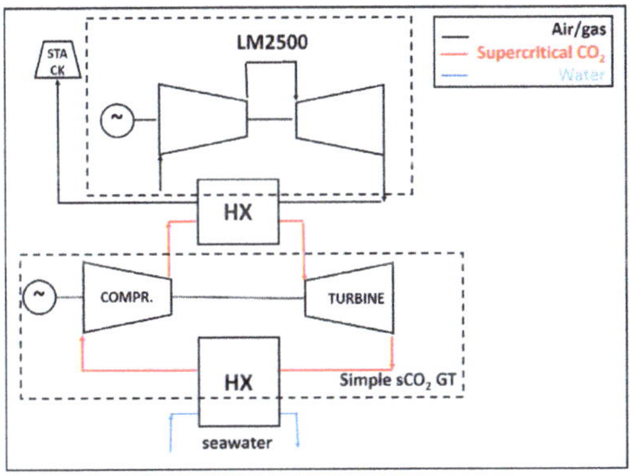

Figure 1. Layout 1, simple sCO_2 GT as WHR system.

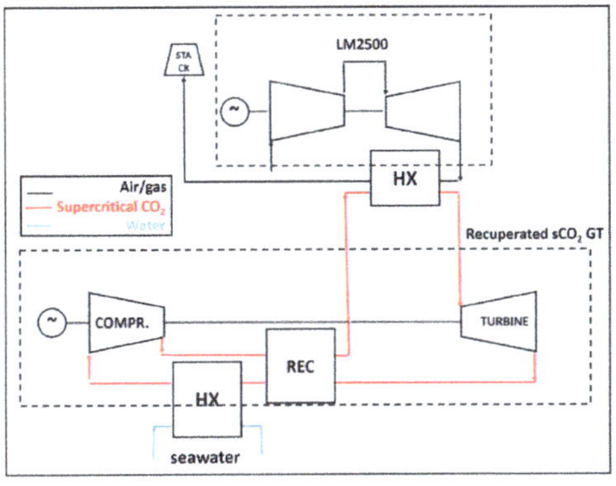

Figure 2. Layout 2, recuperated sCO_2 GT as WHR system.

The third scheme, shown in Figure 3, is a preheated and recuperated one. The additional heat exchanger can be useful to increase the thermal power addressed to the bottoming cycle and to reduce the exhaust gas temperature, with the stack temperature limited to 100 °C. In this way, it is possible to further heat of the working fluid before entering into the recuperator.

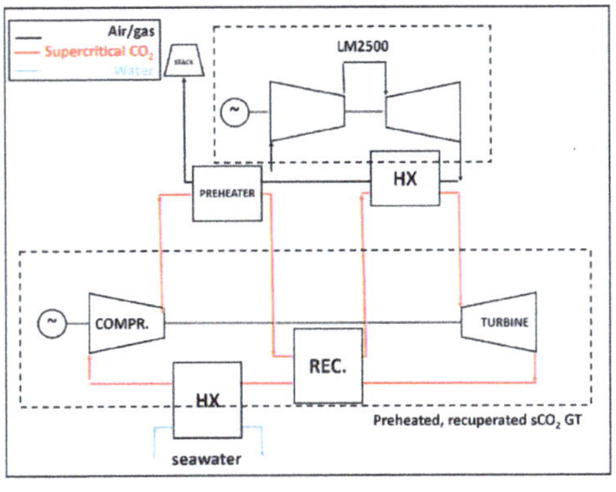

Figure 3. Layout 3, preheated and recuperated sCO_2 GT as WHR system.

In addition, a partial preheating can be considered by the splitting of the compressed working fluid between the recuperator and preheater. This scheme is depicted in Figure 4. In this layout, the sCO_2 stream is separated at the compressor outlet in the following two different streams: the first enters in a low-temperature heater while the second goes to a recuperator. Then, the two streams enter together in a high-temperature heater [28].

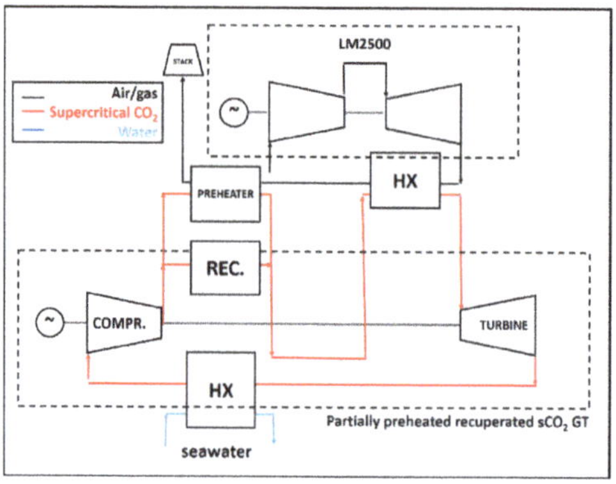

Figure 4. Layout 4, partially preheated, recuperated cycle as WHR system.

The fifth sCO_2 GT layout is reported in Figure 5 and regards the dual heated cascade cycle described in [28] and originally proposed in [29] as follows: a low-temperature turbine and a further heat exchanger are added with respect to the previous scheme. The layout provides two recuperators in series, namely, a low-temperature recuperator (LTR) and a high-temperature recuperator (HTR). The hot exhausts are used to heat the high-pressure sCO_2 before entering in both the turbines.

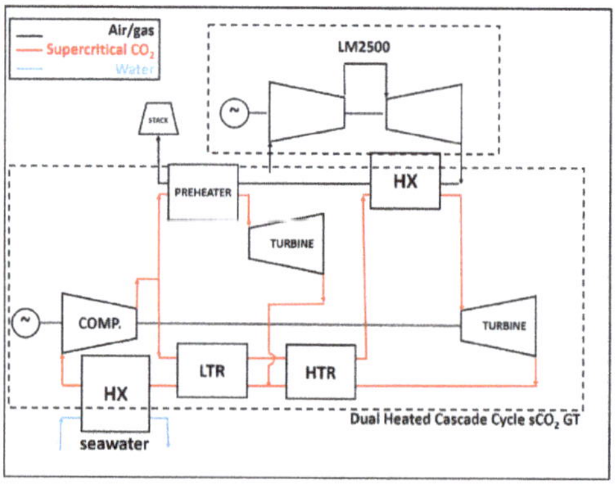

Figure 5. Layout 5 dual-heated cascade cycle sCO_2 GT as WHR system.

Supercritical CO_2 exiting the turbine has a temperature that allows to consider a further waste heat recovery, and the latter scheme (Figure 6) regards a hybrid energy system in which a simple ORC system receives thermal power by the sCO_2 Brayton cycle. The simple Rankine cycle has been individuated to limit the complexity of the layout and to consider that the sCO_2 temperature at the compressor inlet should be close to the supercritical temperature. The cis-1,1,1,4,4,4-hexaflouro-2-butene, R1336mzz (Z), a hydrofluorolefin (HFO), is selected as working fluid of the ORC plant because of its low ODP and GWP values, low toxicity and flammability and good thermodynamic properties [30], as reported in Table 1, in which A1 denotes that R1336mzz (Z) is low toxicity (A) and presents no flame propagation (1) when tested as per the standard.

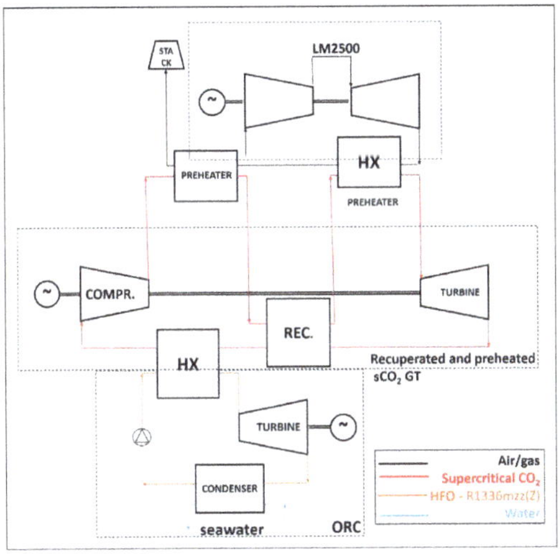

Figure 6. Layout 6, Organic Rankine Cycle coupled to the preheated and recuperated sCO_2 Brayton cycle.

Table 1. R1336mzz (Z) thermo-physical properties.

Chemical Formula	CF3CHCHCF3
ODP	0
GWP 100 years	2
ASHRAE Standard 34 Safety Class	A1
Atmospheric life time, year	0.060274
Molar weight, kg/kmol	164
Critical pressure, bar	29.4
Critical temperature, K	444.45

3. Integrated Energy System Modeling and Validation

The steady-state behavior of the hybrid energy system is simulated using the commercial software Thermoflex of the Thermoflow suite [31], which enables the REFPROP–NIST property function for the gas stream to be used as required [32]. Thermoflex is a thermodynamic modeling software based on mass and energy balancing, and its database offers several modules and subsystems, which can be used as a black box to build complex energy system layouts.

The software allows the simulation of both design and off-design conditions. In this study, Thermoflex has been used in "thermodynamic design" conditions as follows: the software solves energy and mass balances to evaluate the thermodynamic performance. The main gas turbine has been modeled using a black-box approach, considering that the software embedded a wide commercial gas turbine library for design data and part load performance maps. The bottoming cycles have been modeled considering and connecting each single component (e.g., compressor, pump, heat exchanger, turbine, condenser) with the others. Some characteristic parameters (e.g., efficiency, pressure losses, thermal losses) have been taken by the literature. The software allows the use of control loops and to carry out parametric analysis.

The ideal gas equation of state (EoS) is not the most suitable in the case of carbon dioxide in supercritical conditions. To this aim, commercial codes such as Aspen Plus or Ebsilon Professional allow the selection of another, more suitable, equation of state such as the Peng-Robinson EoS [33] or the Lee–Kesler–Plöcker EoS [34]. Thermoflex, instead, allows us to use the REFPROP program, distributed through the Standard Reference Data program of NIST, which provides thermophysical properties of pure fluids and mixtures over a wide range of fluid conditions, including liquid, gas, and supercritical phases. The NIST REFPROP database provides the most accurate thermophysical property models for a variety of industrially important fluids and fluid mixtures, including accepted standards [35]. White et al. [36] conducted a comparative study on the evaluation of property methods for model sCO_2 power cycles, even if mainly focused on direct-fired cycles, as follows: they concluded that REFPROP is a good performing property method when the working fluid is pure carbon dioxide in supercritical conditions, as also highlighted in [37].

In recent years, Thermoflex has been used in various research papers on hybrid energy systems based on gas turbines (see ref. [38] as regards the numerical models used for the small-scale plants based on micro gas turbines) and has been used to simulate sCO_2 power cycles both in case of direct or indirect fired cycles, e.g., in [27,39,40].

This study is addressed to compare the layouts in the case of full-load conditions, and all parameters characterizing each component are considered to be constant (e.g., efficiencies of the rotating components and the heat exchangers) and are defined by the literature.

The main gas turbine is the aero-derivative GE LM2500 [41], which is installed on many ships (e.g., the Italian Navy PPA offshore patrol ships [29] and the V/STOL aircraft carrier Cavour [42], the RMS Queen Mary 2 cruise ship [43] and Millennium class of Celebrity Cruises [44]). Thermoflex includes performance and exhaust data within its software library for several configurations of this gas turbine model [31]; that data have been used by the literature in simulations and model comparisons, e.g., in [45,46]. Table 2 shows the LM2500+ reference data at full load condition, as reported by Thermoflow. The declared maximum model error in the test range is >0.5% for power, exhaust temperature, and mass flow.

Table 2. LM2500 reference data.

Model	GE LM2500 + RB (G4)
Shaft	2
Pressure Ratio	23
Air Flow	88 kg/s
TOT	511 °C
Gen Power	32,686 kW
LHV eta	39.7%

The version of the turbine, which has been chosen for the study, is the natural gas-fueled LM2500+ G4. This solution is actually provided mainly in the case of stationary applications, but it represents an emerging solution in the case of liquefied natural gas (LNG).

The identification of the design parameters of sCO_2 gas turbines has to consider the lack of experimental information. The experimental data regarding mainly prototypes and several characteristic parameters (e.g., isentropic compressor efficiency) can strongly differ with respect to the values considered as references in the theoretical or numerical studies by the literature [47].

The parameters of the sCO_2 power cycle and ORC system models are defined by data published in the literature and presented in Table 3.

Table 3. sCO_2 GT/ORC model parameters.

	Model Parameter
Compressor efficiency	80%
Turbine/expander efficiency	85%
Heat exchanger efficiency	90%
sCO_2 GT min/max pressure	76.3/260 bar
ORC min/max pressure	0.8/25 bar

The isentropic efficiency range of the compressor usually varies between 76% [18] and 80% [20,48,49], while the turbine efficiency values are usually between 85% [49] and 90% [17]. These values are mainly used by the literature. Thermoflex uses the polytropic efficiency, and the compressor efficiency is set at 80%, while the turbine efficiency is set at 85% to consider the isentropic efficiency in the range of the values considered by the literature.

The pressure range of both sCO_2 GT and ORC has been defined considering the values usually published in the literature. The sCO_2 gas turbine has to operate with a minimum pressure greater than the critical pressure, which is close to 73.8 bar. The upper limit is often fixed at 300 bar [12,40]. Regarding the ORC, the pressure corresponding to a condensing temperature of 25 °C for the chosen working fluid is 0.8 bar, while the critical pressure is 29.4 bar, as shown in Table 1. In this way, the working fluid operates always in subcritical conditions.

Once the parameters of the turbomachinery have been fixed, the main issues concern the heat exchangers' characterization. In fact, the recuperator has the fundamental role of recovering waste heat and increasing the system performance, but they also present several issues related to the supercritical conditions of the working fluid that cannot be ignored.

The design and, consequently, the modeling of heat exchangers represent a critical issue related to the study of this type of plant due to the significant variations of the thermodynamic properties of carbon dioxide in the proximity of the critical point. The issue regards mainly the recuperators since in both the high- and low-temperature lines, the fluid is in supercritical conditions, and the different heat capacities of sCO_2, could lead to internal pinch-points near to zero. This problem mainly affects the low-temperature recuperators, while at high temperatures and pressures, the difference in the specific heat is not significant [16].

This issue has to be considered within the numerical modeling, and, among the others, the authors have followed the solution proposed by Scaccabarozzi et al. [50], modeling the recuperator as two heat exchangers in series. In this way, it is possible to limit the issues related to the internal pinch-point. The authors chose to fix in both the heat exchanger the temperature difference at the pinch point at 10 K.

The recuperative heat exchangers' efficiencies vary in the literature between 85% and 95%. In the numerical simulation, the heat exchanger efficiency has been fixed at 90%. An increase in the efficiency of the heat exchangers leads to an increase in the thermal efficiency of the system, but typically it corresponds to an increase in the size of the recuperator, and, consequently, its volume and weight. For example, in a recompression layout composed of two recuperators and a preheater, the increase in efficiency of the recuperators from 85% to 95% increases the total volume and weight by 1.5 m^3 and 4.46 t [21].

3.1. sCO_2 Gas Turbine Model Validation

In this section, a comparison between a model built in Thermoflex and the model based on an in-house numerical tool developed in Matlab [51] has been carried out. A numerical-experimental validation is not possible because of the lack of experimental data on this topic.

Considering the above-mentioned issues related to the presence of the recuperator, the comparison has been carried out on the simple regenerative cycle (layout no. 2).

To obtain a comparison, the boundary conditions and the characteristic parameters of each component have been set by [51].

The compressor and turbine efficiencies are 80% and 85%, respectively. The setting parameters of the main heat exchanger and of the recuperator are shown in Table 4. The comparison has been carried out considering that the heat exchanger has to operate with a fixed sCO_2 stream exit temperature of 391.8 K since the heat exchanger efficiency is not performed.

Table 4. Characteristic parameters of the model used for the validation.

	Model Parameter
Compressor efficiency	80%
Turbine efficiency	85%
ΔT_{pp} HX, K	25
ΔT_{pp} REC, K	10
Δp_{HX}, bar	2
$\Delta p/p_{in}$ REC cold side	0.5%
$\Delta p/p_{in}$ REC cold side	1%
Heat Source Mass flow, kg/s	50
Heat Source in Temp, °C	550
Heat Source Specific Heat, kJ/kgK	1.15
CO_2 mass flow, kg/s	73.13
Cooling water mass flow, kg/s	462.12

Results of the comparison have been reported in Table 5. It is clear that the differences are related to the different modeling approach, mainly regarding the heat exchangers. The net power shown in Table 5 is defined as the difference between turbine and compressor power.

Table 5. Model comparison with Ref. [51].

	Ref. [51]	Results	Variation, %
Inlet Comp temp, °C	33	33	0.00
Inlet Comp pressure, bar	79.19	79.19	0.00
Out Comp pressure, bar	250	250	0.00
TIT, °C	391.8	391.8	0.00
Thermal Power recoverable, MW	19.02	19.78	4
Turbine Power, MW	7.48	7.54	0.8
Compressor Power, MW	2.37	2.31	−2.45
Net Power, MW	5.11	5.23	2.3

The variations in terms of power can be considered acceptable since it is a comparison between two different zero-dimensional thermodynamic models. To obtain more conservative results, the pressure and thermal losses at the heat exchangers have been set greater with respect to those chosen in this section as follows: the normalized heat loss is equal to 5.73%, while the $\Delta p/p_{in}$ has been set at 1.05% and 2.77% respectively for the cold and hot side. The temperature difference at the pinch point has been set at 10 K.

3.2. Identification of the Mass Flow Rate of Each Layout

After setting the model parameters, the working fluid mass flow has been identified through a parametric analysis in which the sCO_2 mass flow has been varied in a wide range, between 60 and 130 kg/s, in order to identify the value able to maximize the net power. In all the cases, a control loop has been used to identify the coolant mass flow rate to guarantee a supercritical carbon dioxide compressor inlet temperature of 32 °C. As shown in Figure 7, the results of the parametric analysis highlighted that the working fluid mass flow to maximize the net power varies from 64 kg/s for the simple cycle (layout 1) to 118 kg/s for the dual heated cascade cycle scheme (layout 5).

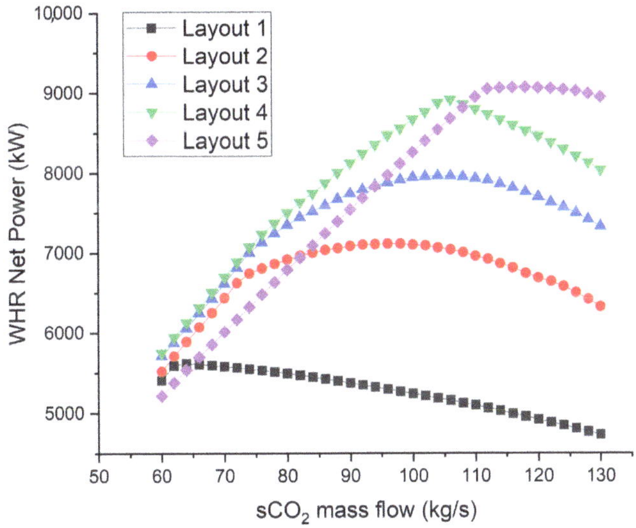

Figure 7. WHR net power variation with sCO_2 mass flow.

4. Energetic and Exergetic Analysis

The net power of the energy system considers the mechanical and electrical losses and the power consumption of the auxiliaries. P_{GT} is the net power of LM2500 GT, while $P_{sCO2\ GT}$ and P_{ORC} are the net power of the bottoming systems, defined as the difference between the power obtained by the turbine and the power consumed by the compressor and/or pumps.

The global efficiency η_{hyb} of the hybrid system is defined in Equation (1), while the efficiency of the waste heat recovery η_{WHR} is defined in Equation (2) as the ratio between the net power of the bottoming cycles and the maximum value of the thermal power available at the exhaust gases as follows:

$$\eta_{hyb} = \frac{P_{GT} + P_{sCO2\ GT}}{\dot{Q}_{FUEL}} \tag{1}$$

$$\eta_{WHR} = \frac{P_{sCO2\ GT} + P_{ORC}}{\dot{m}_{gas} \times \Delta H_{HX_{GT}}} \tag{2}$$

where $\Delta H_{HX_{GT}}$ is the difference between enthalpies at the inlet of the heat exchanger and the limit value at the stack, corresponding to a stack temperature of 100 °C.

The exergy in input to the waste heat recovery system is the exergy transferred to supercritical carbon dioxide from the exhausts of the gas turbine and is equal to $\Delta\xi_{HX} = \dot{m}_{gas} \times (\Delta H_{HX_{GT}} - T_0 \times \Delta s_{HX_{GT}})$, considering the ambient temperature as the reference condition (T_0 = 288.15 K).

The exergetic efficiency is reported in Equation (3).

$$\eta_{ex_{WHR}} = \frac{P_{sCO2\ GT} + P_{ORC}}{\Delta\xi_{HX}} \tag{3}$$

Figure 8 compares the net power of the hybrid schemes analyzed, highlighting the contribution of the WHR system. The introduction of the preheater leads to a slight reduction in the main gas turbine power because of the backpressure at the exhaust. The increase in available power, with respect to the referring value of 32,686 kW shown in Table 1, ranges between 17% and 29% without any additional fuel. However, only for layout 1, this increase remains below 22%. The overall net efficiency increases to about 45%

in the case of the Simple WHR scheme and up to 49% in the hybrid scheme with ORC and sCO_2 GT (Figure 9), with an increase in the overall efficiency between 18% and 30%.

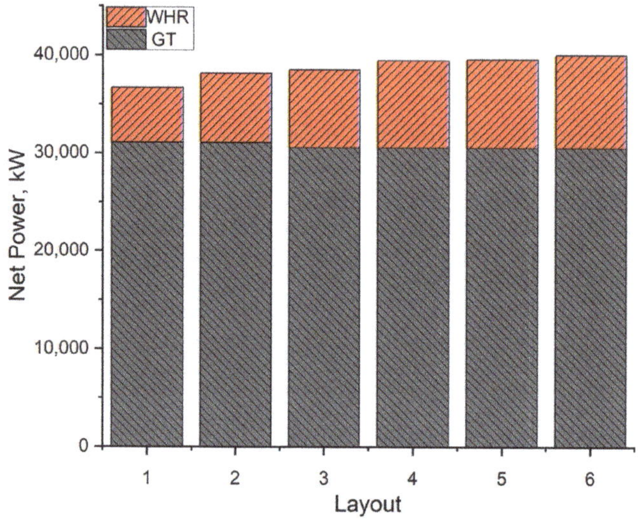

Figure 8. Net power variation with WHR layout.

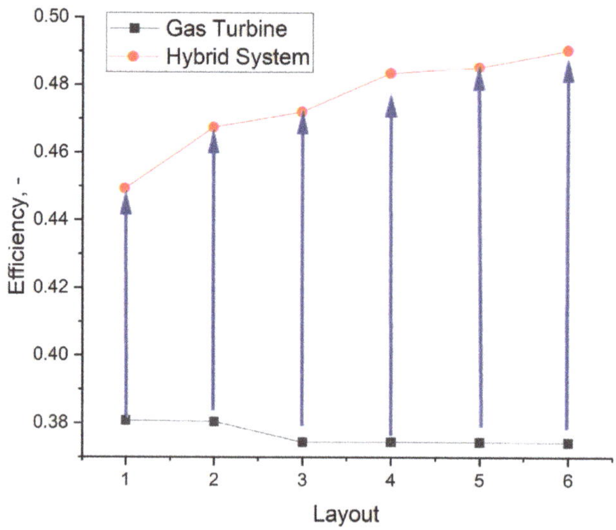

Figure 9. Global efficiency variation with WHR layout.

Figure 10 shows the electrical efficiency and the exergetic efficiency of the bottoming system. All the chosen layouts ensure a satisfactory efficiency, also considering the simplicity of the first cases. At the same time, the exergetic efficiency highlights a good waste heat recovery from the exhausts of the gas turbine.

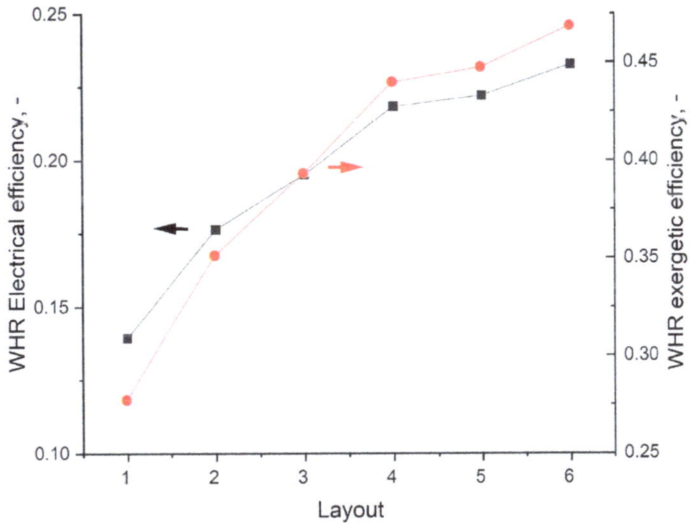

Figure 10. WHR electrical (square) and exergetic efficiencies (circles).

A secondary effect of WHR is the reduction of the exhaust temperature to values between 200 °C and 100 °C with respect to the LM2500+ turbine outlet temperature of 511 °C (Figure 8).

Figure 11 also shows that at the end of the bottoming cycle, the relatively high temperature of sCO_2 and, consequently, the available residual heat allows us to consider the introduction of a further bottoming cycle. ORC systems, for instance, are suitable for heat recovery from medium-low enthalpy sources. Case 6 is an example of this possibility.

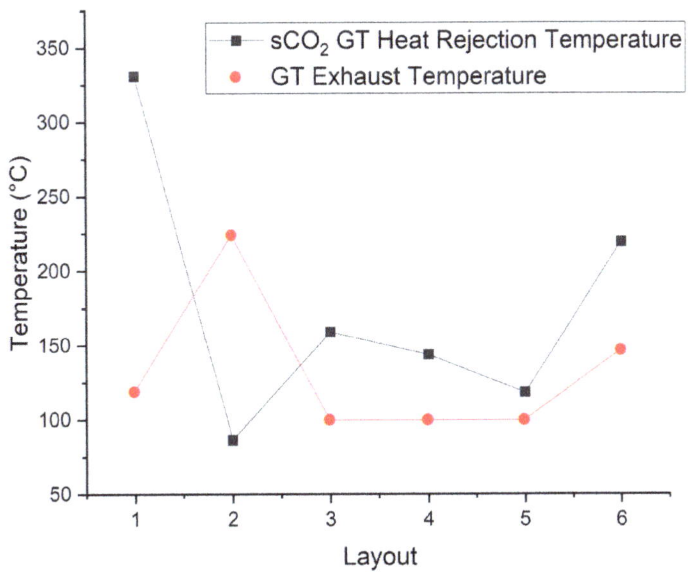

Figure 11. Temp. of sCO_2 at cold sink HX (square) and GT exhaust temperature (circles).

The introduction of WHR implies a significant increase in the global efficiency of the energy system. WHR benefits in marine applications are clearly stated, also considering the most recent IMO index for assessing the energy efficiency and environmental impact of ships, the EEXI index [3]. In fact, this index is defined as follows [52]:

$$EEXI = \frac{CF \times SFC \times P_{ME}}{f_i \times f_C \times f_l \times Capacity \times V_{REF}} \tag{4}$$

where P_{ME} is the power of main engines in kW, SFC the specific fuel consumption for different loads at ship, measured in t_{fuel}/kW, V_{ref} is the speed of the sheep at 75% MCR corresponding to the capacity; CF-Capacity (Deadweight for container vessels, 70% of deadweight) and f_i are correction factors. The introduction of WHR leads to an increase both in P_{ME} and SFC.

However, such schemes introduce complex additional components in the energy scheme, and this may lead to some issues related to plant engineering, maintenance, and management costs. Thus, in order to evaluate the real cost/benefits of the adopted scheme, the analysis of the performance of such systems becomes relevant as the operating conditions vary.

It Is well known that any energy system is affected by ambient conditions. In particular, gas turbines are sensitive to variations in ambient temperature, while any closed-loop thermodynamic cycle, as in the case of sCO_2 and ORC, strongly depends on the temperature of both hot and cold heat sources. In the case of marine applications, seawater is the cold sink whose temperature can vary significantly according to the route followed by the ship. Thus, in the next section, the influence of seawater temperature on the overall energy system performance is analyzed.

5. Effects of Seawater Temperature on Overall Performance

The seawater temperature can be locally considered constant, but container ships, bulk carriers, or oil tankers need to cross throughout different seas/oceans and seasons also in a single travel. Thus, the effects of seawater temperature on the overall performance of the energy system are important parameters in the evaluation of the WHR benefits that cannot be neglected, being that seawater is the cold sink of the cycle. In the specific literature, the water temperature close to the sea/ocean surface, considering also the first 20 m of depth, is called sea surface temperature (SST). SST evolution is an essential climate variable, which is fundamental in any climate regulation analysis [53].

For example, ships long beyond 300 m have to pass the Horne Cape to reach the Pacific Ocean starting from the Atlantic one, since the constraints on the Panama Canal, thus passing, in May, from 20 to 38 °C of the US east coast to 10–18 °C of the US west coast, through the 4–8 °C of the Horne Cape [54]. At the same time, several ships from the East Coast of the US and Canada usually reach China or the Far East passing through both the Mediterranean Sea [53,54] and the Red Sea [53–55]. The difference in annual mean SST between the Mediterranean Sea and the Red Sea is about 8 °C but can be greater as follows: between 14.9 °C and 25.1 °C in winter, 18.2 °C and 27.8 °C in spring, 25.1 °C and 30.3 °C in summer and 20.3 °C and 28.3 °C in autumn.

A parametric analysis by varying seawater temperature has been carried out, in order to consider this aspect. In the analysis, even if ambient conditions influence the main gas turbine operation and, consequently, the amount of thermal power which is available for the bottoming cycle, the ambient air temperature has been considered constant. This is for both the sake of simplicity and to highlight the effect of seawater temperature variations.

In scheme 6, in the first analysis, the design condensing pressure of the ORC has been fixed at 0.8 bar, as in the previous simulations. It is clear that the condensation pressure and, consequently, the temperature have to be defined considering the cold heat source temperature.

The results of the parametric analysis are shown in Figure 12, where it is reported the sCO_2 gas turbine net power variation with sea surface temperature.

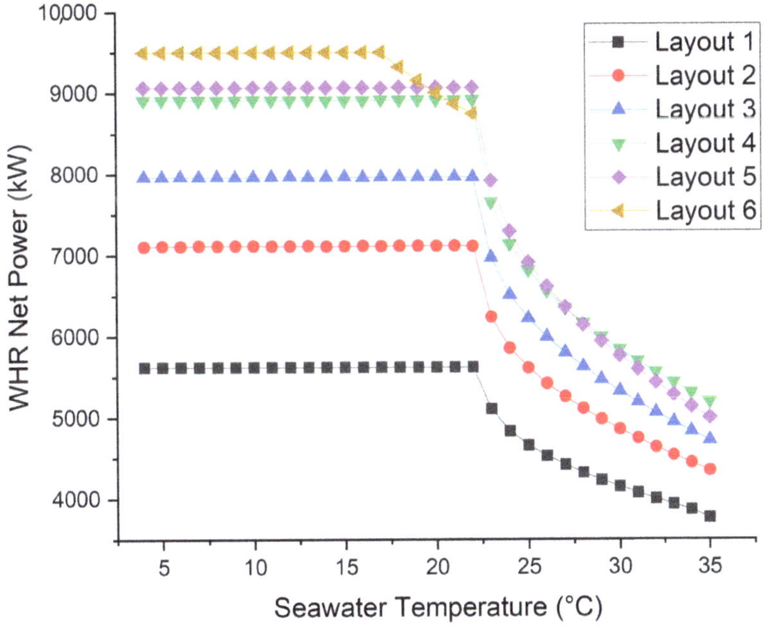

Figure 12. sCO_2 GT net power variation with seawater temperature.

The effect of SST on the bottoming cycle is relevant for high seawater temperatures, and it is due to the peculiar behavior of carbon dioxide in proximity to its supercritical temperature at 31 °C. For sCO_2 temperature close to this value, the cycle presents its best performance, while higher seawater temperatures lead to higher sCO_2 temperature and, consequently, a different behavior in correspondence of the compressor. The WHR system continues to operate fairly, but the power and efficiency gain results are reduced.

As expected, the effect is more relevant for the following sixth case: the condensation temperature of the organic working fluid in the analyzed pressure conditions is within the range of variability of the seaside temperature and, in these conditions, the WHR system could operate only in a reduced range of temperatures. Thus, the use of the latter layout is possible only with a redesign of the pressure levels of the ORC system, and a further parametric analysis has been carried out to consider this point by varying the condensing pressure of the bottoming cycle.

In the analyzed temperature range of the cold source (4–35 °C), the lowest pressure at which the ORC can work in all the conditions is 1.3 bar. The increasing of the minimum pressure of the ORC leads to a decrease in its net power and, consequently, of the WHR net power. As shown in Figure 13, at 1.3 bar, the WHR net power is about 20% below the reference case, but the system can operate also in case of high seawater temperature. Seeing Figures 12 and 13, layout 5 appears to be the most suitable since it guarantees the highest net power in the wider SST temperature range.

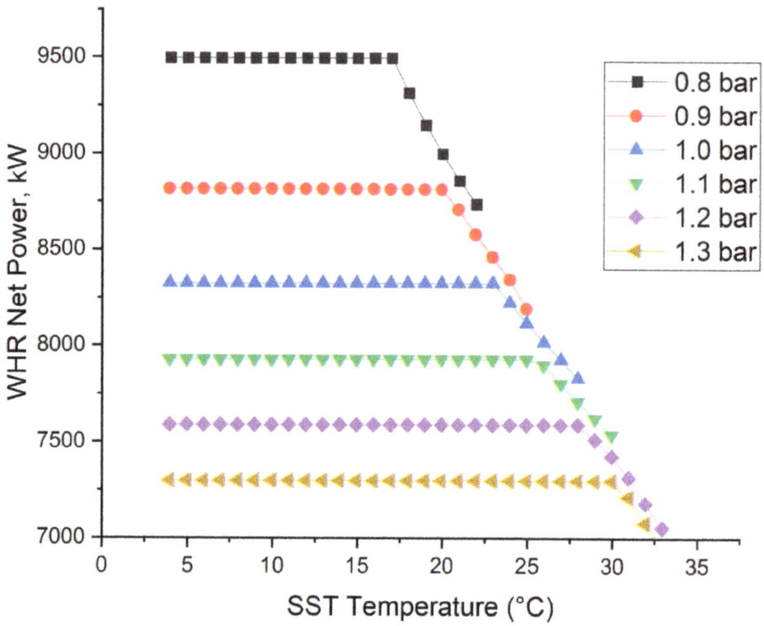

Figure 13. layout 6: WHR net power variation with seawater temperature and condensing pressure.

6. Conclusions

The paper investigates the effects on the performance of waste heat recovery in marine applications when the main engine is a gas turbine, and the bottoming cycle is a supercritical CO_2 closed Brayton cycle. The subject of study is the GE LM2500+. Six different configurations have been evaluated to individuate the best compromise between the increase in power and efficiency and the increase in complexity and weight of the hybrid energy system. Five of these consider sCO_2 gas turbine layouts, from the simplest with just two heat exchangers, to a complex system with five heat exchangers and an additional turbine. Finally, a hybrid system in which an ORC is also present is discussed.

Results of numerical simulations showed that depending on the layout of the WHR, the increase in available power is between 17% and 29%. The overall net efficiency increases to about 45% in the case of the Simple WHR scheme and up to nearly 49% in the hybrid scheme with ORC and sCO_2.

The effect of seawater temperature on the WHR performance has been also analyzed. In the case of seawater temperatures greater than 30 °C, the efficiency of supercritical CO_2 gas turbines decreases by more than 30%, while the efficiency of the cascade layout sCO_2 GT-ORC is also limited by the higher condensing pressure levels.

Author Contributions: Conceptualization, F.R. and P.M.; methodology, F.R.; writing—original draft, F.R.; writing—review and editing, R.C. and P.M. All authors have read and agreed to the published version of the manuscript.

Funding: This research received no external funding.

Data Availability Statement: Data sharing not applicable.

Conflicts of Interest: The authors declare no conflict of interest.

Nomenclature

COGAS	Combined Gas Turbine and Steam turbine
COGES	Combined Gas Turbine Electric and Steam
EEXI	Energy Efficiency Existing Ship Index
EoS	Equation of State
HFO	Hydrofluorolefin
HX	Heat Exchanger
HTR	High Temperature Recuperator
GHG	Greenhouse Gas
GWP	Global Warming Potential
GT	Gas Turbine
IMO	International Maritime Organization
LNG	Liquefied Natural Gas
LTR	Low Temperature Recuperator
MARPOL	"MARitime POLlution", MARPOL 73/78, International Convention for the Prevention of Pollution from Ships
ODP	Ozone Depletion Potential
ORC	Organic Rankine Cycle
sCO_2	Supercritical Carbon Dioxide
SFC	Specific Fuel Consumption
SST	Sea Surface Temperature
TIT	Turbine Inlet Temperature
TOT	Turbine Outlet Temperature
WHR	Waste Heat Recovery
WHRS	Waste Heat Recovery System
Greek	
ΔT	Difference of Temperature [K]
ε	Heat Exchanger Efficiency
η	Efficiency
ξ	Exergy
Subscripts	
Ex	Exergy
Max	Maximum
Pp	Pinch Point
In	Inlet

References

1. IMO. Cutting GHG Emissions from Shipping—10 Years of Mandatory Rules. 2021. Available online: https://www.imo.org/en/MediaCentre/PressBriefings/pages/DecadeOfGHGAction.aspx (accessed on 25 May 2022).
2. Zhu, S.; Zhang, K.; Deng, K. A review of waste heat recovery from the marine engine with highly efficient bottoming power cycles. *Renew. Sustain. Energy Rev.* **2020**, *120*, 109611. [CrossRef]
3. IMO. EEXI and CII—Ship Carbon Intensity and Rating System. Available online: https://www.imo.org/en/MediaCentre/HotTopics/Pages/EEXI-CII-FAQ.aspx (accessed on 13 May 2023).
4. Sing, D.V.; Pedersen, E. A review of waste heat recovery technologies for maritime applications. *Energy Convers. Manag.* **2016**, *111*, 315–328. [CrossRef]
5. Haglind, F. A review on the use of gas and steam turbine combined cycles as prime movers for large ships. Part II: Previous work and implications. *Energy Convers. Manag.* **2018**, *49*, 3468–3475. [CrossRef]
6. Altosole, M.; Benvenuto, G.; Campora, U.; Laviola, M.; Trucco, A. Waste Heat Recovery from Marine Gas Turbines and Diesel Engines. *Energies* **2017**, *10*, 718. [CrossRef]
7. Haglind, F. A review on the use of gas and steam turbine combined cycles as prime movers for large ships. Part I: Background and design. *Energy Convers. Manag.* **2008**, *49*, 3458–3467. [CrossRef]
8. Badeer, G.H. *GE Aeroderivative Gas Turbines-Design and Operating Features*; GER-3695E; GE Power Systems: Evendale, OH, USA, 2000.
9. Wang, Z.; Jiang, Y.; Han, F.; Yu, S.; Li, W.; Ji, Y.; Cai, W. A thermodynamic configuration method of combined supercritical CO_2 power system for marine engine waste heat recovery based on recuperative effects. *Appl. Therm. Eng.* **2022**, *200*, 117645. [CrossRef]

10. Marchionni, M.; Bianchi, G.; Tassou, S.A. Review of supercritical carbon dioxide (sCO_2) technologies for high-grade waste heat to power conversion. *SN Appl. Sci.* **2020**, *2*, 611. [CrossRef]
11. Ahn, Y.; Bae, S.J.; Kim, M.; Cho, S.K.; Baik, S.; Lee, J.I.; Cha, J.E. Review of supercritical CO_2 power cycle technology and current status of research and development. *Nucl. Eng. Technol.* **2015**, *47*, 647–661. [CrossRef]
12. Crespi, F.; Gavagnin, G.; Sanchez, D.; Martinez, G.S. Supercritical carbon dioxide cycles for power generation: A review. *Appl. Energy* **2017**, *195*, 152–183. [CrossRef]
13. Xu, J.; Liu, C.; Sun, E.; Xie, J.; Li, M.; Yang, Y.; Liu, J. Perspective of $S-CO_2$ power cycles. *Energy* **2019**, *186*, 115831. [CrossRef]
14. Feher, E.G. The Supercritical Thermodynamic Power Cycle. *Energy Convers.* **1968**, *8*, 85–90. [CrossRef]
15. Angelino, G. Carbon dioxide condensation cycles for power production. *J. Eng. Gas. Turbines Power* **1968**, *90*, 287–295. [CrossRef]
16. Gotelip, T.; Gampe, U.; Glos, S. Optimization strategies of different SCO_2 architectures for gas turbine bottoming cycle applications. *Energy* **2022**, *250*, 123734. [CrossRef]
17. Wright, S.A.; Davidson, C.S.; Scammel, W.O. Thermo-Economic Analysis of Four sCO_2 Waste Heat Recovery Power Systems. In Proceedings of the 5th Int Symp—Supercrit CO_2 Power Cycles, San Antonio, TX, USA, 28–31 March 2016; pp. 1–16.
18. Sakalis, G.N. Design and partial load operation optimization of integrated ship energy system based on supercritical CO_2 waste heat recovery cycle. *Sustain. Energy Technol. Assess.* **2022**, *51*, 101965. [CrossRef]
19. Hou, S.; Zhang, F.; Yu, L.; Cao, S.; Zhou, Y.; Wu, Y.; Hou, L. Optimization of a combined cooling, heating and power system using CO_2 as main working fluid driven by gas turbine waste heat. *Energy Convers. Manag.* **2018**, *178*, 235–249. [CrossRef]
20. Hou, S.; Wu, Y.; Zhou, Y.; Yu, L. Performance analysis of the combined supercritical CO_2 recompression and regenerative cycle used in waste heat recovery of marine gas turbine. *Energy Convers. Manag.* **2017**, *151*, 73–85. [CrossRef]
21. Du, Y.; Hu, C.; Wang, H.; Dong, W. Size optimization of heat exchanger and thermoeconomic assessment for supercritical CO_2 recompression Brayton cycle applied in marine. *Energy* **2021**, *239*, 122306. [CrossRef]
22. Pan, P.; Yuan, C.; Sun, Y.; Yan, X.; Lu, M.; Bucknall, R. Thermo-economic analysis and multi-objective optimization of $S-CO_2$ Brayton cycle waste heat recovery system for an ocean-going 9000 TEU container ship. *Energy Convers. Manag.* **2020**, *221*, 113077. [CrossRef]
23. Hu, K.; Liu, X.; Chen, Y.; He, D. Influence of marine conditions on thermal efficiency of SCO_2 Brayton cycle for waste heat recovery. *Appl. Therm. Eng.* **2023**, *227*, 120391. [CrossRef]
24. Guo, J.; Du, J.; Li, M.; Zhang, Z. Performance study of a supercritical CO_2 Brayton cycle coupled with a compressed CO_2 energy storage system for waste heat recovery of ship gas turbines under variable load conditions. *Sustain. Energy Fuels* **2022**, *6*, 5557. [CrossRef]
25. Sahoo, A.; Sanjay, R.; Jain, M. *Analysis of Recompression-Regeneration sCO_2 Combined Cycle Utilizing Marine Gas Turbine Exhaust Heat: Effect of Operating Parameters*; SAE Technical Paper 2022-01-5059; SAE International: Warrendale, PA, USA, 2022. [CrossRef]
26. Chacartegui, R.; Muñoz de Escalona, J.M.; Sánchez, D.; Monje, B.; Sánchez, T. Alternative cycles based on carbon dioxide for central receiver solar power plants. *Appl. Therm. Eng.* **2011**, *31*, 872–879. [CrossRef]
27. Reale, F.; Iannotta, V.; Tuccillo, R. Numerical Study of a Micro Gas Turbine Integrated with a Supercritical CO_2 Brayton Cycle Turbine. In Proceedings of the ASME Turbo Expo 2018: Power for Land, Sea and Air; Paper no. GT2018-76656, Oslo, Norway, 11–15 June 2018. [CrossRef]
28. Bonalumi, D.; Giuffrida, A.; Sicali, F. A case study of cascade supercritical CO_2 power cycle for waste heat recovery from a small gas turbine. *Energy Convers. Manag. X* **2022**, *14*, 100212. [CrossRef]
29. Kimzey, G. *Development of a Brayton Bottoming Cycle Using Supercritical Carbon Dioxide as the Working Fluid*; Electric Power Research Institute Report; University Turbine Systems Research Program: Palo Alto, CA, USA, 2012; pp. 1–31.
30. Navarro-Esbrí, J.; Molés, F.; Peris, B.; Mota-Babiloni, A.; Kontomaris, K. Experimental study of an Organic Rankine Cycle with HFO-1336mzz-Z as a low global warming potential working fluid for micro-scale low temperature applications. *Energy* **2017**, *133*, 79–89. [CrossRef]
31. Thermoflow. Thermoflex v. 30. Available online: www.thermoflow.com (accessed on 10 March 2023).
32. Lemmon, E.W.; Bell, I.H.; Huber, M.L.; McLinden, M.O. *NIST Standard Reference Database 23: Reference Fluid Thermodynamic and Transport Properties-REFPROP, Version 10.0*; National Institute of Standards and Technology, Standard Reference Data Program: Gaithersburg, MD, USA, 2018.
33. Peng, D.Y.; Robinson, D.B. A New Two-Constant Equation of State. *Ind. Eng. Chem. Fundam.* **1976**, *15*, 59–64. [CrossRef]
34. Plocker, U.; Knapp, H.; Prausnitz, J. Calculation of High-Pressure Vapor-Liquid Equilibria from a Corresponding States Correlation with Emphasis on Asymmetric Mixtures. *Ind. Eng. Chem. Process Des. Dev.* **1978**, *17*, 324–332. [CrossRef]
35. Reference Fluid Thermodynamic and Transport Properties Database (REFPROP). Available online: https://www.nist.gov/programs-projects/reference-fluid-thermodynamic-and-transport-properties-database-refprop (accessed on 20 May 2023).
36. White, C.W.; Weiland, N.T. Evaluation of Property Methods for Modeling Direct-Supercritical CO_2 Power Cycles. *J. Eng. Gas Turbines Power* **2018**, *140*, 011701. [CrossRef]
37. Zhao, Q.; Mecheri, M.; Neveux, T.; Privat, R.; Jaubert, J.-N. 2016 Thermodynamic Model Investigation for Supercritical CO_2 Brayton Cycle for Coal-Fired Power Plant Application. In Proceedings of the Fifth International Supercritical CO_2 Power Cycles Symposium, San Antonio, TX, USA, 29–31 March 2016; Paper No. 93.
38. Reale, F.; Sannino, R. Numerical Modeling of Energy Systems Based on Micro Gas Turbine: A Review. *Energies* **2022**, *15*, 900. [CrossRef]

39. Colleoni, L.; Sindoni, A.; Ravelli, S. Comprehensive Thermodynamic Evaluation of the Natural Gas-Fired Allam Cycle at Full Load. *Energies* **2023**, *16*, 2597. [CrossRef]
40. Ancona, M.A.; Bianchi, M.; Branchini, L.; De Pascale, A.; Melino, F.; Peretto, A.; Torricelli, N. Systematic Comparison of ORC and s-CO_2 Combined Heat and Power Plants for Energy Harvesting in Industrial Gas Turbines. *Energies* **2021**, *14*, 3402. [CrossRef]
41. GE Aerospace Website. Available online: https://www.geaerospace.com/sites/default/files/lm2500plusg4-ppa-case-study.pdf (accessed on 14 March 2023).
42. Fincantieri Website. Available online: https://www.fincantieri.com/it/prodotti-servizi/navi-militari/classe-cavour/ (accessed on 14 March 2023).
43. GE Aerospace Website. Available online: https://www.geaerospace.com/press-release/marine-industrial-engines/ges-lm2500-gas-turbines-power-queen-mary-2-worlds-largest (accessed on 14 March 2023).
44. GE Aerospace Website. Available online: https://www.geaerospace.com/press-release/marine-industrial-engines/ges-gas-turbine-engines-celebrity-cruises-millennium-log (accessed on 14 March 2023).
45. De Andrade Cruz, M.; de Queiroz Fernandes Araújo, O.; de Medeiros, J.L. Deep seawater intake for primary cooling in tropical offshore processing of natural gas with high carbon dioxide content: Energy, emissions and economic assessments. *J. Nat. Gas Sci. Eng.* **2018**, *56*, 193–211. [CrossRef]
46. Reis, M.M.L.; Gallo, W.L.R. Study of waste heat recovery potential and optimization of the power production by an organic Rankine cycle in an FPSO unit. *Energy Convers. Manag.* **2018**, *157*, 409–422. [CrossRef]
47. Vidoza, J.A.; Andreasen, J.G.; Haglind, F.; dos Reis, M.M.L.; Gallo, W. Design and optimization of power hubs for Brazilian off-shore oil production units. *Energy* **2019**, *176*, 656–666. [CrossRef]
48. Martin, T.; White, M.T.; Bianchi, G.; Chai, L.; Tassou, S.A.; Sayma, A.I. Review of supercritical CO_2 technologies and systems for power generation. *Appl. Therm. Eng.* **2021**, *185*, 1164447.
49. Li, B.; Wang, S.S.; Zu, Y.; Song, L. Study on the off-design performance of supercritical carbon dioxide power cycle for waste heat recovery of gas turbine. *Energy Convers. Manag.* **2021**, *233*, 113890. [CrossRef]
50. Scaccabarozzi, R.; Gatti, M.; Martelli, E. Thermodynamic Analysis and Numerical Optimization of the NET Power Oxy-Combustion Cycle. *Appl. Energy* **2016**, *178*, 505–526. [CrossRef]
51. Alfani, D.; Binotti, M.; Macchi, E.; Silva, P.; Astolfi, M. sCO_2 power plants for waste heat recovery: Design optimization and part-load operation strategies. *Appl. Therm. Eng.* **2021**, *195*, 117013. [CrossRef]
52. Inanova, F. Analysis of the Specifics in Calculating the Index of Existing Marine Energy Efficiency EEXI in Force since 2023. In Proceedings of the 13th Electrical Engineering Faculty Conference, Varna, Bulgaria, 8–11 September 2021. [CrossRef]
53. García-Monteiro, S.; Sobrino, J.A.; Julien, Y.; Sòria, G.; Skokovic, D. Surface Temperature trends in the Mediterranean Sea from MODIS data during years 2003–2019. *Reg. Stud. Mar. Sci.* **2022**, *49*, 102086. [CrossRef]
54. OSPO NOAA. Sea Surface Temperature (SST) Contour Charts. Available online: https://www.ospo.noaa.gov/Products/ocean/sst/contour/ (accessed on 29 January 2023).
55. Shaltout, M. Recent sea surface temperature trends and future scenarios for the Red Sea. *Oceanologia* **2019**, *61*, 484–504. [CrossRef]

energies

MDPI

Article

CFD-Based Analysis of Installed Fuel Consumption and Aerodynamics of Transonic Transport Aircraft during Cruise Flight

Andrea Magrini [1,*], Denis Buosi [2], Francesco Poltronieri [1], Elena De Leo [1] and Ernesto Benini [1,*]

[1] Dipartimento di Ingegneria Industriale, Università Degli Studi di Padova, via Venezia 1, 35131 Padova, Italy
[2] Hit09 S.r.l., Piazzetta Bettiol 15, 35137 Padova, Italy; d.buosi@hit09.com
* Correspondence: andrea.magrini@unipd.it (A.M.); ernesto.benini@unipd.it (E.B.)

Abstract: Gas turbine fuel burn for an aircraft engine can be obtained analytically using thermodynamic cycle analysis. For large-diameter ultra-high bypass ratio turbofans, the impact of nacelle drag and propulsion system integration must be accounted for in order to obtain realistic estimates of the installed specific fuel consumption. However, simplified models cannot fully represent the complexity of installation effects. In this paper, we present a method that combines thermodynamic cycle analysis with detailed Computational Fluid Dynamics (CFD) modelling of the installation aerodynamics to obtain the fuel consumption at a given mission point. The flow field and propulsive forces arising in a transport aircraft powered by an ultra-high bypass ratio turbofan at cruise are first examined to characterise the operating conditions and measure the sensitivity to variations of the incidence at transonic flight. The proposed methodology, in which dynamic balance of the vehicle is achieved at each integration point, is then applied along a cruise segment to calculate the cumulative fuel burn and the change in the specific fuel consumption.

Keywords: installation effects; specific fuel consumption; ultra-high bypass ratio; NASA Common Research Model; propulsion system integration

Citation: Magrini, A.; Buosi, D.; Poltronieri, F.; De Leo, E.; Benini, E. CFD-Based Analysis of Installed Fuel Consumption and Aerodynamics of Transonic Transport Aircraft during Cruise Flight. *Energies* **2023**, *16*, 3323. https://doi.org/10.3390/en16083323

Academic Editors: Michele Pinelli, Alessio Suman and Nicola Casari

Received: 17 March 2023
Revised: 3 April 2023
Accepted: 6 April 2023
Published: 8 April 2023

1. Introduction

Increasing operational costs and regulatory constraints on environmental impact have been pushing aircraft performance, emissions, and pollution towards continuous improvement over the time. Current generation single-aisle transport aircraft in the 200PAX class are able to achieve a gain of around 20% in fuel efficiency per Available Seat Kilometre (ASK) relative to legacy early-2000s counterparts [1]. The next technological target in terms of aircraft propulsion is the implementation of Ultra-High Bypass Ratio (UHBPR) turbines in the 2025+ fleet, as major engine manufacturers are actively developing and testing this propulsor concept. As reviewed by Epstein [2], new design spaces are being opened to tackle ambitious twenty-first century commercial aviation sustainability targets, including smaller and more efficient cores, geared fan architectures, and large-BPR low pressure ratio fans. With these features, the presence of an upper limit on the effective improvements for ducted fans has been readily recognised owing to the counterbalancing effect of detrimental installation penalties in terms of excessive drag and weight of engine and airframe for overlarge fan diameters [3–5]. The optimal single-aisle transport aircraft, therefore, results from the contemporary improvement in propulsion technology and balance of airframe weight and efficiency, considering the entire flight envelope requirements.

In the last twenty years, a large number of parametric studies has been conducted regarding the impact of UHBPR turbofans on overall aircraft performance. Hughes [6] summarized the collaborative NASA work on UHBPR engines technology. Relevant research was conducted by Dagget [7] as well as by Guynn and Berton [8–10]. Dagget carried out an integration study of different high-BPR turbofans on a modified Boeing 777-200ER with composite wings, finding a potential reduction in Thrust Specific Fuel Consumption (TSFC) of up to 16%. This analysis considered several aspects of engine integration, including propulsor weight, nacelle drag and wing interference, ground clearance and engine placement, and engine-out performance. Guynn and Berton conducted a trade-off study with a direct-drive or geared fan applied to an advanced subsonic single-aisle transport aircraft of the B7373/A320 class for Entry Into Service in the 2015–2020 time frame. Using a series of NASA codes for engine thermodynamic modelling and aircraft sizing, they examined the impact of design objective on high-BPR engine layout and aircraft performance.

As highlighted above, apart from the specific gas turbine arrangement and cycle, the essential step to provide realistic aircraft performance in terms of fuel consumption is to incorporate installation effects at the design exploration level. These include the influence of the weight and drag of the propulsor as well as its airframe integration. Considering the aerodynamics, establishing the correlation between a given gas turbine layout and the propulsive forces is mandatory. The available thrust is diminished by the bulk component losses, which are modelled in the thermodynamic cycle analysis through coefficients such as inlet pressure recovery, bypass duct loss, or nozzle velocity and thrust coefficients. The drag is determined by the nacelle shape, which is driven by fan diameter and engine length. At the conceptual design level, given the absence of detailed geometric knowledge, this is typically estimated using semi-empirical models based on flat plate analogies or experimental data. For instance, Alexiou [11] expressed the nacelle drag as a factor decreasing net thrust proportional to flight velocity and specific thrust. Larsson [12] and Thoma [13] employed the ESDU 81024 code [14] and built on NACA 1-series axisymmetric cowl tests at no incidence. Bijewitz [15] derived a correlation between fan diameter and nacelle axial and radial dimensions, scaling the drag to the nacelle wetted area.

Having estimated the installed thrust and the corresponding installed TSFC, the block fuel can then be derived by integrating the mission profile. For large UHBPR turbofans, however, additional interference drag coming from engine/wing interaction is present and is typically not explicitly included. This is a relevant component, amounting to 10-15% of airframe drag, with the nacelle drag alone being around half that of the wings [16].

Aiming to provide a more accurate estimation of the real propulsive performance for a given aircraft model, in this paper we present a method that corroborates thermodynamic cycle analysis with high-fidelity Computational Fluid Dynamics (CFD) simulations to integrate the fuel consumption along a flight segment. The complexity of installation effects, in fact, prevents accurate decomposition into single contributions that can be modelled separately and superimposed at early design [17]. Moreover, even when they can be estimated globally, they impact the engine operation and the resulting turbomachinery status as a result of inlet distortion and nozzle suppression effects. Therefore, it is necessary to simulate both the propulsor and airframe in order to account for their coupling and accurately evaluate the installed performance.

Thus, this paper first reviews the aerodynamics of engine/airframe interference at cruise condition over a range of incidence close to the nominal one, illustrating the sensitivity of the flow field and the aerodynamic forces to the angle of attack. After that, the proposed method is applied to a cruise segment, where 0D thermodynamic data are matched to CFD simulations in order to obtain the fuel burn and installed TSFC. Before presenting the outcome of the study in Section 3, the following Section 2 illustrates the aircraft and engine model, the numerical methods, and the proposed algorithm.

2. Materials and Methods

2.1. Aircraft and Turbofan Model

The aircraft considered in the study is the NASA Common Research Model (CRM), a tube+wing transonic transport configuration designed for a cruise Mach number of M = 0.85 with a nominal lift coefficient C_l = 0.50 and a reference chord Reynolds number of Re = 40 M [18]. This geometry has become a standard for computational model validation and experimental activities, being used as part of the AIAA Drag Prediction Workshop (DPW) 4 to 6, with a huge amount of data having been collected [19–21]. Within the framework of the IVANHOE project, the previous flight conditions were set as the Aerodynamic Design Point (ADP) for the synthesis of a podded UHBPR turbofan designed to meet thrust requirements at cruise and take-off while minimising the TSFC. The engine thermodynamic model was carried out using the commercial GasTurb software [22] considering a current technology level. The resulting engine layout featured a three-spool geared architecture with ADP values of BPR > 15, fan pressure ratio (FPR) < 1.4, and Overall Pressure Ratio (OPR) = 50.

The gas turbine was installed under the wing in a short and slim nacelle designed by incorporating indications reported in the open literature with sensitivity studies on the effect of the main geometric parameters on drag and thrust [23]. The geometry has been described in other papers by these authors regarding engine installation studies; its main features are reported here for clarity. The scarfed and drooped non axisymmetric nacelle had a nominal mass flow capture ratio around 0.7 and a cowl length to highlight diameter of 1.6. The underwing installation position has a major effect on the interference drag and the thrust losses [16]. The nacelle trailing edge was set to have a wing chord-normalised offset from the wing leading edge of -0.05 in the horizontal direction and -0.11 in the vertical direction (with the minus sign indicating upstream and down from the wing leading edge, respectively) to mimic a close-coupled installation. Overall, the whole airplane should be representative of a long-range transport vehicle with a UHBPR turbofan in the 2030 timeframe.

2.2. CFD Modelling

The previously described test case was simulated by solving steady-state Reynolds-Averaged Navier–Stokes (RANS) equations using a commercial finite volume solver (Ansys Fluent [24]) with second order discretisation and $\kappa - \omega$ SST turbulence closure. The computational method was developed in the IVANHOE project using data from the DPW. Grid resolution for the Wing Body Nacelle Pylon (WBNP) configuration of the NASA CRM was first specified in accordance with workshop guidelines [25,26] and assessed for the current geometry with a powered-on nacelle. A detailed description of the numerical model validation is reported in [16,27]. Here, a summary of the mesh sensitivity analysis at the ADP is reported for completeness in Table 1. Five mesh levels with decreasing grid factor $N^{-2/3}$ were generated using an unstructured topology with tetrahedral elements in the external flow region, quadrilateral dominant surface mesh, and a wall-normal grown boundary layer having a slightly decreasing wall $y^+ \approx 1$ starting from the coarsest level. Figure 1 reports three views of a sample mesh, illustrating the multiple refinements block built to better resolve the flow past the nacelle, especially in the jet and pre-entry region (Figure 1a), a closer detail of the area around the nacelle with the boundary layer mesh connected to the tetrahedral zone (Figure 1b), and the surface grid (Figure 1c).

Refinement from that level was achieved by reducing the surface grid size and the expansion ratio according to the specified grid refinement ratio. The table shows a convergent behaviour of the drag coefficient at constant lift. The net force acting on the airplane, NVT (Equation (7)), has a non-monotonic convergence due to a lower value for the fine mesh, although from the medium level onwards it shows stabilization with small amplitude variation. Thus, the medium mesh was employed in the following analyses to avoid the excessive computational cost of finer discretisations without penalising the accuracy.

The representation of powered-on conditions in the numerical simulations was obtained by specifying boundary conditions found from thermodynamic cycle analysis at the engine boundaries in the computational domain. The fan face was modelled as a pressure outlet, with the floating static pressure adjusted to match the inlet mass flow rate. The fan and core outlet sections were modelled as a pressure inlet condition, with the stagnation temperature and pressure specified. The balance of mass fluxes through these boundaries was enforced within the calculation procedure by matching the inlet mass flow to the nozzle discharged mass flow.

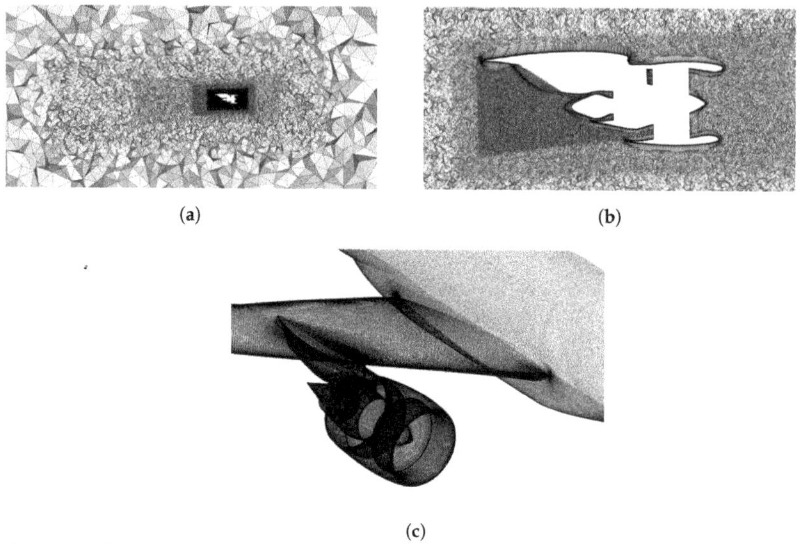

Figure 1. Different views of a sample mesh. (a) Multiple refinement blocks to better resolve the flow around the powered-on nacelle. (b) Detail of the mesh around the nacelle with the wall-normal grown boundary layer. (c) Surface mesh on the model walls.

Table 1. Grid convergence analysis at ADP.

Mesh Level	XCoarse	Coarse	Medium	Fine	XFine
No. of cells [$\times 10^6$]	27.0	33.8	39.1	55.2	74.8
Grid Factor [$\times 10^5$]	1.1111	0.9566	0.8681	0.6898	0.5633
Target Wall Y^+	1.15	1.09	1.03	0.98	0.93
C_l	0.4999	0.4999	0.4999	0.4998	0.4999
$C_d - C_{d,XFine}$	0.0006	0.0005	0.0000	0.0001	0.0000
NVT/NVT_{XFine}	0.6069	0.6759	1.0138	0.9655	1.0000

2.3. Thrust–Drag Bookkeeping

Propulsive forces acting on the vehicle were separated into thrust and drag contributions in order to analyse the propulsive effect, which is a typical approach [28]. The methodology is consistent with previously published papers by these authors. The definition of force components acting on different parts is provided in Figure 2. By summing the drag (ϕ) and thrust (θ) terms, different force metrics can be formed. The following quantities are considered, all expressed as the dot product of the force vector along a chosen direction $\hat{a}$ parallel to the undisturbed velocity:

$$\text{GPT} = FG_{17} + FG_7 + \theta_{fn} + \theta_{cn} + \theta_{cc} + \theta_{plug} + \theta_{pylon} \tag{1}$$

$$F_n = \text{GPT} + FG_0 \tag{2}$$

$$\begin{aligned}\text{NPT} &= FG_{17} + FG_7 - FG_2 + \theta_{fn} + \theta_{cn} + \theta_{cc} + \theta_{plug} + \theta_{int} + \theta_{sp} + \phi_{nac} + \phi_{pylon} = \\ &= F_n + D_{nac}\end{aligned} \tag{3}$$

$$\text{INT} = FG_{17} + FG_7 - FG_2 \tag{4}$$

$$D_{nac} = \phi_{pre} + \phi_{nac} \quad (5)$$

$$D_{tot} = D_{nac} + D_{a/f} \quad (6)$$

$$\text{NVT} = F_n + D_{tot} = \text{NPT} + D_{a/f} \quad (7)$$

where $FG = \dot{m} + (p - p_0)A$ is the gauge stream force. The Gross Propulsive Thrust (GPT) (Equation (1)) represents the gross thrust generated by the engine, accounting for the forces acting on the exhaust surfaces. The net thrust F_n (Equation (2)) is obtained by summing the ram drag FG_0. The Net Propulsive Thrust (NPT) (Equation (3)) amounts to the net force acting on the nacelle delivered as propelling force to the vehicle, which is equivalent to F_n diminished by the absolute nacelle drag D_{nac} (Equation (5)). The sum of the gauge stream forces on the engine boundaries is called the Intrinsic Net Thrust (INT) (Equation (4)), and derives from the specified thermodynamic status obtained from cycle analysis. The total drag D_{tot} (Equation (6)) is provided by the sum of the airframe drag $D_{a/f}$ on the wing and fuselage, with D_{nac} and $C_d = D_{tot}/0.5\rho_0 V_0^2 A_{ref}$. Finally, the Net Vehicle Thrust (NVT) (Equation (7)) is the net force acting on the vehicle along the direction $\hat{a}$ parallel to the flight velocity, provided by the sum of total drag and net propulsive thrust. When positive, it indicates a drag excess that decreases the vehicle speed over time, with the converse being the case for a negative thrust excess, accelerating the aircraft.

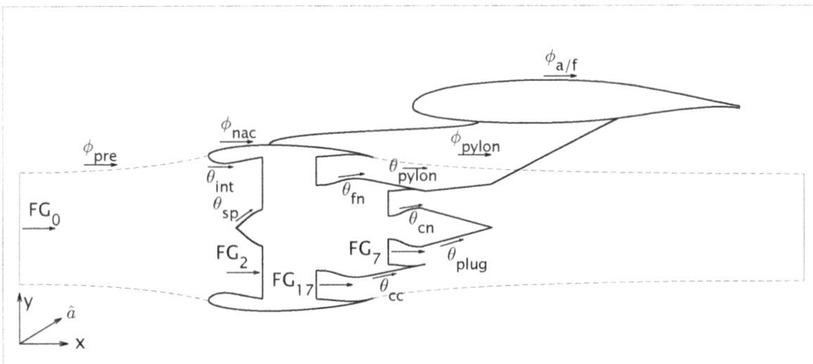

Figure 2. Forces acting on the nacelle and airframe for thrust–drag bookkeeping.

2.4. Mission Point Evaluation

Having established the forces acting on the vehicle, the evaluation of a generic mission point using a CFD model is now discussed. The motion of an aircraft is described by the equation

$$\frac{d(W/g\ V)}{dt} = T + D - W\,sin(\gamma) \quad (8)$$

where V is the flight velocity, is W the aircraft weight, g is the gravitational acceleration, T and D are the thrust and drag, respectively, and γ is the glide angle from the horizontal direction. Assuming a constant mass M over the time period Δt, the equation reduces to

$$\frac{\Delta V}{\Delta t} = \frac{1}{W/g}(T + D - W\,sin(\gamma)) \quad (9)$$

In steady-state CFD simulations, flight conditions are supposed to be constant, which would require a zero net force on the right-hand side for dynamic translational equilibrium to be achieved; otherwise, the calculation of the propulsive forces is biased by the presence of a thrust excess or defect, which in practice results in a deviation from the prescribed boundary conditions. In [29], a method of obtaining a dynamic equilibrium in a steady CFD simulation was presented, and was employed in [16] to compare the present reference nacelle with an optimised version. Here, the procedure is extended to the integration of the aircraft mission over the cruise segment for fuel consumption evaluation. This method, however, is generic, and can be applied to other mission legs as well. Here, it is reviewed and illustrated in the flowchart shown in Figure 3. As anticipated in the introduction, the rationale here is to couple a 0D thermodynamic cycle model used to design and represent the engine with high-fidelity force data from installed CFD simulations in order to obtain the real working conditions of the engine under a prescribed thrust level. In fact, the installed thrust, here meaning the net vehicle thrust from Section 2.3, depends on the sum of engine net thrust and airframe and nacelle drag from the complex interaction of the large podded propulsor and the wing flow. As such, a realistic estimation is only available with an installed CFD simulation. Having computed the net force acting on the vehicle, the proposed approach is to apply engine throttling to reach the resultant prescribed force value, which in the case of steady level flight (where $\gamma = 0$) and assuming constancy of mass over a time interval would be null (NVT = 0).

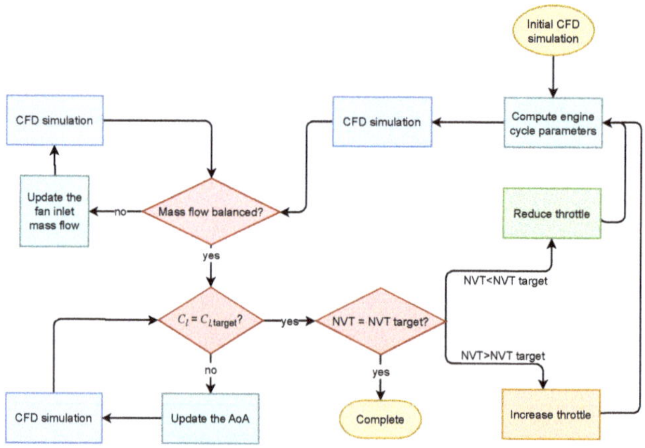

Figure 3. Flowchart of the methodology applied to compute a given flight point with specified lift and net translational force by combining CFD and thermodynamic cycle analysis.

The flowchart shows the iterative procedure by which energy input through fuel mass flow, i.e., engine throttling, is changed in the thermodynamic model. Thus, the CFD boundary conditions are iterated according to the new thermodynamic state until dynamic equilibrium is reached. In this way, control of the mass flow conservation from the fan inlet to the nozzle exhaust is explicitly set. Note that the assumed duct losses in the thermodynamic cycle are consistent with CFD values. Moreover, the angle of attack is updated to match the target C_l.

In this paper, the methodology was applied over four equal time segments starting from the ADP condition, where a constant speed was supposed to hold. Therefore, the aircraft lift was varied according to the fuel burnt in the flight segment, as follows:

$$\left(\frac{W}{g}\right)_{i+1} = \left(\frac{W}{g}\right)_i - \dot{m}_{f,i}\Delta t_i \tag{10}$$

$$C_{l,i+1} = C_{l,i} - (W_{i+1} - W_i)/\left(0.5\rho V_\infty^2\right) \tag{11}$$

where the fuel flow $\dot{m}_{f,i}$ is provided by the thermodynamic cycle model; the total burnt fuel is the sum of the segment consumed fuel, $\sum_{i=1}^{n} \dot{m}_{f,i}\Delta t_i$.

3. Results

3.1. Flow Characterisation

The first part of the analysis was devoted to characterising the flow field at cruise over a range of incidence $\alpha = \{2°, 2.23°, 3°, 4°\}$ in order to understand the transonic aerodynamics resulting from engine/wing interference and the corresponding variation of lift, drag, and thrust forces.

The phenomenology of the interaction between a podded engine and the wing has been described qualitatively and quantitatively in several works based on experimental measurements or CFD analyses [17]. With reference to UHBPR turbofans, the main effect of engine/airframe coupling has been shown to occur in the wing lower side due to jet/surface interaction and flow channelling between the nacelle, pylon, and wing [30]. However, the effect of angle of attack variation at highly subsonic cruise speed is less well documented. This regime has been investigated primarily for wing buffeting with no nacelle or a throughflow nacelle [31,32]. Here, while not capturing any unsteady phenomena, we focus on the steady flow features, in particular those arising nearby the powered-on engine.

The variation of the wing pressure coefficient at different span sections with the angle of attack is illustrated in Figure 4. Case $\alpha = 2.23^\circ$ corresponds to a wing $C_l = 0.50$ equal to the nominal cruise condition. On the suction surface, a normal shock wave is normally present at high subsonic cruise; it originates on the wing root and looses strength as the wing tip is approached. The effect of an increased angle of attack can be seen as a progressive shift in the wave location towards the trailing edge in the two sections surrounding the engine, along with an increase of the pressure jump, particularly in the outboard section. Farther from the nacelle, midway between it and the wing tip (see Figure 4c), the pressure remains flat past the leading edge peak. The effect of the incidence increase is an anticipation of the pressure recovery location, which happens more smoothly at $\alpha = 4^\circ$.

On the lower side, the general behaviour is an enlarged pressure recovery; there is a noticeable difference in the inboard section close to the engine, where a strong shock can be appreciated in Figure 4a. This is the result of nacelle–wing interaction producing a flow discontinuity that detaches from the pylon. Increasing the incidence attenuates the phenomenon by decreasing the diffusion in the pressure side, up to a point where a mild shock wave will be triggered, and eventually no shock wave at all.

A better analysis of the shock pattern near the engine region can be carried out with the help of the numerical Schlieren technique, showing the magnitude of the density gradient in three planes normal to the wing span (straddling the symmetry plane of the nacelle), as reported in Figure 5. On the inboard side, the first row of the figure highlights the position of the shock waves on the wing suction surface as well as in the channel between the wing lower surface and the fan stream boundary. In this area, a strong normal shock that penetrates into the jet is formed. As the incidence is increased, it moves slightly upstream and loses strength; the change in the nacelle flow field with the incidence is less evident in the other planes. The wing shock wave in the upper wall is displaced downstream, as already observed, whereas on the nacelle cowl the discontinuity in the forebody flow remains almost unchanged.

The flow field past the nacelle cowl is analysed in Figure 6 in terms of the pressure coefficient distribution. On both sides, the strong suction peak is terminated by the already observed shock wave. However, on the inboard side the shock position and the minimum pressure do not change, whereas the post-shock flow exhibits a reducing decompression as the incidence grows. On the contrary, the outboard flow has an increasing suction peak and a stronger shock, around $x/c = 25\%$ at $\alpha = 2^\circ$ and shifting downstream at larger incidences. Overall, the nacelle pressure distribution results are considerably less sensitive to the incidence change in the simulated regime as compared to the wing flow shown in Figure 4.

The inboard side remains more critical because of the appearance of the shock wave under the wing, which impinges upon the pylon. This can be clearly appreciated in Figure 7, showing C_p contours and its variation from the 4° case, i.e., $\Delta C_p = C_p - C_p^{\alpha=4^\circ}$. It can be noted that the shock structure is quite articulated, with a branch detaching from the core cowl trailing edge and the foot well inside the bypass stream. The pressure distribution on the core cowl itself is discontinuous because of the shock/expansion sequence in the under-expanded choked fan nozzle, thereby impacting the pressure force exerted there.

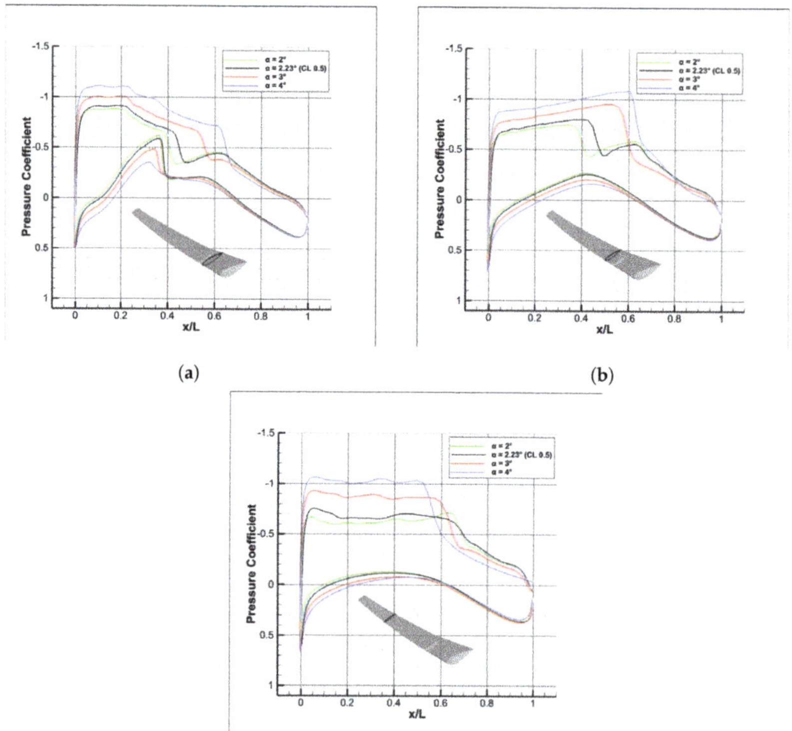

(a) (b) (c)

Figure 4. Wing pressure coefficient distribution at different span locations: (**a**) engine inboard., (**b**) engine outboard, and (**c**) wing outboard.

The relation between the angle of attack and the propulsive forces is summarised in Figure 8. The first chart in Figure 8a reports the gross thrust GPT, net thrust NPT, and net force acting on the whole vehicle NVT, all of which are normalised by a reference net thrust at the ADP. It is interesting to note how both the engine gross thrust and the net propulsive thrust delivered to the wing body increase with the incidence. This effect is caused by the installation, and is owing to the increased backpressure on the cowl exhaust and afterbody under the wing. The pressure growth on the nacelle, core cowl, and plug can be appreciated in Figures 6a and 7. In the latter, $\Delta C_p = C_p - C_p^{\alpha=4^\circ}$ is negative towards the trailing edge, indicating a lower pressure as the angle of attack decreases. Such behaviour must be contrasted with the drag trend reported in Figure 8b, indicating that while D_{nac} exhibits a similar tendency, the airframe and the total drag grow with α as expected. In fact, NVT, being the net balance between the D_{tot} and NPT, changes from a slightly negative value (thrust excess) at $\alpha = 2^\circ$ to a positive one (drag excess).

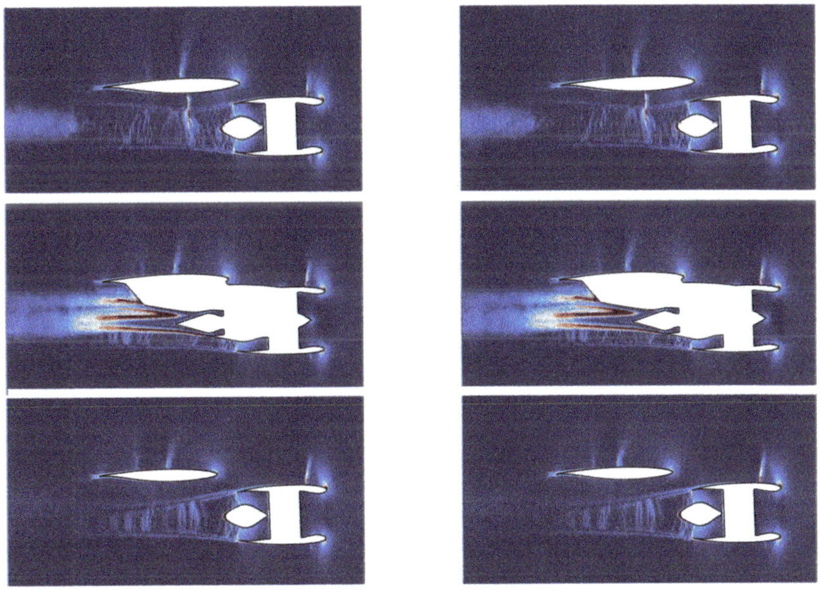

(a) (b)

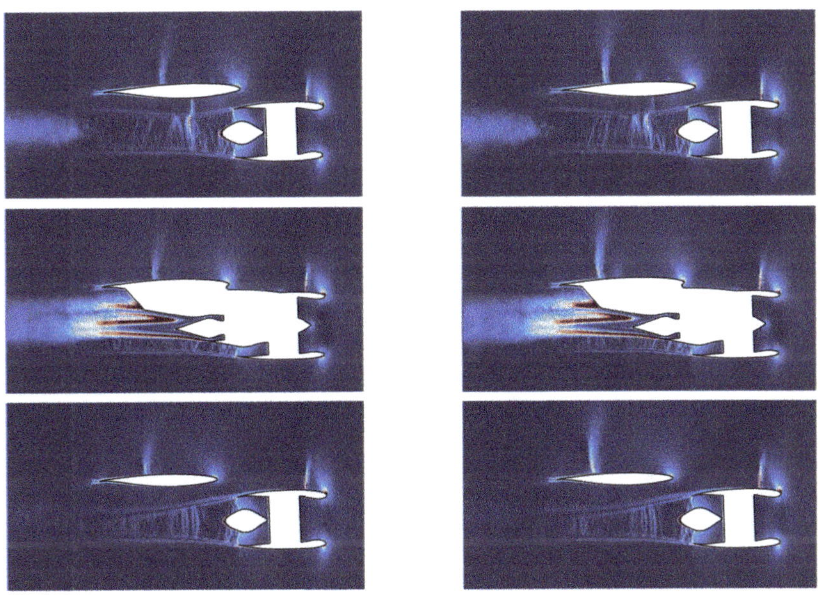

(c) (d)

Figure 5. Shock wave visualisation at different spanwise planes near the engine at increasing angle of attack by means of density gradient contours. From top to bottom: inboard, nacelle symmetry plane, outboard. (**a**) $\alpha = 2°$; (**b**) $\alpha = 2.23°$; (**c**) $\alpha = 3°$; (**d**) $\alpha = 4°$.

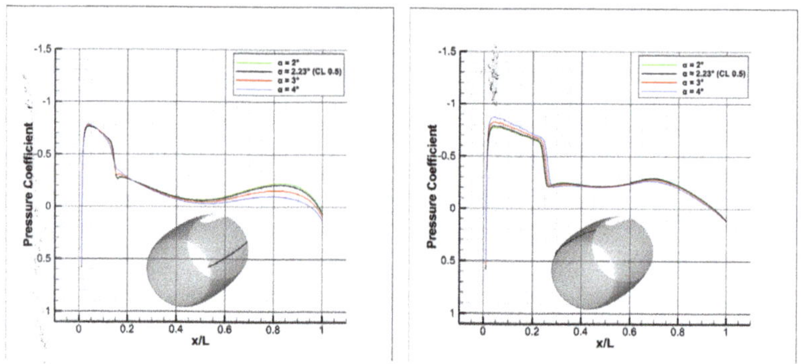

(a) (b)

Figure 6. Pressure coefficient distribution on inboard and outboard nacelle external cowl at increasing incidence: (**a**) inboard and (**b**) outboard.

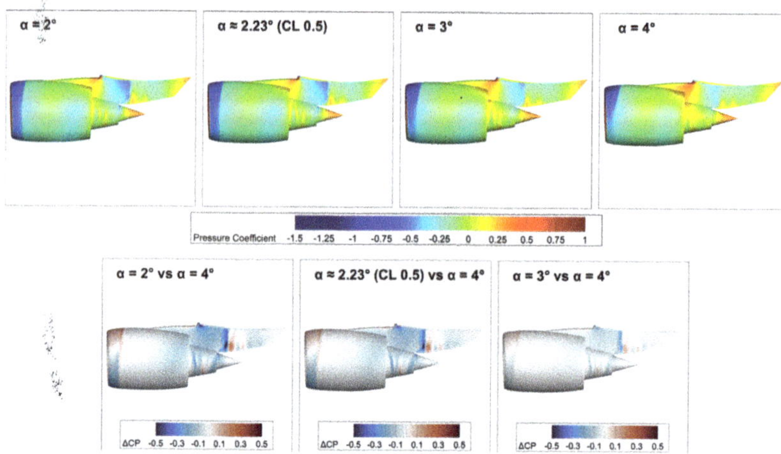

Figure 7. Pressure coefficient distribution on the nacelle and pylon on the inboard side.

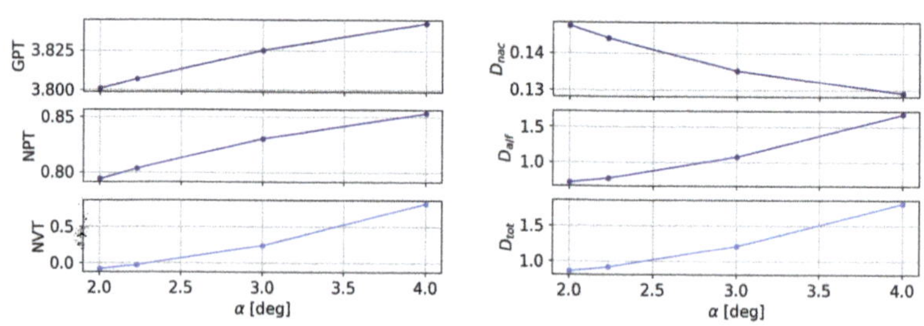

(a) (b)

Figure 8. Variation of thrust and drag forces with the angle of attack. All data are normalised by a reference net thrust. (**a**) Thrust; (**b**) drag.

3.2. Integration of the Mission Profile along the Cruise

The flow field analysed in the previous section presents several interaction phenomena that trigger a complex pattern of shock waves in the exhaust region, which is in addition to those normally found at transonic conditions past a suction peak on the leading edge. The shear stress and pressure distribution arising here have a direct impact on the propulsive forces, generating a penalty on the thrust delivered by the propulsor as well as in terms of the nacelle and airframe drag as a consequence of non-linear interference effects. Starting from this consideration, the procedure presented in Section 2.4 is applied here to integrate the mission profile along a cruise segment coincident with the ADP from the start of cruise condition. Because sub-iterations are needed at each step of the methodology in order to calculate the streamwise dynamic balance of the net vehicle force, to reduce the overall computational time we limited the integration to four Operating Points (OP) along the cruise equispaced by 60 min of flight.

The results of the flight segment analysis are summarised in Table 2. For each point, the table indicates the angle of attack corresponding to the target lift coefficient, drag coefficient variation ΔC_d, and several thrust definitions. Note that the thrust quantities have all been normalised by the net thrust F_n value at point 1.

Table 2. Summary of evaluated mission points along the cruise segment. Thrust quantities are normalised by F_n at OP1.

Quantity	OP1	OP2	OP3	OP4
α [deg]	2.260	2.146	2.025	1.962
C_l	0.5008	0.4876	0.4742	0.4627
$C_d - C_d^{OP1}$ [dc]	0.0	−35.26	−41.48	−47.49
F_n	1.0000	0.9711	0.9467	0.9248
GPT	3.8947	3.8498	3.8121	3.7783
NPF	0.8625	0.8301	0.80258	0.7773
NVF [$\times 10^{-3}$]	−0.0851	0.9432	−0.6344	−0.576
m_f [kg]	3224.05	3131.16	3052.18	2986.68

The variation of the reported parameters can be better appreciated in Figure 9. Figure 9a illustrates the two contributions to the aircraft drag from the nacelle (D_{nac}) and the airframe ($D_{a/f}$). With the cruise speed kept constant, $D_{nac} = D_{nac}(MFCR, \alpha)$, $MFCR = \dot{m}_0/(\rho_0 V_0 A_{hl})$ is the ratio between the far upstream capture area and the highlighted area. Because the inlet mass flow $\dot{m}_0$ is continuously decreased, the spillage drag grows because the pre-entry component is not completely balanced by the lip thrust generated by the leading edge suction. Moreover, as illustrated in Figure 8b, D_{nac} tends to slightly increase as α lowers. Conversely, the airframe drag is reduced almost linearly with α considering the small involved incidence change. The overall variation of the drag coefficient reaches −47.5 drag counts (1 dc = 0.0001).

Looking instead at the thrust components, their trend is uniform and monotonic; lowering the engine throttle diminishes all thrust terms. The overall variation of net thrust is −7.5%, as can be seen from Table 2, while the net propulsive thrust delivered by the engine, which additionally accounts for the external cowl force, is reduced by 8.5% relative to the reference net thrust at OP1.

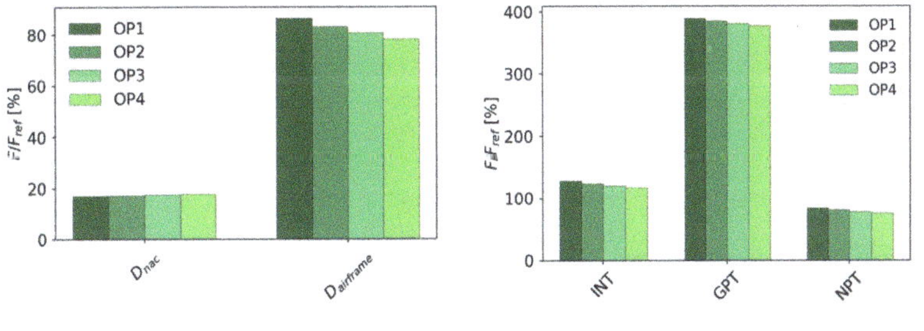

(**a**) (**b**)

Figure 9. Variation of propulsive forces along the integrated cruise segment. All data are normalised by the net force at OP1. (**a**) Drag; (**b**) thrust.

Figure 10 illustrates the absolute variation of force components acting on the nacelle surfaces relative to the OP1 value and normalised with respect to F_n^{OP1}. The nacelle cowl force ϕ_{nac} is negative because of the lip suction, and is further reduced as the inlet mass flow dims, adding more thrust. However, as already noted, this term is balanced by the augmented pre-entry drag ϕ_{pre}. The intake force θ_{int} has a very limited change, at less than 0.1% of F_n^{OP1}. Exhaust wall forces exhibit a larger variability. While the internal fan nozzle force provides an almost 1% larger thrust contribution, on the core cowl the force tends towards higher drag. That region has been shown in Section 3.1 to be influenced by the shock/expansion pattern in the bypass stream and the interaction with the inboard shock wave attached to the wing and pylon. The core nozzle duct force θ_{cn} trend resembles that of θ_{fn}. Lastly, θ_{plug} is increased slightly, by 0.25% in total. This exposed surface is sensitive to the core flow and to the external pressure under the wing, which varies with the angle of attack.

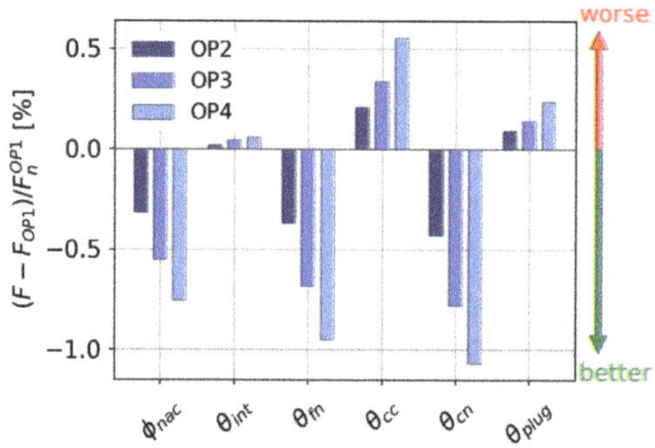

Figure 10. Absolute percentage variation from OP1 of forces acting on the nacelle walls, normalised to F_n^{OP1}.

As a final outcome of the procedure, the relation between the normalised net force F_n/F_n^{OP1} and the installed TSFC is depicted by the chart in Figure 11. With decreasing engine throttle and incidence, the TSFC from the start to the end of cruise is reduced by 0.04 [g/kNs]. According to the proposed algorithm, this value, computed using the thermodynamic cycle model, is derived by the throttle setting that guarantees the contemporary achievement of the target NVT, the target lift coefficient, and the mass flux balance through the engine boundaries. As such, it incorporates the installation effects, at the same time altering the drag, thrust, and lift; thus, it represents a more accurate estimate of the amount of fuel needed to sustain the examined flight condition.

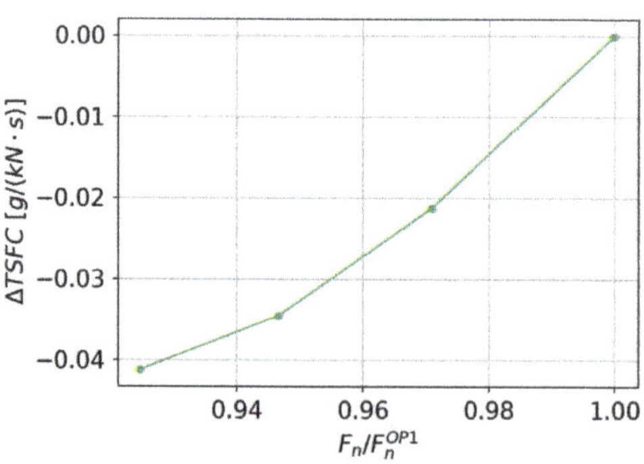

Figure 11. Relation between net force and TSFC change along the cruise segment.

4. Conclusions

The installed performance of a reference transport aircraft (the NASA CRM) with a powered UHBPR turbofan has been investigated at highly subsonic cruise, matching a thermodynamic cycle model with detailed CFD simulations. The initial assessment of small lift variations from the nominal flight highlighted the strong interactional effect caused by the engine/airframe coupling, resulting in flow displacement in the wing pressure side and shock discontinuities in the pylon inboard side. The sensitivity of surface pressure to the angle of attack was larger in the wing than in the nacelle, where the increased pressure in the cowl afterbody and exhaust surfaces generated a net force growth with α. However, this effect was balanced by the augmented airframe drag, leading to a shift in the net vehicle force towards an excess of drag.

The flow characterisation results have been used to explain the outcome of the simulated cruise segment, where each of the four operating points were computed according to the proposed procedure, achieving dynamic balance along the lift and drag directions. Because of the lower lift coefficient as fuel is burnt, airframe drag is reduced despite the limited growth of nacelle drag favoured by spillage. Thrust forces, on the contrary, were all reduced in absolute value, as the required propelling force followed the drag trend, with their sum assumed to be zero in the quasi-steady flight discretisation.

Overall, the proposed approach is able to compute the installed fuel consumption and associated fuel burn by modelling engine–airframe interaction through CFD in a more realistic way compared to semi-empirical models, in which the drag of single components is superimposed linearly. As such, it is expected to offer increased accuracy in capturing the installation penalties of integrated powerplants. Clearly, the increased fidelity comes at the cost of detailed CFD simulations; thus, the proposed method is suitable as a post-design verification or as an off-design assessment. In fact, it might be applied to integrate the flight trajectory in mission legs different from cruise as well, where fewer analytical models are available to estimate the installation penalties.

Author Contributions: Conceptualization, E.B., D.B. and A.M.; Methodology, D.B.; Software, D.B. and A.M.; Validation, D.B. and A.M.; Formal analysis, A.M. and D.B.; Investigation, F.P. and E.D.L.; Resources, E.B.; Data curation, A.M.; Writing—original draft preparation, A.M.; Writing—review and editing, E.B.; Visualization, A.M. and D.B.; Supervision, E.B.; Project administration, E.B.; Funding acquisition, E.B. All authors have read and agreed to the published version of the manuscript.

Funding: This research was conducted within the Clean Sky 2 project IVANHOE (InstalledadVAnced Nacelle uHbr Optimisation and Evaluation) funded by the European Union's Horizon2020 research and innovation programme under grant agreement number 863415.

Data Availability Statement: Not applicable.

Acknowledgments: The authors acknowledge University of Padova Strategic Research Infrastructure Grant 2017: "CAPRI: Calcolo ad Alte Prestazioni per la Ricerca e l'Innovazione" for providing computing infrastructure for the study.

Conflicts of Interest: The authors declare no conflict of interest.

Nomenclature

The following symbols and acronyms are used in this manuscript:

a/f	airframe
CFD	Computational Fluid Dynamics
C_d	Drag coefficient
C_l	Lift coefficient
C_p	Pressure coefficient
cc	core cowl

cn	core nozzle
FG	Gauge stream force
F_n	Net Force
fn	fan nozzle
g	Gravitational acceleration
GPT	Gross Propulsive Thrust
INT	Intrinsic Net Thrust
int	intake
$\dot{m}$	Mass flow rate
m_f	Fuel mass
M	Mach number
NPT	Net Propulsive Thrust
NVT	Net Vehicle Thrust
nac	nacelle
p	Pressure
OP	Operating Point
ϕ, θ	Surface aerodynamic force
pre	pre-entry
ρ	Density
RANS	Reynolds-Averaged Navier–Stokes
sp	spinner
TSFC	Thrust Specific Fuel Consumption
V	Velocity
W	Weight

References

1. NLR; SEO Amsterdam Economics. *Destination 2050—A Route to Net Zero European Aviation;* Report NLR-CR-2020-510; NLR—Royal Netherlands Aerospace Centre: Amsterdam, The Netherlands, 2021.
2. Epstein, A.H. Aeropropulsion for Commercial Aviation in the Twenty-First Century and Research Directions Needed. *AIAA J.* **2014**, *52*, 901–911. [CrossRef]
3. Borradaile, J. Towards the optimum ducted UHBR engine. In Proceedings of the 24th Joint Propulsion Conference, Boston, MA, USA, 11–13 July 1988. [CrossRef]
4. Zimbrick, R.A.; Colehour, J.L. Investigation of very high bypass ratio engines for subsonic transports. *J. Propuls. Power* **1990**, *6*, 490–496. [CrossRef]
5. Peacock, N.J.; Sadler, J.H. Advanced propulsion systems for large subsonic transports. *J. Propuls. Power* **1992**, *8*, 703–708. [CrossRef]
6. Hughes, C.E. NASA Collaborative Research on the Ultra High Bypass Engine Cycle and Potential Benefits for Noise, Performance, and Emissions. In Proceedings of the ISABE 2009, Indianapolis, IN, USA, 29 September–1 October 2009.
7. Daggett, D.L.; Brown, S.T.; Kawai, R.T. *Ultra-Efficient Engine Diameter Study;* Technical Report NASA/CR-2003-212309; NASA: Washington, DC, USA, 2003.
8. Guynn, M.; Berton, J.; Fisher, K.; Haller, W.; Tong, M.; Thurman, D. Analysis of Turbofan Design Options for an Advanced Single-Aisle Transport Aircraft. In Proceedings of the 9th AIAA Aviation Technology, Integration, and Operations Conference (ATIO), Hilton Head, SC, USA, 21–23 September 2009. [CrossRef]
9. Berton, J.J.; Guynn, M.D. Multi-Objective Optimization of a Turbofan for an Advanced, Single-Aisle Transport. *J. Aircr.* **2011**, *48*, 1795–1805. [CrossRef]
10. Guynn, M.D.; Berton, J.J.; Fisher, K.L.; Haller, W.J.; Tong, M.T.; Thurman, D.R. *Refined Exploration of Turbofan Design Options for an Advanced Single-Aisle Transport;* Technical Report NASA/TM–2011-216883; NASA: Hampton, VA, USA, 2011.
11. Alexiou, A.; Aretakis, N.; Roumeliotis, I.; Kolias, I.; Mathioudakis, K. Performance modelling of an ultra-high bypass ratio geared turbofan. In Proceedings of the ISABE 2017, Manchester, UK, 3–8 September 2017.
12. Larsson, L.; Grönstedt, T.; Kyprianidis, K.G. Conceptual Design and Mission Analysis for a Geared Turbofan and an Open Rotor Configuration. In Proceedings of the ASME 2011 Turbo Expo: Turbine Technical Conference and Exposition, Vancouver, BC, USA, 6–10 June 2011. [CrossRef]
13. Thoma, E.M.; Grönstedt, T.; Zhao, X. Quantifying the Environmental Design Trades for a State-of-the-Art Turbofan Engine. *Aerospace* **2020**, *7*, 148. [CrossRef]
14. ESDU. *Drag of Axisymmetric Cowls at Zero Incidence for Subsonic Mach Numbers;* Technical Report 81024; ESDU International: London, UK, 1981.
15. Bijewitz, J.; Seitz, A.; Hornung, M.; Luftfhart Bauhaus, E. Architectural Comparison of Advanced Ultra-High Bypass Ratio Turbofans for Medium to Long Range Application. In Proceedings of the Deutscher Luft- und Raumfhahrtkongress DLR, Augsburg, Germany, 16–18 September 2014; p. 12.

16. Magrini, A.; Buosi, D.; Benini, E. Analysis of installation aerodynamics and comparison of optimised configuration of an ultra-high bypass ratio turbofan nacelle. *Aerosp. Sci. Technol.* **2022**, *128*, 107756. [CrossRef]
17. Magrini, A.; Benini, E.; Yao, H.D.; Postma, J.; Sheaf, C. A review of installation effects of ultra-high bypass ratio engines. *Prog. Aerosp. Sci.* **2020**, *119*, 100680. [CrossRef]
18. Vassberg, J.; Dehaan, M.; Rivers, M.; Wahls, R. Development of a Common Research Model for Applied CFD Validation Studies. In Proceedings of the 26th AIAA Applied Aerodynamics Conference, Honolulu, HI, USA, 18–21 August 2008. [CrossRef]
19. Vassberg, J.; Tinoco, E.; Mani, M.; Rider, B.; Zickuhr, T.; Levy, D.; Brodersen, O.; Eisfeld, B.; Crippa, S.; Wahls, R.; et al. Summary of the Fourth AIAA CFD Drag Prediction Workshop. In Proceedings of the 28th AIAA Applied Aerodynamics Conference, Chicago, IL, USA, 28 June–1 July 2010. [CrossRef]
20. Levy, D.W.; Laflin, K.R.; Tinoco, E.N.; Vassberg, J.C.; Mani, M.; Rider, B.; Rumsey, C.L.; Wahls, R.A.; Morrison, J.H.; Brodersen, O.P.; et al. Summary of Data from the Fifth Computational Fluid Dynamics Drag Prediction Workshop. *J. Aircr.* **2014**, *51*, 1194–1213. [CrossRef]
21. Tinoco, E.N.; Brodersen, O.P.; Keye, S.; Laflin, K.R.; Feltrop, E.; Vassberg, J.C.; Mani, M.; Rider, B.; Wahls, R.A.; Morrison, J.H.; et al. Summary Data from the Sixth AIAA CFD Drag Prediction Workshop: CRM Cases. *J. Aircr.* **2018**, *55*, 1352–1379. [CrossRef]
22. Kurzke, J.; Halliwell, I. *Propulsion and Power*; Springer International Publishing: Cham, Switzerland, 2018. [CrossRef]
23. Magrini, A.; Buosi, D.; Benini, E.; Sheaf, C. Ultra-high bypass nacelle geometry design space exploration. In Proceedings of the AIAA Scitech 2021 Forum, Virtual, 11–15 & 19–21 January 2021. [CrossRef]
24. ANSYS. *ANSYS Fluent 19.2 Theory Guide*; ANSYS, Inc.: Canonsburg, PA, USA, 2018.
25. Mavriplis, D.J.; Vassberg, J.C.; Tinoco, E.N.; Mani, M.; Brodersen, O.P.; Eisfeld, B.; Wahls, R.A.; Morrison, J.H.; Zickuhr, T.; Levy, D.; et al. Grid Quality and Resolution Issues from the Drag Prediction Workshop Series. *J. Aircr.* **2009**, *46*, 935–950. [CrossRef]
26. Vassberg, J. A Unified Baseline Grid about the Common Research Model Wing/Body for the Fifth AIAA CFD Drag Prediction Workshop (Invited). In Proceedings of the 29th AIAA Applied Aerodynamics Conference, Honolulu, HI, USA, 27–30 June 2011. [CrossRef]
27. Magrini, A.; Buosi, D.; Benini, E. Maximisation of installed net resulting force through multi-level optimisation of an ultra-high bypass ratio engine nacelle. *Aerosp. Sci. Technol.* **2021**, *119*, 107169. [CrossRef]
28. Ministry-Industry Drag Analysis Panel (MIDAP) Study Group. *Guide to In-Flight Thrust Measurement of Turbojets and Fan Engines*; AGARD: Neuilly-sur-Seine, France, 1979.
29. Buosi, D.; Magrini, A.; Benini, E. Installed performance of ultra-high bypass turbofans: Estimation of power saving in optimised configurations at steady flight. In Proceedings of the AIAA Scitech 2022 Forum, San Diego, CA, USA, 3–7 January 2022. [CrossRef]
30. Magrini, A.; Benini, E. Aeropropulsive assessment of engine installation at cruise for UHBPR turbofan with body force fan modelling. *Aerosp. Sci. Technol.* **2023**, *132*, 108048. [CrossRef]
31. Balakrishna, S.; Acheson, M. Analysis of NASA Common Research Model Dynamic Data. In Proceedings of the 49th AIAA Aerospace Sciences Meeting including the New Horizons Forum and Aerospace Exposition, Orlando, FL, USA, 4–7 January 2011. [CrossRef]
32. Sebastian, S.; Ralf, R. Experimental assessment of wing lower surface buffet effects induced by the installation of a UHBR nacelle. *CEAS Aeronaut. J.* **2022**. [CrossRef]

energies

MDPI

Article

Simulation of Particle Trajectories in Gas Turbine Components and Assessment of Unsteady Effects Using an Efficient Eulerian-Lagrangian Technique

Stefano Oliani [1,*], Nicola Casari [1], Michele Pinelli [1] and Mauro Carnevale [2]

[1] Department of Engineering, University of Ferrara, Via Saragat 1, 44122 Ferrara, Italy; michele.pinelli@unife.it (M.P.)
[2] Department of Mechanical Engineering, University of Bath, Claverton Down, Bath BA2 7AY, UK
* Correspondence: stefano.oliani@unife.it

Abstract: In recent years, CFD has proven to be a very useful asset to help with predicting complex flows in a wide range of situations, including multiphase and gas-particle flows. On this track, numerical modelling of particle-laden flows in multistage turbomachinery has become an important step in helping to analyse the behaviour of a discrete phase in gas turbines. Furthermore, unsteady effects due, for example, to rotor–stator interaction may have an effect on trajectories and capture efficiencies of the discrete phase. Unfortunately, computational times for transient simulations can be exceedingly high, especially if a discrete-phase needs also to be simulated. For this reason, this work reports a new method for the efficient and accurate simulation of particle-laden flows in gas turbine engines components. The Harmonic Balance Method is exploited to gain orders of magnitude speedup exploiting the idea that once the flow field has been embedded in the spectral basis, it can be reconstructed at any desired time. In this way, not only can the computational time needed to reach convergence of the flow field be dramatically reduced, but there is also no need to keep simulating the flow field during particle tracking. On the contrary, the continuous phase field can be retrieved at any desired time through flow reconstruction. This technique is conceptually simple, but, to the authors' knowledge, has never been applied so far in particle-laden flow simulations and represents a novelty in the field. First, the implementation of the method is described, and details are given on how phase-lagged boundary conditions can be applied to flow and particles to further speed up the calculation. Then, some relevant case studies are presented to highlight the performance of the method.

Keywords: OpenFOAM; Lagrangian tracking; particle deposition; harmonic balance; turbomachinery; compressible flows

Citation: Oliani, S.; Casari, N.; Pinelli, M.; Carnevale, M. Simulation of Particle Trajectories in Gas Turbine Components and Assessment of Unsteady Effects Using an Efficient Eulerian-Lagrangian Technique. *Energies* **2023**, *16*, 2810. https://doi.org/10.3390/en16062810

Academic Editor: Jae Su Kwak

Received: 21 February 2023
Revised: 13 March 2023
Accepted: 15 March 2023
Published: 17 March 2023

1. Introduction

The requirements for efficiency and safe operation of gas turbines have become of paramount importance in modern society. One of the main causes of performance degradation in gas turbines is solid particle ingestion, causing erosion and deposition on aerodynamic surfaces. Land-based units can ingest soot as a consequence of operation in highly polluted industrial environments, or sand particles during operation in desertic places. Filtration systems tend to limit the particles entering the machinery, but they are unable to completely prevent their ingestion [1]. Aircraft engines can ingest particles at takeoff and landing during sandstorms, or encounter ash clouds transported by volcanic eruptions to cruise altitude. Dunn et al. [2] showed that it is possible to consume the surge margin very quickly when the engine operates in a dust cloud. Clearly, investigation of fouling and erosion in gas turbines has had an important relevance for academic researchers and engine manufacturers. Many numerical and experimental investigations about these phenomena can be found in the literature. The interested reader is referred to [3–5] for comprehensive reviews.

In recent years, CFD has proven to be a very useful asset to help with predicting complex flows in a wide range of situations, including multiphase and gas-particle flows. Consequently, numerical modelling of particle-laden flows in turbomachinery has become an important step in analyzing the behaviour of a discrete phase in gas turbines. On the basis of this premise, the main driver of this work was the necessity of efficiently simulating particle-laden flows in gas turbine components. For turbomachinery problems, Eulerian-Lagrangian methods are the most commonly employed due to their accuracy and simplicity. On the other hand, such method requires a large number of particles to be tracked, so that its major drawback is the computational cost. This could translate into a massive deployment of resources for complex turbomachinery domains. Several authors used different methods to model solid particles in gas turbine passages. Ghenaiet [6] and Hamed et al. [7] studied erosion in an axial flow turbine using a steady simulation with a frozen—rotor interface to couple the different stages. It must be said that this method did not include unsteady effects and, in addition, the erosion pattern and the flow depended on the clocking between the turbine rows. Tabakoff et al. [8] studied erosion in a turbine stage. They also employed steady-state simulations, but they introduced random particles circumferential redistribution at the stator/rotor interface. This allowed taking into consideration the uniforming effect of reciprocal motion on time-averaged particle distribution at stator outlet. In a similar fashion, Mustafa [9] investigated droplet trajectories during online washing of a multistage compressor with the commercial code CFX-TASCflow. Yang and Boulanger [10] simulated the full annulus of an axial fan and compared the erosion rates obtained for steady and transient simulations. Zagnoli et al. [11] and Prenter et al. [12] used ANSYS FLUENT to simulate particle-laden flow in an axial turbine stage with steady and unsteady simulations. A different methodology was adopted by Aldi et al. [13] for analyzing particle ingestion in an axial compressor stage. In this case, the analysis was carried out with separate particle injections for the isolated rotor and stator rows. The results of the isolated rotor and stator characterisations were post-processed to include the interaction between the rotor and the discrete phase while analyzing deposition on the stator. This technique avoids the explicit treatment of particles crossing stator/rotor interfaces but includes only a first-order approximation of the effect of the upstream blade row. Recent advances in the simulation of particle-laden flows in multistage turbomachinery were carried out by Oliani et al. [14] by developing a unified and simple treatment for each type of turbomachinery interface. The baseline methodology is the same employed in this work.

Following the previous studies, the aim of this work is to propose a general methodology for the simulation of particle-laden flows in gas turbine components. We highlight some important features regarding particle trajectories and impacts on aerodynamic surfaces. Moreover, we investigate the impact of the flow unsteadiness caused by rotor–stator interaction on particle trajectories. This topic has still not been addressed in detail in the literature. Unfortunately, computational times to carry out fully transient simulations can be exceedingly high. This is all the more so when a discrete-phase needs also to be simulated. The reason for this is that, even after the convergence, the flow field needs to be continuously updated in order to include the unsteady effects on particle trajectories, often leading to unacceptable computational requirements. For these reasons, we propose an efficient method to overcome this issue. Specifically, we employ the Harmonic Balance Method (HBM) to tackle the problem. This method was devised relatively recently by Hall [15] and is based on casting a set of equations in the frequency domain in order to switch the time derivative with a spectral operator. This technique has gained popularity over the years due to its efficiency compared to fully transient calculations for the continuous phase. To gain orders of magnitude speed up, the idea that once the flow field has been embedded in the spectral basis, it can be reconstructed at any desired time, is exploited. In this way, not only can the computational time needed to reach convergence of the flow field be dramatically reduced, but there is also no need to keep simulating the flow field during particle tracking. On the contrary, the continuous phase field can be retrieved at any desired time through flow reconstruction. This technique is rather straightforward but has never been applied so far in particle-laden flow simulations and represents a novelty in the field.

The paper is organised as follows: In Section 2, the governing equations for the continuous and the discrete phase are described. Then, we briefly review the HBM and its implementation for turbomachinery problems. In Section 3, we introduce the HBM-reconstruction method for the efficient coupling between HBM and Lagrangian tracking of the discrete phase. The method is validated on an ad hoc test case in Section 4. Finally, in Section 5, we investigate the discrete phase behaviour in two representative gas turbine components. Particle trajectories and impacts on aerodynamic surfaces are discussed, and unsteady effects are assessed using the proposed HBM-reconstruction method.

2. Numerical Modelling

2.1. Governing Equations for the Continuous and the Discrete Phase

A new implicit density-based solver was implemented in OF by the authors to carry out the calculations of the continuous phase [16]. The unsteady three-dimensional compressible Reynolds-averaged Navier–Stokes (RANS) equations are written in integral form:

$$\int_V \frac{\partial \boldsymbol{Q}}{\partial t} dV + \int_{\partial V} (\boldsymbol{F_c} - \boldsymbol{F_v}) dS = 0 \tag{1}$$

where V and ∂V are the control volume and the corresponding closed surface, respectively. The conservative variables $\boldsymbol{Q}$, the convective flux $\boldsymbol{F_c}$ and the diffusive flux $\boldsymbol{F_v}$ arrays are equal to

$$\boldsymbol{Q} = \begin{bmatrix} \rho \\ \rho \boldsymbol{u} \\ \rho E \end{bmatrix}, \boldsymbol{F_c} = \begin{bmatrix} \rho \boldsymbol{u} \cdot \boldsymbol{n} \\ (\rho \boldsymbol{u} \otimes \boldsymbol{u}) \cdot \boldsymbol{n} + p \boldsymbol{n} \\ \rho \boldsymbol{u} H \cdot \boldsymbol{n} \end{bmatrix}, \boldsymbol{F_v} = \begin{bmatrix} 0 \\ \boldsymbol{\tau} \cdot \boldsymbol{n} \\ (\boldsymbol{\tau} \cdot \boldsymbol{u} + \boldsymbol{q}) \cdot \boldsymbol{n} \end{bmatrix} \tag{2}$$

where $\boldsymbol{n}$ is the face normal vector, $\boldsymbol{u}$ is the velocity, ρ is the density, p is the static pressure, E is the total internal energy, H is the total enthalpy, $\boldsymbol{\tau}$ is the viscous stress tensor and $\boldsymbol{q}$ is the heat flux vector. To preserve the code structure and keep the freedom to choose among the many built-in OF turbulence models, turbulence equations are solved in a segregated manner. Finally, to relate the pressure and enthalpy to conservation variables, the system of equations is completed using the ideal gas law. Applying a finite-volume discretisation to Equation (1), a semi-discretised form is obtained:

$$V \boldsymbol{D_t} \boldsymbol{Q} = \boldsymbol{R}(\boldsymbol{Q}) \tag{3}$$

where $\boldsymbol{D_t}$ is the time derivative operator and $\boldsymbol{R}(\boldsymbol{Q})$ is the residual term. The system of equation is linearised and solved using the GMRES solver [17] combined with a LU-SGS preconditioner [18]. In OF, the built-in solvers and the underlying code structure are based on SIMPLE formulation. For this reason, the new ICSFoam library [16] has been implemented to carry out the implicit solution of the system of equations using a density-based formulation. Now, it is possible to assemble an arbitrary number of blocks and equations, thanks to the generalisation of the structure for block-coupled matrices. As pointed out in Section 2.2, this is necessary for the solution of a fully implicit version of the HBM. Approximate Riemann solvers are used to calculate the inviscid fluxes at cell interfaces. Finally, the support for Multiple Reference Frame (MRF) solution in rotating frames has been included into the solvers. Second order accuracy in space is obtained with the MUSCL approach. The Van Leer limiter is applied to primitive variables for the reconstruction of the solution from cell average values to the faces of control volumes. A classic second-order accurate central difference formula is used for viscous fluxes.

For what concerns the discrete phase, we solve the dynamic equation of motion for each particle in the domain:

$$\frac{dx_p}{dt} = u_p \tag{4}$$

$$m_p \frac{du_p}{dt} = f_I + f_{NI} \tag{5}$$

where f_I and f_{NI} are the inertial and non-inertial forces, respectively. Inertial forces can include different actions such as drag, gravity, Saffman lift and virtual mass. For turbomachinery applications in which a steady-state approach is adopted (use of MRF), f_{NI} includes centrifugal and Coriolis forces, defined as

$$f_{centr} = -m_p \boldsymbol{\Omega} \times \boldsymbol{\Omega} \times \boldsymbol{x}_p \tag{6}$$

$$f_{Cor} = -2m_p \boldsymbol{\Omega} \times \boldsymbol{u}_p \tag{7}$$

where $\boldsymbol{\Omega}$ is the shaft angular speed in rotating domains and zero elsewhere. Depending on whether the continuous phase field is updated or not after each time-step of Equations (4) and (5), an unsteady or a steady Eulerian-Lagrangian algorithm is obtained, respectively. To track particles across the domain, we also need to know the current cell of the mesh where they reside. In OF, each cell is decomposed into tetrahedrons and the tracking is performed in barycentric coordinates. This significantly improves the accuracy and the robustness of the algorithm. At each step, the current cell is updated using a tracking routine; see, e.g., [19].

The treatment for particles crossing mixing plane interfaces in steady-state simulations is carried out according to the technique described in [20]. Particles are redistributed in a random manner in the tangential direction to simulate the reciprocal motion of stator and rotor. The velocity vector is rotated according to the redistribution angle to maintain the radial and tangential components. Since mixing planes usually divide one frame of reference from another, particle velocity is finally updated correspondingly. In addition, to ensure mass conservation in a full-annulus sense, the particle transfer function has been modified by adding or deleting particles basing on the coverage in the tangential direction of the two rows.

2.2. The Harmonic Balance Method

A detailed description of the implementation of a fully implicit HBM in OF can be found in [21]. In general, spectral techniques are based on a frequency-domain formulation of periodic unsteadiness. The main idea is that a nonlinear solution with N_f harmonics is equivalent to $2N_f + 1$ coupled steady flow problems. Therefore, the HBM leverages the signal sparsity in the Fourier basis to substitute an unsteady solution with a coupled set of steady-state calculations. In this work, we follow the mathematical formulation proposed in the time-spectral approach of Hall et al. [15].

Let us suppose that the vector $\boldsymbol{Q}$ of conservative variables evolves with a known set of K frequencies ω_k in time. Now, we expand the vector $\boldsymbol{Q}_l$ (representing $\boldsymbol{Q}$ in the control volume l) and the residual term $\boldsymbol{R}(\boldsymbol{Q}_l)$ in Fourier series

$$\boldsymbol{Q}_l \approx \hat{\boldsymbol{Q}}_{l,0} + \sum_{k=-K}^{K} \hat{\boldsymbol{Q}}_{l,k} e^{-im\omega_k t}, \qquad \boldsymbol{R}(\boldsymbol{Q}_l) \approx \hat{\boldsymbol{R}}_{l,0} + \sum_{k=-K}^{K} \hat{\boldsymbol{R}}_{l,k} e^{-im\omega_k t} \tag{8}$$

where $\hat{\boldsymbol{Q}}_{l,0}, \hat{\boldsymbol{Q}}_{l,k}$ are the $N_T = 2K + 1$ Fourier coefficients of the time-average and the positive and negative frequencies terms, respectively. These coefficients are not known a priori. Substituting Equation (8) into Equation (1), we obtain, for the control volume l,

$$iV\boldsymbol{A}\widehat{\mathcal{Q}_l} = \widehat{\mathcal{R}_l} \tag{9}$$

where

$$\widehat{\mathcal{Q}_l} = \begin{Bmatrix} \hat{Q}_{l,0} \\ \hat{Q}_{l,1} \\ \vdots \\ \hat{Q}_{l,K} \\ \hat{Q}_{l,-K} \\ \vdots \\ \hat{Q}_{l,-1} \end{Bmatrix}, \widehat{\mathcal{R}_l} = \begin{Bmatrix} \hat{R}_{l,0} \\ \hat{R}_{l,1} \\ \vdots \\ \hat{R}_{l,K} \\ \hat{R}_{l,-K} \\ \vdots \\ \hat{R}_{l,-1} \end{Bmatrix}, \text{and} A = \begin{bmatrix} 0 & \cdots & \cdots & \cdots & \cdots & \cdots & 0 \\ \vdots & \omega_1 & & & & & \\ \vdots & & \ddots & & & & \\ \vdots & & & \omega_K & & & \\ \vdots & & & & -\omega_K & & \\ \vdots & & & & & \ddots & \\ 0 & & & & & & -\omega_1 \end{bmatrix} \tag{10}$$

Equation (9) is a set of equations coupled through the nonlinear term $\widehat{\mathcal{R}_l}$. As reported by Hall et al. [15], the system of equations is more easily solved by modelling these terms implicitly, transforming them back into the time domain. This allows for working with time-domain solutions stored at $2N_f + 1$ time levels, and the flux term can be computed as usual for each time level. The choice of the sampling points for the time levels is non trivial and influences the entries of the Fourier matrix $\boldsymbol{E}$ that relates the Fourier coefficients vectors $\widehat{\mathcal{Q}}$ and $\widehat{\mathcal{R}}$ with the time-domain counterparts:

$$\widehat{\mathcal{Q}_l} = \boldsymbol{E}\mathcal{Q}_l, \quad \text{and} \quad \widehat{\mathcal{R}_l} = \boldsymbol{E}\mathcal{R}_l \tag{11}$$

where $\mathcal{Q}$ contains the variables $\boldsymbol{Q}$ stored at N_T subtime levels.

If the spectrum is composed of a single fundamental frequency ω and its harmonics (periodic flow), the Fourier matrix is always well-conditioned. Indeed, in such a scenario, a uniform sampling of the base period $T = 2\pi/\omega$ leads to a condition number equal to 1 because of the orthogonality of the Fourier basis. On the contrary, if the frequency set is arbitrary, finding a proper set of time instants corresponding to a low condition number of the matrix $\boldsymbol{E}$ is way more difficult.

Indeed, in many applications of industrial interest, almost-periodic signals are considered, where multiple discrete dominant frequencies exist. These frequencies need not be integral multiples of each other, a typical example being multi-rows turbomachinery where the number of blades and vanes are not multiples. When this is the case, we can still project the flow variables onto a set of non-harmonically related frequencies. It must be emphasised that now the frequency set does not form, in general, an orthogonal basis anymore. Additionally, there is no analytical expression for the matrix $\boldsymbol{E}$. Instead, we define its inverse $\boldsymbol{E}^{-1}$ analytically by its components as $E^{-1}_{n,k} = e^{i\omega_k t_n}$, and then compute numerically $\boldsymbol{E}$ [22,23]. The correct selection of the time levels is of paramount importance for the convergence of the simulation. Indeed, the higher is the condition number of the matrix $\boldsymbol{E}$, the more errors arising from the HB term will be amplified during the iterative solution [22].

In the present paper, we follow the OptTP approach introduced by Nimmagadda et al. [24]. This method has negligible performance drop with respect to other methods, is robust, and is easier to implement. The basic idea is to compute the condition number for many possible periods and then uniformly sample the time period T^* that minimizes the condition number of matrix $\boldsymbol{E}$. Using Equation (11), and multiplying on the left Equation (9) by the inverse matrix $\boldsymbol{E}^{-1}$, we have

$$V\boldsymbol{D}\mathcal{Q}_l = \mathcal{R}_l \tag{12}$$

where $\boldsymbol{D} = \boldsymbol{E}^{-1} i A \boldsymbol{E}$. We can observe that Equation (12) is a system of steady-state equations. A comparison with Equation (3) shows that this procedure has allowed us to find an approximation of the time derivative operator $\boldsymbol{D}_t \boldsymbol{Q}_{l,j} \approx (\boldsymbol{D}\mathcal{Q}_l)_j$ in terms of the flow field snapshot j in each control volume l. The acceleration methods for steady-state simulations can be used (e.g., multigrid and local time-stepping), greatly reducing the computational cost of the solution.

Now, we must solve a set of equations coupling all the subtime levels through the HB term $V\boldsymbol{D}\mathcal{Q}$. We can directly discretise this term implicitly since the operator $\boldsymbol{D}$ is linear: $V\boldsymbol{D}\mathcal{Q}^{n+1} = V\boldsymbol{D}\mathcal{Q}^{n} + V\boldsymbol{D}\Delta\mathcal{Q}^{n}$. Including this linearisation into Equation (12), one obtains:

$$\boldsymbol{M}\Delta\mathcal{Q}^{n} = \mathcal{R}^{n} - V\boldsymbol{D}\mathcal{Q}^{n} \tag{13}$$

where $M_{ii} = \left(-\frac{\partial \boldsymbol{R}_i^n}{\partial \boldsymbol{Q}_i^n} + VD_{i,i}\right)$ and $M_ij = VD_{i,j}\boldsymbol{I}$. In the present work, the GMRES linear solver [17] with an LU-SGS preconditioner [18] are employed for the solution of the linearised system. Hence, the flow variables are updated for all the snapshots at the same time. The block-coupled solver structure described in [16] allows for a convenient assembly of the matrix $\boldsymbol{M}$ for the linear system.

2.3. Turbomachinery Boundary Conditions

In multi-stage turbomachinery, flow unsteadiness can be expressed as a linear combination of the blade passing frequencies of the neighbouring rows [25]. The most common method that can be found in the HBM literature to deal with multi-stage turbomachinery is to resolve different time instants for each blade row. In this way, of course, the snapshots solved in each row do not, in general, match each other. To solve this problem, the data between adjacent zones are exchanged through spectral interpolation of the flow quantities from the donor side to the time levels of the receiving side [26], or via matching of the spatial Fourier coefficients [27]. The drawback of these techniques is that, if the interpolation is not combined with non-reflecting conditions [27] or oversampling [23,26], they can generate spurious wave reflections which may in turn corrupt the solution. In addition, if Lagrangian tracking must be coupled with the HBM, it is much easier to deal with synchronised time instants. Therefore, in this work, we follow the synchronised HBM strategy proposed by Crespo and Contreras [28]. This method ensures the flow continuity between stator and rotor interfaces by using the same time instants in all the blade rows.

Another important aspect concerns the exploitation of spatial periodicity in the azimuthal direction, as often carried out in turbomachinery simulations. It must be said, however, that, in unsteady calculations, it is necessary to have the same pitch for all different rows. For this reason, often half or even the full annulus of the machine needs to be simulated. The usage of phase-shifted boundary conditions [29] is a way to resolve the problem by exploiting the known spatio temporal periodicity of the flow [30]. The method allows for reducing the computational domain to just one passage per row, independently of the blade number ratio. Indeed, there is a correspondence between the flow field inside a blade passage at a certain time t and the adjacent passage at a different time $t+\mathcal{T}$

$$\boldsymbol{Q}(r,\theta+\Delta\theta,z,t) = \boldsymbol{Q}(r,\theta,z,t+\mathcal{T}) \tag{14}$$

where $\mathcal{T}$ is called the time lag between the two passages, $\Delta\theta$ is the row pitch and θ is the tangential coordinate. Let us consider for simplicity a single stage, the results being generalizable to multiple stage. The time lag can be calculated as the phase of a wave traveling at a rotational speed $\omega_k = 2\pi k f_{BP}$ in the azimuthal direction: $\mathcal{T} = k\sigma/\omega_k$, where σ is the interblade phase angle (IBPA) and f_{BP} is the blade passing frequency of the opposite row. To enforce phase-lagged conditions, one needs to update the Fourier coefficients on the boundaries at each time-step. Since HBM is intrinsically related to the Fourier decomposition of the solution, phase-shifted boundary conditions are well-suited and much simpler to implement in frequency-domain solvers. Equivalently, Equation (14) can be expressed in the frequency domain as

$$\sum_{k=-K}^{K} \hat{\boldsymbol{Q}}_{\boldsymbol{k}}(r,\theta+\Delta\theta,z)e^{i\omega_k t} = \sum_{k=-K}^{K} \hat{\boldsymbol{Q}}_{\boldsymbol{k}}(r,\theta,z)e^{i\omega_k t}e^{i\omega_k \mathcal{T}} \tag{15}$$

With this in mind, we can reformulate the relation between the time instants on the two periodic boundaries as [23]

$$\mathcal{Q}(r,\theta+\Delta\theta,z) = \boldsymbol{E}^{-1}\boldsymbol{S}\boldsymbol{E}\mathcal{Q}(r,\theta,z) \tag{16}$$

where S is diagonal with components $S_{kk} = e^{ik\upsilon}$. The implementation of phase-lagged boundary conditions in the HBM solver is detailed in [21].

3. Coupling the HBM and Lagrangian Tracking

Previous studies showed that unsteady effects may have a more or less pronounced impact on the predictions of erosion and deposition in turbomachinery [10,12,31]. It is difficult to know a priori if the discrete phase will show an influence from rotor–stator interactions and to what extent for a particular case. Therefore, in general, one must know and take into account that steady simulations can experience a certain degree of error. The alternative is, of course, to perform unsteady Lagrangian tracking calculations, with the associated high computational cost. The term unsteady Lagrangian tracking denotes the calculations in which the continuous phase fields are updated after each Lagrangian step for the discrete phase. Therefore, after each particle p_i in the domain is advanced of Δx_{p_i} in space, the continuous phase fields are advanced from time t to $t + \Delta t$, where Δt is the physical time-step of the calculation. The entire process is then repeated for as long as needed to track all the particles throughout the domain.

To gain orders of magnitude speed up the simple idea that once the flow field has been embedded in the spectral basis, it can be reconstructed at any desired time is exploited. In this way, not only can the computational time needed to reach convergence for the flow field be dramatically reduced, which is the advantage of using HBM over transient calculations, but there is also no need to keep simulating the flow field during particle tracking. On the contrary, the continuous phase field can be retrieved at any desired time through flow reconstruction. Therefore, the continuous phase quantities needed to calculate the forces acting on solid particles can be directly retrieved, without actually computing the flow field. This new approach will be called from now on the HBM-reconstruction method. An implicit assumption of this method is that particles do not affect the fluid flow (one-way coupling), but their trajectory is influenced by the continuous phase through fluid-dynamic forces. This is a typical assumption in turbomachinery flows since usually the particle's volume fraction is very low (<<10%). This technique is rather straightforward but has never been applied so far in particle-laden flow simulations and represents a novelty in the field.

Some authors [10] have adopted a different approach by saving many timesteps of the converged flow field across one fundamental period of the current blade row. Then, after solid particles' injection, the snapshot corresponding to the actual instant is loaded from the memory, instead of computing it. In this way, particle tracking can proceed as usual using the saved timesteps. However, this approach is not efficient and has several drawbacks. Firstly, it requires a lot of memory because the entire flow field needs to be stored for all the desired instants of tracking (which are typically tens or one hundred for one passing period). For larger simulations, even modern computers can not keep so many timesteps in the RAM due to memory limitations. Therefore, a continuous process of loading and unloading of the saved timesteps from the disk storage is needed. This, though much cheaper than recomputing the flow field at each timestep, results anyway in a very cumbersome and time-consuming process. Indeed, even with modern SSD, the memory bandwidth is limited compared to RAM, resulting in a pronounced access latency during the process. Just to give an idea, the memory bandwidth of RAM is typically around 10 GB s^{-1}, depending on the computer, while SSD bandwidth is about 500 MB s^{-1}. Things are even worse when a HDD storage is used, for which load/write speed is about 100 MB s^{-1}. On the contrary, embedding the flow field with a ROM such as the HBM allows for keeping all the necessary data in the main memory and therefore there is no need to load data from the storage drive. Therefore, memory and computational efficiency are achieved. Of course, if the Fourier basis is chosen as the basis for the ROM, one could compress the data after performing a conventional transient calculation and reconstruct the field at any desired time. During the Lagrangian tracking step, this would be equivalent to the method presented above. Anyway, the HBM allows for obtaining accurate results in a reduced turnaround time also during the continuous phase calculation, and naturally embeds the results in the spectral basis without need for further post-processing. Hence, this strategy is preferable and has been used in this work.

Another major disadvantage of saving and loading the snapshots for the tracking is that the timestep of the calculation can not be modified. Indeed, since the snapshots of the flow field used for the tracking are saved and loaded from the storage, once the timestep has been selected, the only way to modify the snapshots is to recompute them. Instead, with the HBM, the flow field can be reconstructed at any time and one can therefore choose and modify the timestep used for Lagrangian tracking as preferred, with no restrictions. This is particularly useful if a suitable timestep for particle tracking can not be established a priori.

Of course, with the HBM being a reduced order model, accuracy losses can be expected if the number of harmonics retained in the spectrum is not sufficient. Indeed, errors arising from inaccurate reconstruction of the flow field can potentially accumulate along particle trajectory, leading to an increased computational error. A sensitivity analysis on the number of harmonics retained in the spectrum is carried out in Section 4 to further investigate this aspect.

To fully exploit the potential of the HBM, it is also desirable for the new method to allow particle tracking in reduced domains of a single passage per row. This can be achieved if the concept of phase-shift between periodic boundaries is extended to the discrete phase, and represents a further novelty in the methodology. The idea is based on the observation that, when a particle crosses a periodic boundary with a phase-shift boundary conditions imposed on it, another particle must reappear on the other side with the same velocity after a time lag:

$$\mathcal{T} = \sigma_k / \omega_k \quad \forall k \tag{17}$$

where σ_k is the IBPA associated with a travelling wave of rotational speed ω_k. Remember that, even if multiple fundamental frequencies are present in the same row, Equation (17) must hold for each of them. In other words, the time-lag is a constant and can be calculated a priori for each row.

Please notice that, physically, the particle reappearing on the other side of the cyclic can not be the same particle that exits from the first side. Despite this fact, for computational purposes, the same particle is used on both halves of the cyclic since this is perfectly equivalent and easier to implement. Another way to see this is considering that, if a particle exits from a cyclic at a time t and is injected from the other half at time $t + \mathcal{T}$, then this is equivalent to continuing to track the same particle in the adjacent passage of the row at time t. Indeed, the flow fields of two adjacent passages are exactly the same after the time lag $\mathcal{T}$. This allows for taking advantage of single passage reduction also during the Lagrangian tracking step, thus avoiding reconstructing the flow field in adjacent passages. The underlying hypothesis in doing this is that particles' injection inside the domain has the same periodicity of the flow. As a matter of fact, there is hardly any reason why this is not true in turbomachinery flows. In other words, this means that it is very unlikely that particles are injected with a temporal frequency different from the one of the flow field. In addition, particles are typically injected uniformly (in space and time) from the domain inlet, in which case this assumption is certainly true.

The implementation is as follows. For phase-lag cyclics, it is sufficient to transfer the particle from one side to the other and deactivate the particle for a time corresponding to the time lag. After a time $\mathcal{T}$ has passed, the particle is reactivated and the tracking proceeds as usual. It is also emphasised that, necessarily, the two halves of the phase-lag cyclic have IBPAs of the opposite sign (if half A is in advance with respect to half B, then half B is in delay of the same amount with respect to A). Although negative phase angles do not represent a problem, negative time lags are of no use for particle tracking. For this reason, it is always necessary to convert negative IBPAs into the corresponding positive time lags. For phase-lag AMI interfaces, things are slightly more complex. A strategy similar to the one described in [20] is used to find the correct position and cell on the other side of the interface. The particle is deactivated for a time $N_{Cp}\mathcal{T}$, where N_{Cp} is the number of the copy on the receiving side and $\mathcal{T}$ is the time lag of the receiving blade row. Once again, $\mathcal{T}$ is always positive, depending on whether N_{Cp} corresponds to a copy transformed in the forward or backward direction. The schematic procedure for particle transfer between phase-shifted interfaces is depicted in Figure 1.

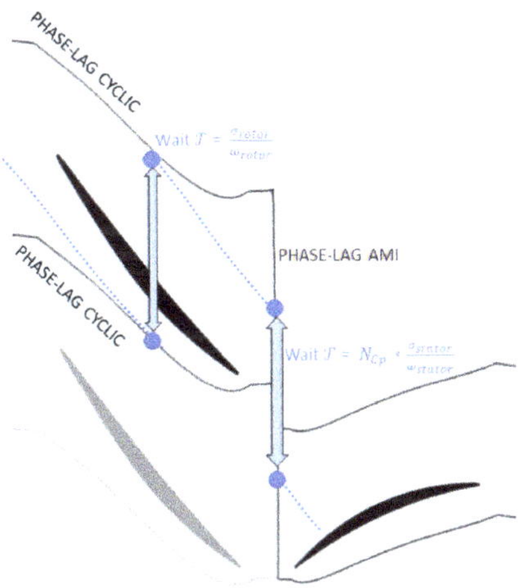

Figure 1. Schematic procedure for particle transfer between phase-shifted interfaces. Particle trajectory is represented with a dashed blue line.

4. Validation of the Method

The validation of the method is carried out on the rotor blade of the first stage of an E^3 axial turbine [32]. A fully transient simulation is compared with the results obtained with the HBM-reconstruction technique. The geometry is two-dimensional, and an artificial wake is generated and imposed at the domain inlet, according to:

$$p_0(y) = \bar{p}_0[1 - \Delta p \cdot e^{-0.693(2\frac{y}{L})^2}] \tag{18}$$

$$T(y) = \bar{T}[1 + \Delta T \cdot e^{-0.693(2\frac{y}{L})^2}] \tag{19}$$

where $\bar{p}_0$ is the average total pressure and $\bar{T}$ is the average temperature at domain inlet, respectively equal to 240 kPa and 600 K. The outlet total pressure is set equal to 144 kPa. Δp and ΔT are the total pressure and temperature variations equal to of 0.10969 and 0.015, respectively. L is the azimuthal wake width and has been selected as 20% of the blade pitch. The average values of total pressure and temperature, as well as the flow inlet angle, are chosen according to a preliminary steady-state simulation of the entire stage with a mixing plane interface.

The standard k-ε turbulence model with its original constants and parameters is employed for the RANS closure of the turbulent viscosity ν_t [33]:

$$\frac{\partial \rho k}{\partial t} + \nabla \cdot (\rho k \boldsymbol{u}) = \nabla \cdot (\rho \frac{\mu_t}{\sigma_k} \nabla k) + P - \rho \varepsilon \tag{20}$$

$$\frac{\partial \rho \varepsilon}{\partial t} + \nabla \cdot (\rho \varepsilon \boldsymbol{u}) = \nabla \cdot (\rho \frac{\mu_t}{\sigma_k} \nabla \varepsilon) + \frac{C_1 \varepsilon}{k} P - C_2 \rho \frac{\varepsilon^2}{k} \tag{21}$$

$$\nu_t = \frac{C_\mu k^2}{\varepsilon} \tag{22}$$

where P is the turbulent kinetic energy production rate. The model constants employed are set according to the standard values: $C_\mu = 0.09$, $C_1 = 1.44$, $C_2 = 1.92$, $\sigma_k = 1.00$, $\sigma_\varepsilon = 1.30$. The automatic wall treatment switches between the viscous and inertial sublayers values for the turbulent quantities by comparing the local y^+ with the estimated intersection of the viscous and inertial sublayers y^+_{lam}.

The mesh is composed of nearly 20,000 elements. During the simulation, the wake profile described by Equations (18) and (19) is slid in the azimuthal direction at the domain inlet, to recreate an unsteady flow field similar to the one seen by the blade row in its relative frame during a simulation of the entire stage. The wavelength of the wake is twice the pitch of the rotor, to mimic an upstream vane row with a count ratio of 1:2 with respect to the blade row.

The fully transient calculation is performed on a domain including two blade passages (to take into account the wave length of the incoming wake) using periodic boundaries, while HBM simulations are carried out on a single passage domain exploiting phase-lag periodic boundary conditions. Figure 2a shows the instantaneous temperature field in the blade passages for the transient simulation. HBM solutions were obtained considering 4, 5 and 6 harmonics of the wake passing frequency. The number of sample points is obtained with the usual relation $2N_f + 1$ and is respectively equal to: 9, 11, 13. Figure 2b shows instantaneous temperature profiles along the pitch in front of the rotor for the HBM solutions compared with the reference transient simulation. As can be seen, with six harmonics, the HBM results closely match the transient ones, while five harmonics are already sufficient to capture the shape of the profile well. Finally, a steady state calculation is also carried out using a mixed out state at the domain inlet with values of total pressure and temperature of $\bar{p}_0$ and $\bar{T}$.

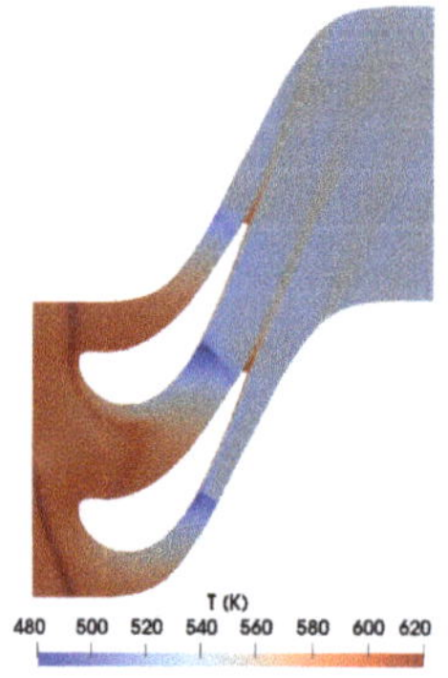

(a)

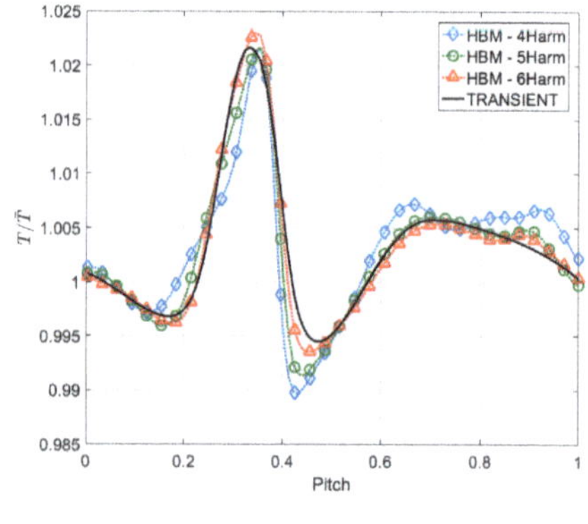

(b)

Figure 2. (**a**) Temperature field in the blade passages with unsteady wake imposed at the inlet; (**b**) temperature profiles in front of two blades obtained with the fully transient and HBM calculations. Reference pitch is twice the single blade pitch.

For the unsteady calculations, particles are injected at the inlet following a probabilistic distribution in which the probability is high inside the incoming wake and very low outside. Particles are injected at the equilibrium velocity with the flow. In this way, a strongly non uniform incoming condition is generated for the particles. Indeed, the aim here is not so much to reproduce a realistic unsteady behaviour in the rotor as to create a test case in which unsteady effects are willingly enhanced to highlight the difference with a steady-state calculation. During the Lagrangian tracking steps of the fully transient and the HBM solutions, a timestep of 1×10^{-6} is employed for the flow field, corresponding to a physical Courant number of 2. For the steady solution, instead, particles are injected uniformly from the domain inlet as if a mixing plane was placed between the stator and the rotor. In order to focus only on the correct implementation of the methodology, turbulent dispersion is not accounted for in the simulations of this case study. Particles are considered spherical and smooth, and the unique force acting on them is viscous drag. Three different diameters are injected to analyse the sensitivity of the results to particle size: 2, 8 and 16 μm. Assuming a density of 2700 kg m^{-3}, the Stokes number of the particles based on the turbine axial chord is: 0.125, 2.00 and 8.00, respectively. A perfectly elastic rebound is assumed when a particle impacts on the blade surface.

Figure 3 shows trajectories of 16 μm particles for transient and HBM-reconstruction methods at four equispaced instants in time. Transient results are shown on two blade passages while, of course, HBM results are available only on a single passage. The lower passage of the transient simulation is synchronised with the passage of the solution reconstructed using HBM. The qualitative comparison of the trajectories clearly shows that the new technique is able to reproduce the discrete phase movement across the domain. In particular, the non uniform injection and the variation in particles concentration due to wake stretching and distortion inside the passage are captured properly. Particles distribution is also very similar after they impact on the blade and exit the domain, meaning that impact angle and velocity are correctly reproduced.

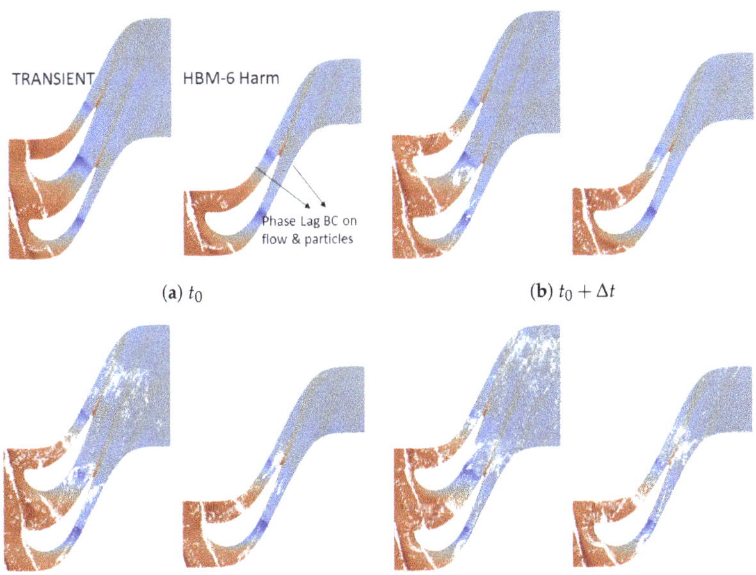

(a) t_0 (b) $t_0 + \Delta t$ (c) $t_0 + 2\Delta t$ (d) $t_0 + 3\Delta t$

Figure 3. Comparison of 16 μm particle trajectories for transient simulation and HBM-reconstruction method at four equispaced instants.

Beside the instantaneous behaviour of the discrete phase, one is typically interested in obtaining reliable prediction of global parameters or time-averaged values quantifying impacts on the blade. Figure 4 reports the impact efficiency along the blade chord, defined as the ratio between the number of particles impacting on the blade and the total number of particles injected. For particles with Stokes 0.125 and 2.00, only the pressure side of the blade is shown, since almost no impacts were found on the suction side. Indeed, as also reported by Jacobs et al. [34], shock–particle interaction may create particle velocity components perpendicular to the main flow, producing transport perpendicular to the flow streamlines. This effect, combined with the curvature of the fluid flow streamlines is responsible for the very small number of impacts found on the suction side. Bigger particles instead tend to be projected towards the blade suction side after they rebound on the pressure side of the blade below (see Figure 3). A very good agreement is found for all diameters between HBM-reconstruction with six harmonics and transient cases (Figure 4a,c,e). Conversely, steady-state solutions are significantly different, showing that an unsteady treatment is necessary in this test case. A sensitivity analysis with respect to the number of harmonics is also carried out. A comparison of Figure 4b,d,f reveals that smaller particles are the most sensitive to the number of harmonics used for the simulation, being more influenced by the surrounding flow field due to the lower inertia. For these particles, six harmonics need to be retained in the spectrum to provide a good resolution of the impact efficiency. For $St = 2.00$ and $St = 8.00$, on the other hand, five harmonics are sufficient to obtain a good accuracy. From these observations, it can be inferred that, if the number of harmonics is adequate for the correct resolution of the flow field, a good approximation of the impact of unsteady effects on the discrete phase will also be obtained.

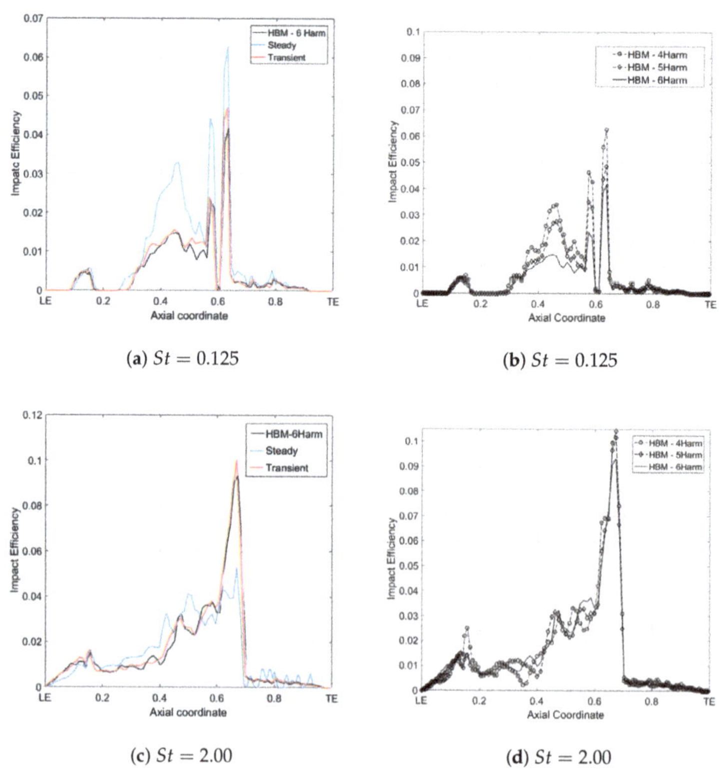

(**a**) $St = 0.125$ (**b**) $St = 0.125$

(**c**) $St = 2.00$ (**d**) $St = 2.00$

Figure 4. *Cont.*

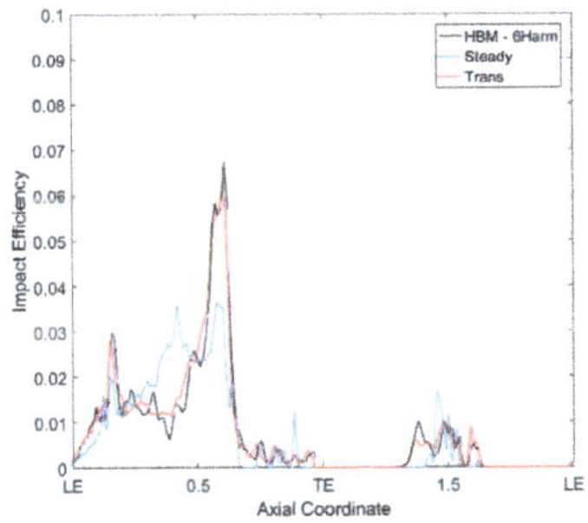

(**e**) $St = 8.00$

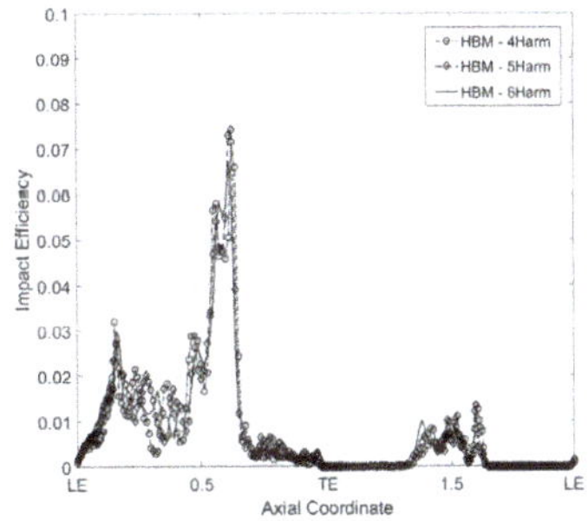

(**f**) $St = 8.00$

Figure 4. (**a**,**c**,**e**) Impact efficiency on the blade surface for the steady, transient, and HBM-reconstruction methods; (**b**,**d**,**f**) impact efficiency for different number of harmonics in the HBM-reconstruction method. Results are shown for the various Stokes numbers: 0.125 (**a**,**b**), 2.00 (**c**,**d**) and 8.00 (**e**,**f**).

Inspection of computational times in Figure 5 reveals the potential of the new method for a remarkable speedup compared to conventional transient simulations. The CPU times are normalised with respect to the total simulation time for the steady case (notice the log scale on the y-axis). Total times are represented by the sum of the flow time (time to reach convergence of the flow field before injecting particles) and the Lagrangian time (time necessary to track all the particles until they exit the domain). It can be noticed that the time for the transient is two orders of magnitude greater than the steady simulation. Moreover, focusing just on the Lagrangian time (yellow bars), this is increased by three orders of magnitude with respect to steady-state. This observation clearly highlights the need for a reduction in the time required by this step of the calculation. This is accomplished by the HBM-reconstruction that reduces it by more than two orders of magnitude, making the bars diagram look very similar to the steady calculation, where the Lagrangian step represents only a minor contribution to the overall time. This is due to the combined effect of HBM and the reduction to a single passage with phase-lag BCs. This second contribution is expected to have a much bigger influence in real turbomachinery cases, where several passages typically have to be simulated. Overall, in this case, the total time for HBM-reconstruction is reduced by one order of magnitude compared to transient.

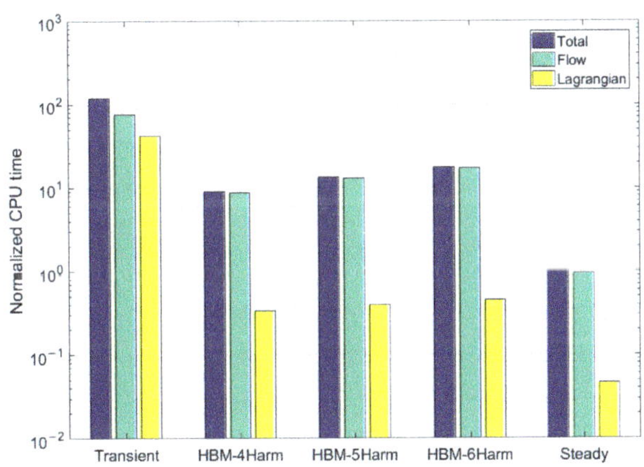

Figure 5. Normalised CPU times (reference is steady-state calculation) for the E^3 rotor case.

5. Results

In this chapter, the behaviour of a discrete phase inside two gas turbine components is analysed: a transonic axial compressor stage and a 1.5 axial turbine stage. Despite the geometries being taken from experimental rigs rather than actual engines' components, they can be considered as representative cases of turbomachinery applications of interest. Steady and unsteady results are presented, where the latter are obtained thanks to the HBM-reconstruction method. Mixing plane interfaces are used to connect different rows in the steady-state simulations. The HBM-reconstruction method is used to include transient effects and compared with steady results to assess their importance. The behaviour of the discrete phase is analysed both in terms of particle trajectories and global/local parameters referring to particle impacts on aerodynamic surfaces.

5.1. *NASA Stage 37*

The 3D geometry of the NASA stage 37 [35] is selected as the first case study. The stage is composed of 36 blades and 46 vanes. The rotor angular velocity is 18,000 rpm. One passage per row is modeled thanks to the single passage reduction technique described in Section 2.3. The computational domain for the stage is composed of a structured hexahedral mesh of 1.3 million elements generated using Ansys Turbogrid. The grid parameters were chosen according to a preliminary grid sensitivity study; see [13]. In the spanwise direction, about 100 grid points are used, and 45 grid points in the circumferential direction. The grid is realised by employing an O-grid around the rotor blade and the stator vane, with local refinements near the hub and shroud regions. In addition, 200 points are used around the rotor blade. The layer of cells on blade walls is such that the y^+ values range from 1 to 10. The tip clearance between the blade and the shroud is resolved with a mesh composed of six nodes across the gap span. At the inlet, an absolute total pressure of 101,325 Pa and an absolute total temperature of 288 K are imposed. A turbulence intensity of 3% and a turbulent length scale of 2×10^{-3} m were also prescribed at the inlet. Two operating conditions are considered for the study, namely choked flow and the near stall. To achieve these conditions, the choking mass flow rate was initially computed by imposing a static pressure outlet boundary condition. The computed choking mass flow rate was 20.42 kg s^{-1}, which is 2.3% below the experimental value of 20.9 kg s^{-1}. The static pressure was then increased until the mass flow rate achieved the desired value, equal to $p/p_{amb} = 1.39$ and $p/p_{amb} = 1.66$ for the choked and near stall conditions, respectively. For a detailed description and validation of this case in OF, see [16,21]. In each row, four harmonics of the passing frequencies are considered in the spectrum. Figure 6 shows the instantaneous pressure and entropy contours at midspan for the choked flow condition. The flow field has been obtained by reconstruction at the desired time. Four adjacent passages are shown, considering the phase-shift between them.

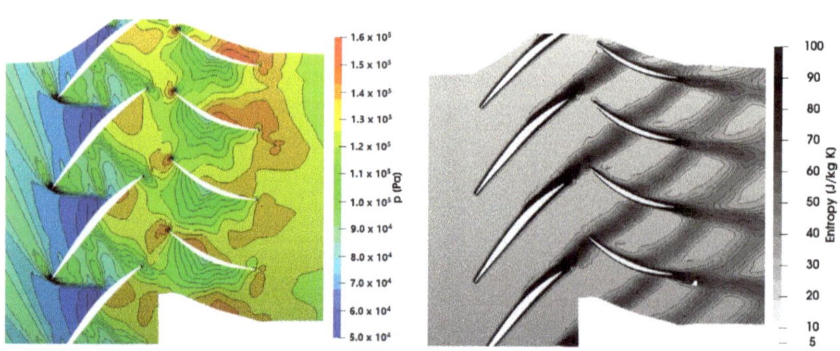

(**a**) Pressure contours (**b**) Entropy contours

Figure 6. Contours of pressure and entropy at midspan for the NASA stage 37.

Spherical drag and Saffman–Mei [36] lift effect are considered to be the only forces acting on the particles. Here, a constant density of $2700\,\mathrm{kg\,m^{-3}}$ for the discrete phase is assumed due to the nature of the air contaminants, made up of a large part of sand and very small particles of soil. Particles are injected at the same velocity as the air flow from the stage inlet, with randomly positioned injection points. Particle diameter d_p varies in the range $0.5 \div 32\,\mu\mathrm{m}$, while the Stokes number, calculated at the rotor and stator inlet sections, are reported in Table 1.

Table 1. Injection data for the NASA stage 37. N_{inj} is the number of particles injected for each diameter.

d_p [μm]	Rotor				
	0.5	**1**	**2**	**8**	**32**
St	2×10^3	8×10^{-3}	3×10^{-2}	0.5	8
τ^+	25	98	394	6.3×10^3	1.0×10^5
N_{inj}	1×10^6	1×10^6	5×10^5	1×10^5	1×10^5
d_p [μm]	Stator				
	0.5	**1**	**2**	**8**	**32**
St	3×10^3	1×10^{-2}	4×10^{-2}	0.7	11
τ^+	19	77	309	4.9×10^3	7.9×10^4
N_{inj}	1×10^6	1×10^6	5×10^5	1×10^5	1×10^5

Small particles ($<2 \div 10\,\mu\mathrm{m}$) are more prone to stick to the blades, causing fouling [37]. Larger ones tend to impact without adhesion, causing the erosion of aerodynamic surfaces. In this work, the attention is focused on particles' impacts and deposition on aerodynamic surfaces. For each diameter, a sufficient amount of particles to obtain a statistically significant number of impacts on the blades surface must be injected. Since smaller particles tend to follow the fluid streamlines better, only a small number hits the walls. For this reason, a smaller number of particles were injected with growing diameter, as reported in the last row of Table 1.

Tian and Ahmadi [38] pointed out that different turbulence models have an impact on the particles velocity deposition in horizontal and vertical ducts. Their sensitivity analysis highlights the influence of turbulence models, wall mesh refinement and particle dimensions. They show that the most important parameter is the non dimensional particle relaxation time defined as

$$\tau^+ = \frac{(\rho_p)/\rho d_p^2 u_\tau^2}{18\nu^2} \tag{23}$$

The turbulence model can be important in the resolution of particle trajectories near the wall. The turbulence model used for the NASA stage 37 calculation is the standard $k - \varepsilon$ model and near-wall effects are modeled by means of scalable wall functions. In a previous work, Aldi et al. [13] demonstrated that the $k - \varepsilon$ model is suitable to model particle dispersion near the walls for the NASA Stage 37 geometry. Their analysis shows that the $k - \varepsilon$ turbulence model overpredicts deposition for particles in Brownian ($\tau^+ < 10^{-2}$) regions, but in the inertial ($\tau^+ > 10$) region, the predicted deposition velocity trend is similar to experimental data. As reported in Table 1, the τ^+ for the compressor stage simulations is in the range corresponding to the inertial region in accordance with Tian and Ahmadi [38].

Particle rebound velocity and direction are computed with a specific particle–wall interaction model imposed on rotor blades, stator vanes and endwalls. The normal e_n and tangential e_t restitution coefficients are defined according to Forder et al. [39], as functions of the particle impact angle α. The correlations for the restitution coefficients are based on impingement testing using AISI 4130 carbon steel and sand. The analysis of particle adhesion is instead performed using the experimental results from Poppe et al. [40]. Based on the trends of sticking probability, we can define representative functions to relate the normal impact velocity u_n and the sticking probability S_p. According to the analysis carried out by Suman et al. [41], the trend can be represented by two equations. The first one refers to lower normal impact velocities (<4 m/s)

$$S_p = -0.112u_n + 0.99 \tag{24}$$

and the second one refers to normal impact velocities in the range (>4 m/s)

$$S_p = -6 \cdot 10^{-5}u_n^2 - 6 \cdot 10^{-4}u_n + 0.545 \tag{25}$$

For the sticking decision, an auxiliary random number in the range $[0, 1]$ is generated and compared with S_p. If greater, the algorithm rejects the sticking, and the restitution coefficients are used to calculate the new particle velocity. Otherwise, the particle sticks, and its velocity is set to zero. Two parameters can be defined in order to quantify the impact and adhesion characteristics. The impact efficiency (IE) has already been defined as the ratio between the number of impacting particles and the number of particles injected for a fixed diameter. The capture efficiency (CE) is defined as the ratio between the number of stuck particles and the number of particles injected for a fixed diameter. A time step of 1×10^{-6} s is used for the flow field reconstruction, corresponding to 100 instants per blade passing period. The computational time necessary to obtain the solution was about one order of magnitude higher than the steady-state case. This is considered as satisfactory, since fully unsteady calculations would have required the simulation of half annulus of the stage with an estimated computational time of two to three orders of magnitude higher than the steady-state solution.

The trends of the impact and capture efficiencies as a function of the Stokes number are shown in Figure 7 for the rotor and the stator. In the case of the rotor, only results for the choked flow condition are shown (Figure 7a) since the results were very similar in the near stall case. Results for both steady and unsteady calculations are shown. Steady results are discussed first (dashed lines). In general, it is possible to observe that the percentage of particles that hit the blade surface increases with the particle diameter for the rotor, while the stator has a maximum for 8 µm particles, more pronounced for the near stall condition. The number of particles that impact the blade and the vane is similar for small diameters. For large diameters, the blade has a greater number of impacts, and this difference increases with the particle diameter. As can be seen, smaller particles are more likely to stick when they make impact on the walls, but they undergo a lower number of impacts because they follow better the fluid streamlines thanks to the lower inertia. For larger particles, the opposite is true. The combination of these factors results in 8 µm particles having the highest capture efficiency (red line). The trends and the values are in line with those presented by Suman et al. [41] for the rotor and by Aldi et al. [13] for the entire stage. The most significant difference between the two operating conditions is represented by the vane impact efficiency. It can be observed that the near stall condition provides a higher number of impacts on the stator for the whole range of diameters, with 8 µm particles reaching a value almost double with respect to the choked condition. The reason for this behaviour can be explained in fluid dynamics terms, as described later in the text.

As expected, unsteady effects due to upstream running pressure disturbances caused by the vane are negligible, and the lines are almost overlapped. The situation for the near stall condition is equivalent and is not reported for the sake of brevity. Figure 7c refers instead to the stator. It can be observed that also for the vane the rotor–stator interaction effects are not extremely important. A slight increase in impact and capture efficiencies is observed in the choked case, but only of a few percentage points, with the maximum difference for $St = 0.7$ (8 µm particles). For the near stall condition, results are almost unchanged, as can be noticed in Figure 7c. These findings are corroborated by the fact that the local impact pattern is also almost unchanged moving to the unsteady calculation, as testified by Figure 8, which compares the two calculations in terms of local impact efficiency on the pressure and suction sides of the blade and vane surfaces. Only

8 µm particles at a choked flow condition are shown since this size resulted in the largest discrepancy in the global IE. Nevertheless, only a small increase of impacts towards the trailing edge on the pressure side of the vane can be observed compared to the steady simulation. Therefore, it can be concluded that, for this geometry, transient effects do not produce important differences on particle behaviour and steady calculations are appropriate to capture the relevant features. This is an important remark that could not be argued a priori, and shows that, for a transonic axial compressor stage, the operating point is a more important parameter influencing deposition. Therefore, we investigate this aspect further.

The near stall condition results in general in higher impact efficiencies, especially for medium-sized particles. This behaviour can be explained by looking at Figure 9, which shows the trajectories inside the rotor and stator passages for the two operating conditions. A relative tangential velocity component is reported for the blade row, while absolute tangential velocity is shown for the vane row. The smallest particles are highly influenced by the flow field, and their velocity is very close to the local flow velocity (Figure 9a). Due to the different shock wave intensity, these particles experience a higher deceleration for the near stall condition and, in much the same way as the continuous phase, their incidence angle at the stator entry is increased (Figure 9a). However, these particles undergo a very small number of impacts on the blades, as reported in Figure 7 and, therefore, their impact efficiency is not significantly altered by the operating condition. On the contrary, larger particles are much less sensitive to the deceleration induced by the shock at the blade leading edge. Indeed, as illustrated in Figure 9c, trajectories are only marginally altered by the operating condition of the compressor. Nevertheless, the high impact efficiency amplifies these effects and results in higher local impacts on the rear part of the vane pressure side for the near stall condition. 8 µm particles are somewhat in between the two situations, since they are large enough to have a high impact efficiency on the blade, but their inertia still allows for experiencing the stronger deceleration caused by the stronger shock in the near stall condition. Indeed, looking at Figure 9b, it can be seen that a stronger deceleration is observed in the area enclosed inside the black ellipse, producing an increased incidence at the entrance of the stator. This is also evidenced by the wider particle "wake" observed in the rear part of the vane with respect to the choked condition. At this point, contrary to 0.5 µm particles, they are not able to follow the flow streamlines and the impact and capture efficiencies are significantly increased for the near stall condition. However, it is important to notice that the higher incidence angle increases the impacts on the pressure side, but in the leading edge area of the suction side, fewer impacts occur. For this reason, it can not be immediately concluded that the increased deposition for near stall condition will cause higher performance losses than the choked condition. Indeed, suction side and leading edge shape modifications are the most impacting on the losses inside the passage and could therefore be more detrimental to the global compressor performance [42,43]. Additional studies including blade surface roughness changes caused by deposition must be performed to assess the importance of these effects.

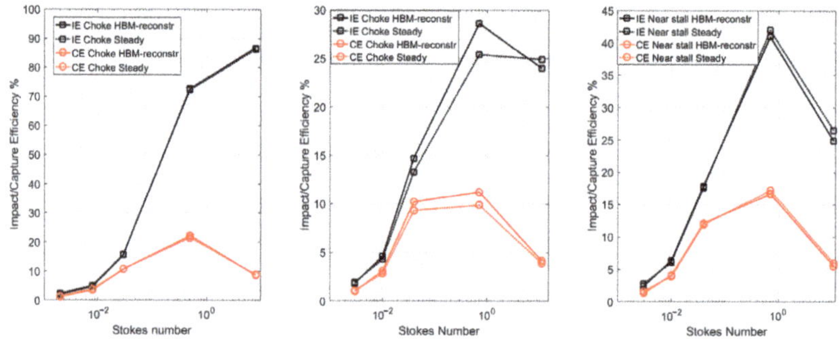

(**a**) Rotor (**b**) Stator—Choked (**c**) Stator—Near stall

Figure 7. Comparison of efficiencies for the NASA stage 37 obtained with steady and HBM-reconstruction methods.

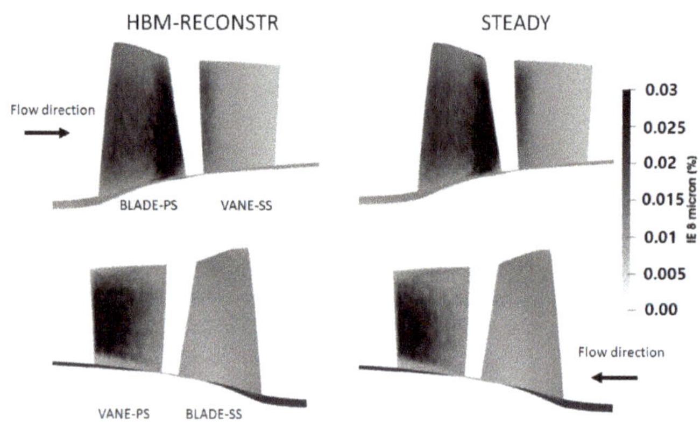

Figure 8. NASA stage 37 at choked flow condition: impact efficiency for 8 µm on the blade and vane surfaces obtained with steady and HBM-reconstruction methods.

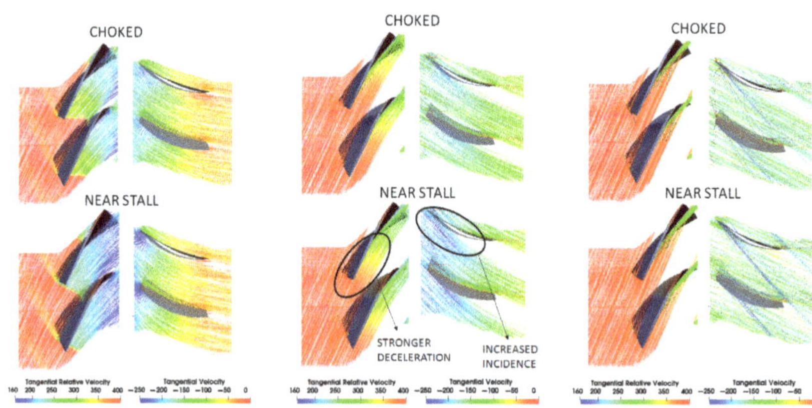

(**a**) 0.5 µm (**b**) 8 µm (**c**) 32 µm

Figure 9. NASA stage 37: particles trajectories for the steady calculations at choked and near stall conditions.

5.2. 1.5 Axial Turbine Stage

The second geometry analysed is the 1.5 axial turbine stage rig of the University of Bath [44]. The rig is composed of 32 upstream vanes, 48 rotor blades and 32 downstream vanes. A structured body-fitted grid of nearly 2.1 million elements was generated using ANSYS ICEM CFD. The mesh size is chosen according to a precedent study [44] showing that the cell number is appropriate for the simulation. The average y^+ value on the blades is about 14, allowing the usage of wall-functions.

The boundary conditions for the calculation are shown in Table 2. The flow coefficient is defined as $C_F = \frac{u_x}{\Omega_d b}$, where u_x is the axial velocity at the inlet section, Ω_d is the rotor angular speed, and b is the turbine hub radius.

The timestep of the HBM-reconstruction calculation was chosen so that 100 instants per blade passing period were resolved. Figure 10a illustrates the time-averaged pressure field on stator and rotor walls and endwalls. Figure 10b shows the validation of the solution, in terms of the comparison between the pressure coefficient for numerical and experimental data. Illustrated is the time-averaged value of C_p over two passages on a probe located on the hub downstream of the first vane.

Table 2. Operating conditions for the 1.5 axial turbine stage.

Parameter	Value
Flow coefficient, C_F	0.41
Inlet total temperature	300 K
Outlet static pressure	99,300 Pa
Disc rotational speed	3000 rpm
Inlet turbulent intensity k	3%
Inlet turbulent length scale	3×10^{-3} m

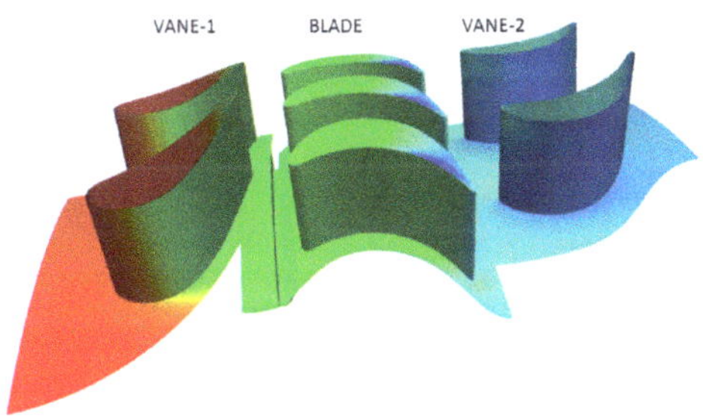

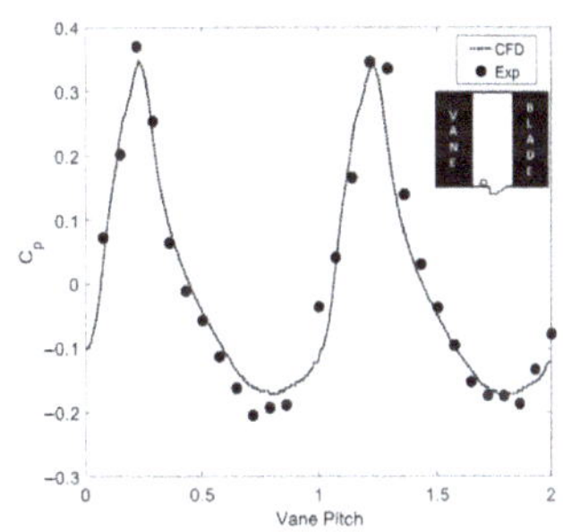

Figure 10. (**a**) Time -averaged pressure contours on the wall of the 1.5 axial turbine stage; (**b**) pitchwise distribution of time-averaged C_p: experimental and numerical results.

The same structure of the previous section is followed here, by presenting a steady analysis to highlight the salient features of the discrete phase behaviour, followed by a comparison with the HBM-reconstruction technique. Spherical drag and Saffman–Mei lift force acting on the particles are considered during the calculation. A constant density of 2700 kg m^{-3} is assumed for the discrete phase. Particles are injected from the domain inlet at the same velocity of the air flow. The variation of the particle diameter d_p is in the range 0.5 ÷ 32 µm, while the Stokes number, calculated at the three rows inlet sections, is reported in Table 3.

Table 3. Stokes numbers for the 1.5 axial turbine stage. N_{inj} is the number of particles injected for each diameter.

d_p [µm]	N_{inj}	St Stator1	St Rotor	St Stator2
0.5	1×10^6	1.7×10^{-3}	7.2×10^{-3}	1.8×10^{-3}
1	1×10^6	6.8×10^{-3}	2.9×10^{-2}	7.3×10^{-3}
2	5×10^5	2.7×10^{-2}	0.12	2.9×10^{-2}
8	1×10^5	0.44	1.85	0.47
32	1×10^5	6.99	29.5	7.47

A mixing plane interface is placed between stators and the rotor for the steady analysis. The global impact and capture efficiencies for the first stator, the rotor and the second stator as a function of particle diameter are reported in Figure 11. It can be noticed that, for the first stator, these properties grow with particle size. This trend resembles the one seen for the rotor in the NASA stage 37 case (Figure 7a). Consequently, it can be deduced that, for the first upstream row of other axial turbomachinery, the trend will be roughly the same. In this case, however, the impact number is higher due to larger turning angle imposed to the flow by the vane row. The monotonic trend is interrupted after $d_p > 8$ µm because, for this size, all particles have already impinged on the vane surface. Indeed, it can be noticed that, for larger particles, the impact efficiency is higher than 100%, meaning that some particles rebound multiple times on the vane. On the contrary, the diagrams for the rotor and the second vane row present a maximum for middle-sized particles. For larger particles, the first stage acts as a "filter" with respect to the second stator. Indeed, despite the fact that the sticking probability is significantly lower, the larger inertia causes a high impact and capture efficiencies in the first stage, with the result that only a small fraction of large particles can reach the downstream vane row. This is testified by the fact that the impact and capture efficiencies of 8 and 32 µm particles for the second stator are orders of magnitude lower than for the other rows. This is not true for smaller particles since only a fraction of them actually make impact on the walls. To a lower extent, the first vane row acts as a filter for the rotor, explaining the decreasing number of impacts for large particles.

To better analyse these features, trajectories for a hundred particles are shown in a meridional plane for three diameters in Figure 12, coloured by absolute velocity magnitude. 1 µm particles undergo a higher acceleration in the first vane row, but their inertia is not sufficiently high and drag force is prevalent along their trajectory. When the Stokes number grows, particles are accelerated less by the flow, and they approach the following row with a lower incidence. Moreover, their inertia begins to be important and the average radial impact locations on the blade is shifted towards the shroud of the turbine. This is in agreement with the findings of Tabakoff et al. [8] who analysed the impact locations for particles of similar size on an axial turbine. Some of them are also able to move through the tip gap of the blade and exit on the suction side. It can be seen that 32 µm particles are centrifuged out so fast that almost no impacts occur on the pressure side of the blade; instead, they impact on the rotor case. As previously described, for larger particles, the almost totality of particles are not able to reach the downstream stator. However, it is important to note that, although Equations (24) and (25) predict a low sticking probability for large particles (due to the higher normal impact velocity), this value could be overestimated compared to other models typically used in the literature. This results in a higher number of particles being able to reach the downstream vane row in the end. It can be also noticed that the particles are not able to infiltrate inside the cavity, and their trajectory is not affected by the purge flow exiting from the rim gap, even for 1 µm. The same simulation (not shown here) was also conducted for the high-purge flow condition, showing almost identical results.

Figure 11 shows the impact and capture efficiencies for the three rows. The only noteworthy differences between steady and transient simulations are due to smaller particles' impacts. Up to 2 µm, the impacts on the blade pressure side are increased and more uniformly distributed, while for 8 and 32 µm, they are almost unmodified. Unsteady effects are negligible on the first vane row. This can be demonstrated also by comparing the global impact and capture efficiency for the upstream stator obtained with steady and HBM simulations (Figure 11a). As expected, the lines are superimposed to each other. For what concerns the rotor blade, Figure 11b shows that some differences are present for the transient solution. The most important are the total impact and capture efficiencies for 2 µm particles that are increased from 30.5% to 40% and from 24.5% to 30.5%, respectively. Once again, rotor–stator interactions are more important for small particles due to the lower inertia, but it is the combination of this effect with the number of impacts that must be accounted for.

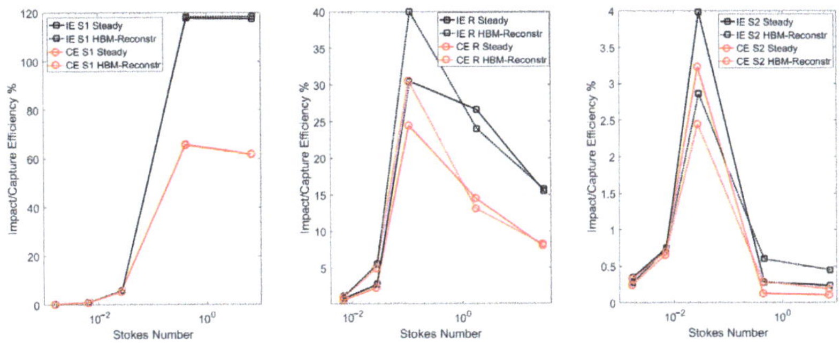

(**a**) First stator (**b**) Rotor (**c**) Second stator

Figure 11. Comparison of efficiencies for the 1.5 axial turbine stage obtained with steady and HBM-reconstruction methods.

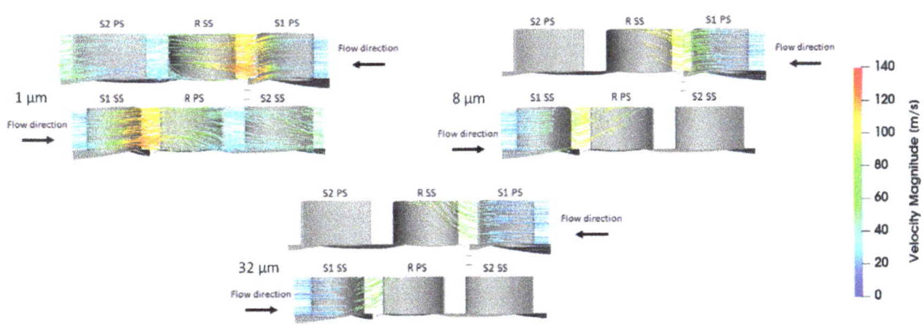

Figure 12. 1.5 axial turbine stage: particle trajectories for three different diameters.

The impact efficiencies for 1 µm and 2 µm are shown in Figure 13. These are the particle sizes that resulted in the larger differences in the impact and capture efficiencies between the steady and transient calculations. Pressure and suction sides are reported for the three rows. The pressure side is the most struck by the discrete phase. It can be also noticed that a small percentage of 1 µm particles impact the suction side of the rotor near the tip. This is due to the tip leakage flow that transports some particles across the tip gap towards the suction side of the blade.

Unsteady effects could be potentially more important in the downstream stator, but the number of impacts is so small that no considerable percentage differences are observed, as illustrated in Figure 11c. As for the NASA stage 37, the very similar patterns found that comparing steady-state and transient approaches allows for arguing that steady-state solutions are a good approximation of the discrete phase physics, at least for the tested geometries and particle sizes. Unfortunately, this is not of much interest from the scientific standpoint, but allows for putting a higher trust in the results obtained with stationary simulations of discrete phases, which are still the most commonly employed in this research field. Please notice that under no circumstances should this remark be directly extended to the continuous phase, as already mentioned several times. Anyway, it is shown that unsteady effects on the discrete phase are small but not negligible for the rotor blade. Differences up to 10% are present for $St \approx 0.1$ as shown in Figure 11b.

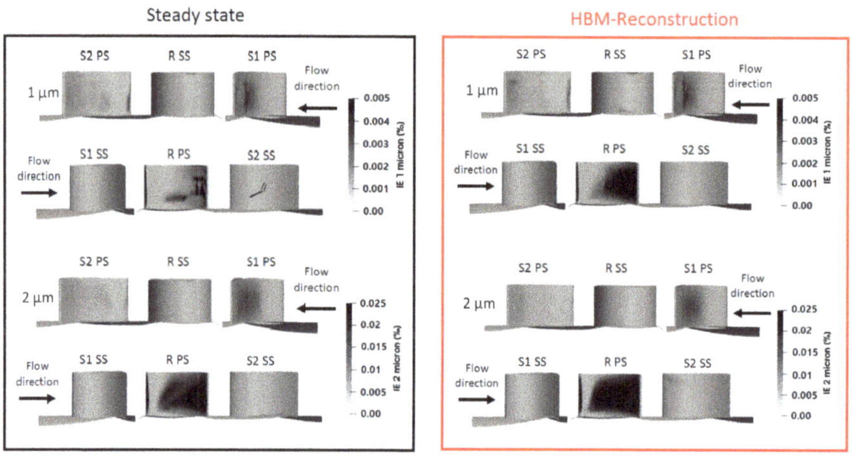

Figure 13. Impact efficiencies comparison between steady and HBM simulations for 1 µm and 2 µm in the three rows of the 1.5 axial turbine stage.

Unsteady analysis is also very useful to visualize the dynamic of particles approaching the rotating passage. To explore this feature, a small stream of particles exiting from the upstream vane is selected. Figure 14 shows the locations of 1 and 8 µm particles at four different time steps, coloured by absolute velocity magnitude. In the 1 µm case, particles are accelerated as they leave the vane passage and travel faster than the rotor blade, avoiding suction surface impacts. Larger particles instead experience a smaller acceleration and tend to concentrate in the wake region after impacting the vane walls. Differently from smaller particles, they travel slower and are chopped by the oncoming rotor blades. Having a smaller sticking probability, they rebound on the rotor blade suction surface as the blade keeps moving toward them.

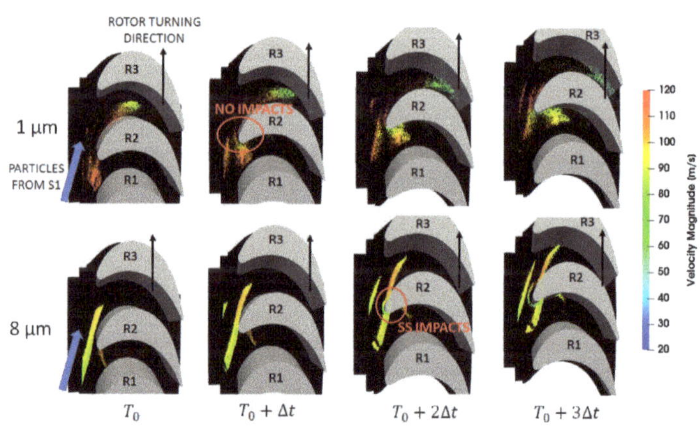

Figure 14. 1.5 axial turbine stage: particle positions at four equispaced time steps for two particle diameters. Particles are coloured by absolute velocity magnitude.

6. Conclusions

This work consisted of analyzing the behaviour of a discrete phase inside gas turbine components. A technique that combines HBM with Lagrangian tracking was presented, allowing for significant speedup during the continuous and discrete phase solution for unsteady calculations. The matter of particles crossing cyclic boundaries with a phase-shift relation was properly addressed. Moving to the results, firstly, a simple test-case has been used to assess the correct implementation of the methodology. Then, two representative applications of the Lagrangian tracking algorithm to gas turbine components were presented, namely a transonic axial compressor stage and a 1.5 axial turbine stage. A deposition model from the literature was used to determine which particles tend to stick on the blade surfaces. Impact statistics and particle trajectories were used to explain the observed trends in global impact and capture efficiencies. For the compressor stage, two operating conditions were investigated. Steady-state calculations with mixing plane interfaces and unsteady simulations with the HBM-reconstruction were performed and compared to each other for both case studies. Generally speaking, significantly different behaviours can be expected for particles with different Stokes number. For a transonic compressor stage, it was demonstrated that different shock intensities caused by different operating conditions have important effects on the vane impact efficiency according to particle size. For the axial turbine, it was demonstrated that impacts and deposition occur almost exclusively in the first stage that acts as a filter for the successive rows. Additionally, due to the significantly higher swirl component of the flow compared to the axial compressor, larger particles are centrifuged towards the casing of the turbine much faster. Interestingly, unsteady effects seem to affect the particle impacts only slightly. The most important differences are observed for the axial turbine case, in which the impact efficiency for 2 µm particles is increased by about 10% on the rotor blade. However, these differences do not dramatically modify the conclusions drawn with the steady models. This is perhaps disappointing from a scientific standpoint but confirms that steady-state solutions are a good approximation of the discrete-phase physics at a reduced computational time, at least for the tested geometries and particle sizes. In any case, the HBM-reconstruction technique allows for obtaining more accurate results at an acceptable computational cost and therefore represents a valuable tool for the prediction of the discrete-phase behaviour.

Author Contributions: Software development and CFD simulations, S.O.; resources, review and supervision, N.C.; resources, review and supervision, M.P.; resources and review, M.C. All authors have read and agreed to the published version of the manuscript.

Funding: This research received no external funding.

Data Availability Statement: Data presented in the paper can be obtained upon reasonable request by contacting the authors.

Conflicts of Interest: The authors declare no conflict of interest.

Nomenclature

The following abbreviations are used in this manuscript:

Ω	Domain angular velocity vector
F	Vector of fluxes
M	Matrix of the equations system
Q	Vector of conservative variables
$R(Q)$	Discretised fluxes residual term
u	Velocity
$\hat{Q}_m$	m-th Fourier coefficient of Q
D_t	Time derivative operator
D	Harmonic balance operator
E	Discrete Fourier transform matrix

ω	Angular frequency corresponding to a known flow periodicity
σ	Interblade phase angle
θ	Azimuthal coordinate
N_B	Number of blades in a row
N_T	Number of snapshots for the HBM simulation
V	Control volume

References

1. Wilcox, M.; Kurz, R.; Brun, K. Technology Review of Modern Gas Turbine Inlet Filtration Systems. *Int. J. Rotating Mach.* **2012**, *2012*, 128134. [CrossRef]
2. Dunn, M.G. Operation of gas turbine engines in an environment contaminated with volcanic ash. *J. Turbomach.* **2012**, *134*, 051001. [CrossRef]
3. Suman, A.; Casari, N.; Fabbri, E.; Pinelli, M.; Di Mare, L.; Montomoli, F. Gas Turbine Fouling Tests: Review, Critical Analysis, and Particle Impact Behavior Map. *J. Eng. Gas Turbines Power* **2019**, *141*, 032601. [CrossRef]
4. Suman, A.; Casari, N.; Fabbri, E.; Di Mare, L.; Montomoli, F.; Pinelli, M. Generalization of particle impact behavior in gas turbine via non-dimensional grouping. *Prog. Energy Combust. Sci.* **2019**, *74*, 103–151. [CrossRef]
5. Hamed, A.; Tabakoff, W.; Wenglarz, R. Erosion and Deposition in Turbomachinery. *J. Propuls. Power* **2006**, *22*, 350–360. [CrossRef]
6. Ghenaiet, A. Prediction of erosion in an axial turbine with initial position of blade. In Proceedings of the 13th European Turbomachinery Conference, ETC2019-111, Lausanne, Switzerland, 8–12 April 2019.
7. Hamed, A.A.; Tabakoff, W.; Rivir, R.B.; Das, K.; Arora, P. Turbine Blade Surface Deterioration by Erosion. *J. Turbomach.* **2004**, *127*, 445–452. [CrossRef]
8. Tabakoff, W.; Hamed, A.; Metwally, M. Effect of Particle Size Distribution on Particle Dynamics and Blade Erosion in Axial Flow Turbines. *J. Eng. Gas Turbines Power* **1991**, *113*, 607–615. [CrossRef]
9. Mustafa, Z.; Pilidis, P.; Amaral Teixeira, J.; Ahmad, K. CFD Aerodynamic Investigation of Air-Water Trajectories on Rotor-Stator Blade of an Axial Compressor for Online Washing. In Proceedings of the Turbo Expo: Power for Land, Sea, and Air, Barcelona, Spain, 8–11 May 2006; pp. 1385–1394.
10. Yang, H.; Boulanger, J. The Whole Annulus Computations of Particulate Flow and Erosion in an Axial Fan. *J. Turbomach.* **2012**, *135*, 011040. [CrossRef]
11. Zagnoli, D.; Prenter, R.; Ameri, A.; Bons, J.P. Numerical Study of Deposition in a Full Turbine Stage Using Steady and Unsteady Methods. In Proceedings of the ASME Turbo Expo 2015: Turbomachinery Technical Conference and Exposition, Montreal, QC, Canada, 15–19 June 2015 ; Paper No: GT2015-43613.
12. Prenter, R.; Ameri, A.; Bons, J.P. Computational Simulation of Deposition in a Cooled High-Pressure Turbine Stage with Hot Streaks. *ASME J. Turbomach.* **2017**, *139*, 091005. [CrossRef]
13. Aldi, N.; Morini, M.; Pinelli, M.; Spina, P.R.; Suman, A. An Innovative Method for the Evaluation of Particle Deposition Accounting for Rotor/Stator Interaction. *J. Eng. Gas Turbines Power* **2016**, *139*, 052401. [CrossRef]
14. Oliani, S.; Friso, R.; Casari, N.; Pinelli, M.; Suman, A.; Carnevale, M. Progresses in Particle-Laden Flows Simulations in Multistage Turbomachinery with OpenFOAM. *J. Turbomach.* **2022**, *144*, 101007. [CrossRef]
15. Hall, K.C.; Thomas, J.P.; Clark, W.S. Computation of Unsteady Nonlinear Flows in Cascades Using a Harmonic Balance Technique. *AIAA J.* **2002**, *40*, 879–886. [CrossRef]
16. Oliani, S.; Casari, N.; Carnevale, M. ICSFoam: An OpenFOAM library for implicit coupled simulations of high-speed flows. *Comput. Phys. Commun.* **2023**, *286*, 108673. [CrossRef]
17. Saad, Y.; Schultz, M.H. GMRES: A generalized minimal residual algorithm for solving nonsymmetric linear systems. *Siam J. Sci. Stat. Comput.* **1986**, *7*, 856–869. [CrossRef]
18. Yoon, S.; Jameson, A. Lower-upper Symmetric-Gauss-Seidel method for the Euler and Navier–Stokes equations. *AIAA J.* **1988**, *26*, 1025–1026. [CrossRef]
19. Macpherson, G.B.; Nordin, N.; Weller, H.G. Particle tracking in unstructured, arbitrary polyhedral meshes for use in CFD and molecular dynamics. *Commun. Numer. Methods Eng.* **2009**, *25*, 263–273. [CrossRef]
20. Oliani, S.; Casari, N.; Pinelli, M.; Suman, A.; Carnevale, M. Numerical study of a centrifugal pump using Harmonic Balance Method in OpenFOAM. In Proceedings of the Taormina, Italy: The 34th International Conference on Efficiency, Cost, Optimization, Simulation and Environmental Impact of Energy Systems, Taormina, Italy, 27 June–2 July 2021.
21. Oliani, S.; Casari, N.; Carnevale, M. A New Framework for the Harmonic Balance Method in OpenFOAM. *Machines* **2022**, *10*, 279. [CrossRef]
22. Guédeney, T.; Gomar, A.; Gallard, F.; Sicot, F.; Dufour, G.; Puigt, G. Non-uniform time sampling for multiple-frequency harmonic balance computations. *J. Comput. Phys.* **2013**, *236*, 317–345. [CrossRef]
23. Gopinath, A.; van der Weide, E.; Alonso, J.; Jameson, A.; Ekici, K.; Hall, K. Three-Dimensional Unsteady Multi-stage Turbomachinery Simulations Using the Harmonic Balance Technique. In Proceedings of the 45th AIAA Aerospace Sciences Meeting and Exhibit, Reno, NV, USA, 8–11 January 2007.
24. Nimmagadda, S.; Economon, T.D.; Alonso, J.J.; da Silva, C.R.I. Robust uniform time sampling approach for the harmonic balance method. In Proceedings of the 46th AIAA Fluid Dynamics Conference, Washington, DC, USA, 13–17 June 2016.

25. Tyler, J.M.; Sofrin, T.G. Axial Flow Compressor Noise Studies. In *Proceedings of the SAE Technical Paper*; SAE International: Warrendale, PA, USA, 1962.
26. Sicot, F.; Dufour, G.; Gourdain, N. A time-domain harmonic balance method for rotor/stator interactions. *J. Turbomach.* **2012**, *134*. [CrossRef]
27. Ekici, K.; Hall, K.C. Nonlinear Analysis of Unsteady Flows in Multistage Turbomachines Using Harmonic Balance. *AIAA J.* **2007**, *45*, 1047–1057. [CrossRef]
28. Crespo, J.; Contreras, J. On the Development of a Synchronized Harmonic Balance Method for Multiple Frequencies and its Application to LPT Flows. In Proceedings of the ASME Turbo Expo 2020: Turbomachinery Technical Conference and Exposition, Virtual, 21–25 September 2020; Paper No: GT2020-14952.
29. Erdos, J.I.; Alzner, E.; McNally, W. Numerical Solution of Periodic Transonic Flow through a Fan Stage. *AIAA J.* **1977**, *15*, 1559–1568. [CrossRef]
30. Gerolymos, G.A.; Michon, G.J.; Neubauer, J. Analysis and Application of Chorochronic Periodicity in Turbomachinery Rotor/Stator Interaction Computations. *J. Propuls. Power* **2002**, *18*, 1139–1152. [CrossRef]
31. Oliani, S.; Friso, R.; Casari, N.; Pinelli, M.; Suman, A.; Carnevale, M. A comparative analysis of particle-mixing plane interaction in multistage turbomachinery simulations. In Proceedings of the European Turbomachinery Conference 14, Gdansk, Poland, 12–16 April 2021.
32. Thulin, R.D.; Howe, D.C.; Singer, I.D. *High-Pressure Turbine Detailed Design Report*; Technical Report CR-165608; NASA Report; Lewis Research Center: Cleveland, OH, USA, 1982.
33. Launder, B.; Spalding, D. The numerical computation of turbulent flows. *Comput. Methods Appl. Mech. Eng.* **1974**, *3*, 269–289. [CrossRef]
34. Jacobs, G.; Don, W.S.; Dittmann, T. High-order resolution Eulerian–Lagrangian simulations of particle dispersion in the accelerated flow behind a moving shock. *Theor. Comput. Fluid Dyn.* **2012**, *26*, 37–50. [CrossRef]
35. Reid, L.; Moore, R.D. *Design and Overall Performance of Four Highly Loaded, High Speed Inlet Stages for an Advanced High-Pressure-Ratio Core Compressor*; Technical Report 1337; NASA Technical Paper; Lewis Research Center: Cleveland, OH, USA, 1978.
36. Mei, R.; Klausner, J. Shear lift force on spherical bubbles. *Int. J. Heat Fluid Flow* **1994**, *15*, 62–65. [CrossRef]
37. Kurz, R.; Brun, K. Degradation in Gas Turbine Systems . *J. Eng. Gas Turbines Power* **2000**, *123*, 70–77. [CrossRef]
38. Tian, L.; Ahmadi, G. Particle deposition in turbulent duct flows–comparisons of different model predictions. *J. Aerosol Sci.* **2007**, *38*, 377–397. [CrossRef]
39. Forder, A.; Thew, M.; Harrison, D. A numerical investigation of solid particle erosion experienced within oilfield control valves. *Wear* **1998**, *216*, 184–193. [CrossRef]
40. Poppe, T.; Blum, J.; Henning, T. Analogous Experiments on the Stickiness of Micron-sized Preplanetary Dust. *Astrophys. J.* **2000**, *533*, 454–471. [CrossRef]
41. Suman, A.; Morini, M.; Kurz, R.; Aldi, N.; Brun, K.; Pinelli, M.; Ruggero Spina, P. Quantitative Computational Fluid Dynamic Analyses of Particle Deposition on a Transonic Axial Compressor Blade—Part II: Impact Kinematics and Particle Sticking Analysis. *J. Turbomach.* **2014**, *137*, 021010. [CrossRef]
42. Morini, M.; Pinelli, M.; Spina, P.R.; Venturini, M. Numerical Analysis of the Effects of Nonuniform Surface Roughness on Compressor Stage Performance. *J. Eng. Gas Turbines Power* **2011**, *133*, 072402. [CrossRef]
43. Kurz, R.; Brun, K. Fouling Mechanisms in Axial Compressors. *J. Eng. Gas Turbines Power* **2012**, *134*, 032401. [CrossRef]
44. Cosmo, G.D.; Scobie, J.A.; Lock, G.D.; Sangan, C.M.; Carnevale, M. Fluid Dynamics of Turbine Rim Seal Structures: A Physical Interpretation Using URANS. *J. Eng. Gas Turbines Power* **2022**, *145*, 031009. [CrossRef]

energies

MDPI

Article

Performance Degradation of a Shell-and-Tube Heat Exchanger Due to Tar Deposition

Nicola Aldi, Nicola Casari, Michele Pinelli, Alessio Suman * and Alessandro Vulpio

Department of Engineering (DE), University of Ferrara, 44122 Ferrara, Italy; nicola.aldi@unife.it (N.A.); nicola.casari@unife.it (N.C.); michele.pinelli@unife.it (M.P.); alessandro.vulpio@unife.it (A.V.)
* Correspondence: alessio.suman@unife.it

Abstract: Biomass represents a programmable renewable energy source that is useful for reducing issues related to the transfer from fossil fuels to the renewable energy era. The exploitation of biomass is strongly related to the development of power technologies that are designed to improve efficiency; however, at the same time, they have to be designed to improve the life cycle of the entire installation—especially in relation to maintenance operations. In this paper, a numerical analysis is proposed to assess the performance of a heat exchanger used for separating condensing tar from syngas generated by the gasification of lignocellulosic wood chips and pellets. The analysis included clean, fouled, and clogged conditions. Flow maldistribution characterized the inlet section of shell-and-tube configurations and was responsible for clogging phenomena. Starting from field detection, analyses of fouled and clogged conditions showed a reduction in the effectiveness of the heat exchanger, causing dangerous conditions for the internal combustion engine used to exploit the syngas flow.

Keywords: biomass gasification; tar deposition; shell-and-tube heat exchanger; CFD; conjugate heat transfer; performance degradation

Citation: Aldi, N.; Casari, N.; Pinelli, M.; Suman, A.; Vulpio, A. Performance Degradation of a Shell-and-Tube Heat Exchanger Due to Tar Deposition. *Energies* **2022**, *15*, 1490. https://doi.org/10.3390/en15041490

Academic Editor: Dimitris S. Manolakos

Received: 17 January 2022
Accepted: 14 February 2022
Published: 17 February 2022

1. Introduction

Negative environmental impacts from fossil fuels have encouraged researchers to look for greener energy sources. Among all renewable energy sources, biomass appears to be one of the most attractive [1]. The gasification process is considered a key technology for the conversion of biomass to biofuels [2] because of its high efficiency and flexibility [3]. The solid feedstock is converted into a gaseous fuel (syngas), whose main components are carbon monoxide (CO), hydrogen (H_2), methane (CH_4), carbon dioxide (CO_2), and nitrogen (N_2) [2]. Cogeneration systems based on biomass gasification can utilize syngas as a fuel gas in internal combustion engines or gas turbines for electricity and heat production [4].

Despite the numerous advantages of biomass gasification, the technology is still in the development stage due to issues related to the process itself. During gasification, undesirable byproducts such as particulate matter, ammonia, sulfur compounds, and tar are unavoidably produced and entrained in the fuel gas. The types of syngas contaminants and the potential problems associated with their presence have been reported by Belgiorno et al. [5]. Since these impurities can cause severe issues in downstream equipment, cleaning syngas is essential before its being utilized [6]. In particular, reducing tar content is one of the most significant challenges in biomass gasification [7]. Tar condensation can indeed occur if the local temperature drops until the tar dew point is reached, leading to fouling and clogging of fuel lines, heat exchangers, filters, and engines [8].

As pointed out by Morf [9], there is no clear and uniform definition for the term "tar" in the literature because of the complexity of the mixture of organic compounds that represent "tar". A comprehensive overview of operational "tar" definitions used by researchers is given in [10]. The tar composition in syngas is strongly dependent on its formation conditions. Evans and Milne [11,12] identified four major tar product classes: primary products derived from cellulose, hemicellulose, and lignin; secondary products, characterized by phenolics and olefins; alkyl tertiary products, which include methyl derivatives of aromatics; and condensed tertiary products, which include the polyaromatic hydrocarbon series without substituents. Primary products are destroyed before the tertiary products appear [13]. Kiel et al. [14] derived a different tar classification based on the behavior of tar components in downstream processes. In this respect, two significant tar properties compounds were considered, i.e., condensation behavior and water solubility. The proposed classification system groups tar components into five classes [14]: GC-undetectable tars (class 1)—namely, the heaviest tars that condense at high temperatures even at very low concentrations; heterocyclic compounds (class 2), which generally exhibit high water solubility due to their polarity; aromatic compounds (class 3), i.e., light hydrocarbons that do not show condensation or water solubility issues; light polyaromatic hydrocarbons (class 4), which condense at relatively high concentrations and intermediate temperatures; and heavy polyaromatic hydrocarbons (class 5), which condense at relatively high temperatures at low concentrations.

As stated above, an essential property for tar condensation is its dew point. The tar dew point is the temperature at which the actual total partial pressure of tar equals its saturation pressure [14]. The Energy Research Center of The Netherlands developed a model for calculating tar dew point T_{dp} for different tar classes as a function of tar-class concentration C [14] (Figure 1). The model includes vapor/liquid equilibrium data for tar components in syngas based on the ideal gas behavior. Raoult's law is applied to calculate the mixture of hydrocarbons, using the vapor pressure data of individual compounds [14].

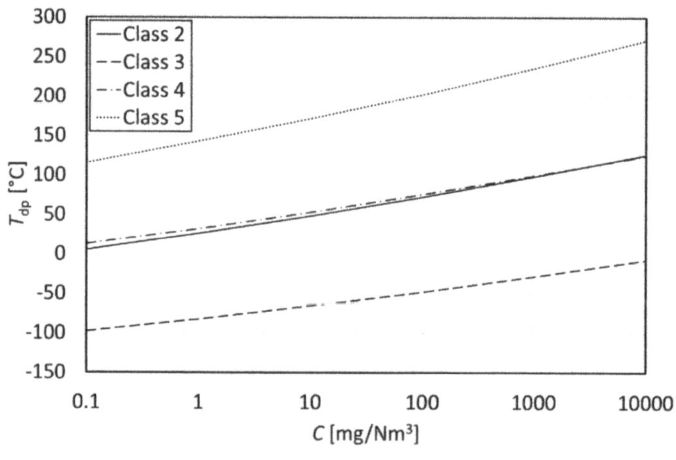

Figure 1. Tar dew point for different tar classes as a function of tar-class concentration.

Several approaches for tar reduction are reported in the literature. Tar removal technologies can be broadly divided into two groups, i.e., primary and secondary measures, depending on the location where the tar is removed. According to Devi et al. [15], primary methods can be defined as all the measures taken in the gasification step to prevent or convert tar formed in the gasifier. The authors provided a thorough overview of the primary methods used for tar elimination during biomass gasification. In particular, they pointed out the importance of the proper selection of operating parameters (temperature, pressure, gasifying medium, residence time, etc.), the use of adequate bed additives or catalysts during gasification, and a proper gasifier design. Secondary measures are conventionally employed as syngas treatments outside the gasifier [15]. As outlined by Anis and Zainal [16] in their review, secondary methods include both chemical (catalytic and thermal cracking) and mechanical/physical treatments—the latter being further classified into dry (cyclones, rotating particle separators, various types of filters, adsorbers, etc.) and wet (wet scrubbers, wet electrostatic precipitators, wet cyclones, etc.) gas cleanup systems.

Among all the wet cleaning techniques, wet scrubbers are necessary devices that use water scrubbing to condense tar and simultaneously remove particulates from syngas [16]. However, crucial issues related to wet scrubbing systems involve expensive wastewater treatment and decreases in both syngas heating values and process net energy efficiency [16]. To overcome the disadvantages of conventional wet cleaning technologies, Thapa et al. [17] developed a cleanup system consisting of a single-tube heat exchanger for syngas cooling followed by a dry biomass-based filter with wood shavings as filter media. The use of an indirect heat exchanger (no contact between hot syngas and chilled water) eliminates the need for wastewater treatment, while filter media can be reused as gasification feedstock. The application of the system for cleaning the syngas produced in a 20 kW downdraft gasifier resulted in a tar reduction efficiency of 61%. More recently, Thapa et al. [18] designed a tar removal technology that uses a vegetable oil bubbler in series with a shell-and-tube heat exchanger. The authors tested the ability of the system to remove tar from the syngas generated in a pilot-scale downdraft gasification plant. About 60% of the tar was condensed in the heat exchanger, which cooled the syngas below the tar dew point, and 96% of the remaining tar was absorbed by the oil bubbler. However, even if syngas cleanup techniques based on indirect heat exchangers for promoting tar condensation appear promising, the accumulation of tar deposits on heat transfer surfaces can lead to extremely short maintenance intervals.

This work presents a numerical investigation of performance degradation due to tar deposits on a shell-and-tube heat exchanger installed in a small-scale biomass gasification plant. The novelty of the present research is related to using a numerical simulation to comprehend the cooling process of the syngas and the consequent tar separation. The use of heat exchangers as a tar trap means that the common rules and methods used in the design of a shell-and-tube heat exchanger must be balanced with the need to condense the greatest amount of tar with the widest overhaul intervals. At first, the flow distribution and heat transfer performance were analyzed for the heat exchanger with clean surfaces. Then, to evaluate if favorable conditions are established for tar condensation and to predict the potential deposition zones, Eulerian–Lagrangian simulations were carried out for the clean heat exchanger. Finally, based on on-field fouling observations, different fouled heat exchanger models were developed to examine the progressive deterioration of heat transfer performance due to solid deposit build-up.

2. Conjugate Heat Transfer Methodology

The heat exchanger under study is part of a commercial-grade system for the gasification of lignocellulosic wood chips and pellets coupled to a cogeneration unit located inside the campus of the University of Parma within the context of project SYNBIOSE, which is aimed at building a knowledge base and the best practices for promoting small-scale gasification and CHP for the tertiary sector [19,20]. The project involved Siram S.p.A., the Center for Energy and Environment of the University of Parma (CIDEA), and the Department of Engineering of the University of Ferrara.

The purpose of the heat exchanger was to cool the syngas stream that feeds a 125 kW internal combustion engine and, at the same time, to clean the fuel gas by promoting tar condensation. As can be seen from Figure 2, it is a vertical shell-and-tube heat exchanger with one pass on both the shell and tube sides. The syngas enters the heat exchanger from the upper header and flows inside 55 carbon steel tubes with an inner diameter of 30 mm, a wall thickness of 2.3 mm, and a length of 1400 mm. The tubes are accommodated in a carbon steel shell, characterized by an inside diameter of 450 mm and a thickness of 9.5 mm, according to a triangular layout with variable tube pitch (44–60 mm). The shell-side fluid is water, which is fed from the bottom and forced to flow across the tube bundle by 11 segmental baffles with a spacing of 110 mm and a baffle cut of 33%. Since tar condensation occurs during the heat exchanger operation, the vertical configuration of the device allows the liquid to flow out of the tubes and collect in a condensate tank connected to the bottom header.

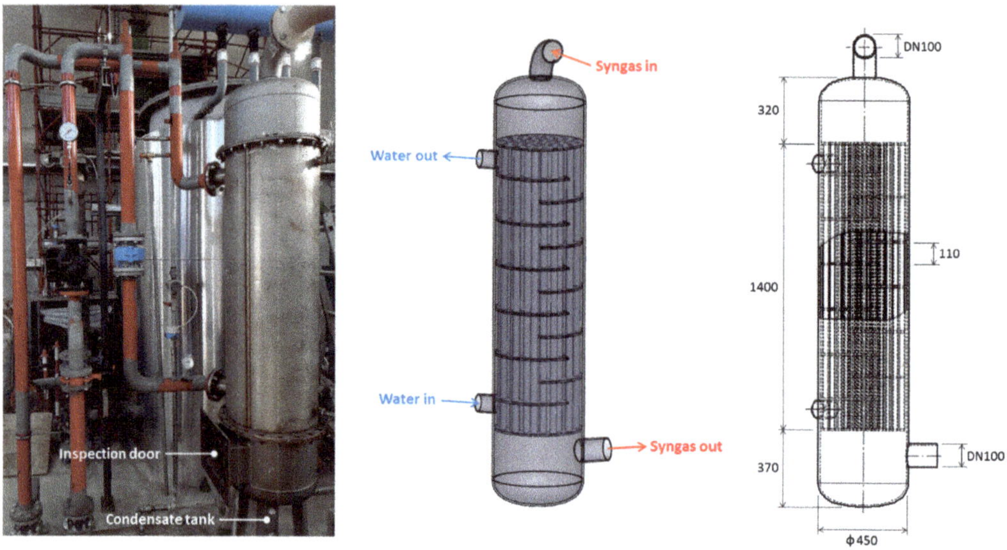

Figure 2. Shell-and-tube heat exchanger for syngas cooling: actual geometry and reconstructed three-dimensional model.

The starting point of the numerical analysis was the creation of a three-dimensional model for the heat exchanger. Since the three-dimensional model of the device was not available, its virtual geometry was reconstructed from technical drawings and on-field measurements (Figure 2). During this phase, the geometry of the heat exchanger was slightly simplified to reduce the computational effort of the subsequent simulations. In particular, the connections of the bottom header to the inspection door and the condensate tank were removed from the three-dimensional model, as well as small discontinuities related to the flanged and welded joints.

Operating Conditions and Model Setup

To characterize the performance of the shell-and-tube heat exchanger, steady conjugate heat transfer simulations were performed with the commercial CFD code ANSYS CFX 20.2 under its nominal operating conditions, which are outlined in Table 1. The computational domain for the analysis of the clean device consisted of two fluid domains (syngas and water) and one solid domain. Since all heat exchanger components (shell, tube bundle, tube plates, baffles, headers) were merged to form a single solid domain, thermal contact resistances at the interfaces between adjacent elements were neglected. A grid sensitivity analysis was performed, checking the variation of the power of the heat exchanger and the pressure losses of the syngas compared to a progressive refinement of the mesh from about 40 million to 70 million elements. The chosen grid was composed of about 61 million elements. The mesh was generated by employing near-wall refinements, with 3 prism layers added to the solid walls (Figure 3).

Table 1. Nominal operating conditions of the heat exchanger.

Quantity	Value
Syngas mass flow rate [kg/s]	0.10
Syngas inlet relative pressure [Pa]	−15,000
Syngas inlet temperature [°C]	350
Water volume flow rate [m^3/h]	11
Water inlet temperature [°C]	80

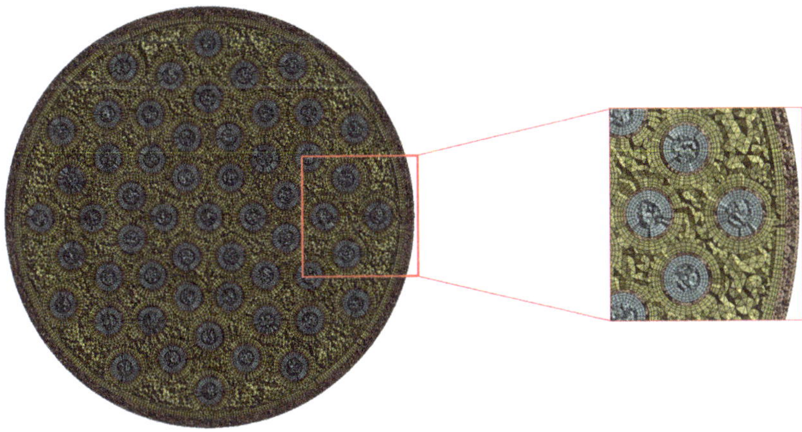

Figure 3. Cross-sectional view of heat exchanger numerical grid (detail of the mesh around the tubes shown in a close-up).

The fuel gas produced in the downdraft gasifier was sampled, and its overall composition is reported in Table 2. Since no reaction is expected within the heat exchanger except for tar condensation (which would involve only a slight change in the gas composition), the composition was considered to be uniform over the corresponding fluid domain. Therefore, the syngas was approximated as an ideal gas mixture, whose thermophysical properties were assumed to be constant and was evaluated by weighted averaging of the properties of the components according to their molar fractions (Table 3).

Table 2. Syngas composition.

Component	Molar Fraction
H_2	0.150
O_2	0.020
N_2	0.420
CH_4	0.015
CO	0.230
CO_2	0.165

Table 3. Thermophysical properties of syngas and water.

Property	Syngas	Water
Density [kg/m^3]	variable	971.8
Specific heat [J/(kg·K)]	1142	4197
Dynamic viscosity [Pa·s]	$1.655 \times e^{-5}$	$3.55 \times e^{-4}$
Thermal conductivity [W/(m·K)]	0.0244	0.670

The boundary conditions for the conjugate heat transfer calculations reflected the previously highlighted nominal operation point of the device. For the tube-side fluid (syngas), the inlet relative pressure and temperature were specified at the inlet section of the domain, while the mass flow rate was imposed at the outlet section. For the shell-side fluid (water), the mass flow rate and inlet temperature were set at the domain inflow boundary, whereas a zero relative pressure was applied at the outflow boundary. Solid walls that were wet by the fluids were treated as hydraulically smooth, and thermal conditions on the external surfaces were specified by imposing a heat transfer coefficient of 10 W/(m^2·K), representative of natural convection over a vertical surface according to [21]. Finally, a reference temperature for the environment was set at 15 °C. The turbulence model used in the simulations was the standard *k-epsilon* model, and near-wall effects were modeled with scalable wall functions [22]. A second-order high-resolution advection scheme was adopted for both the flow and turbulence equations.

3. Clean Heat Exchanger Performance

3.1. Flow Distribution Analysis

A uniform distribution of flow in the tube bundle of shell-and-tube heat exchangers is assumed in conventional designs. Nevertheless, as stressed by Mohammadi and Malayeri [23], in practice, flow maldistribution is an inevitable occurrence, which may have severe implications on heat exchanger performance. Among the different causes of flow maldistribution identified by the authors, gross flow maldistribution is caused by the geometrical characteristics of the device. This type of maldistribution can significantly increase the tube-side flow pressure drop and result in a reduced heat transfer rate [23]. In particular, Kim et al. [24] demonstrated that the configuration of the inlet nozzle and intake header strongly influences the flow distribution in the tube bundle.

Figure 4, which shows a clean heat exchanger under examination, illustrates the velocity field of syngas within the upper header on two section planes orthogonal to each other, passing through the header axis (the left section contains the axis of the inlet nozzle). It is possible to observe that the syngas flow coming from the inlet nozzle was not uniformly distributed among the tubes. In such a configuration of the intake header, the tubes in line with the nozzle are preferentially fed by the gas flow. Moreover, a significant flow recirculation occurred when syngas entered the header, resulting in extended dead zones.

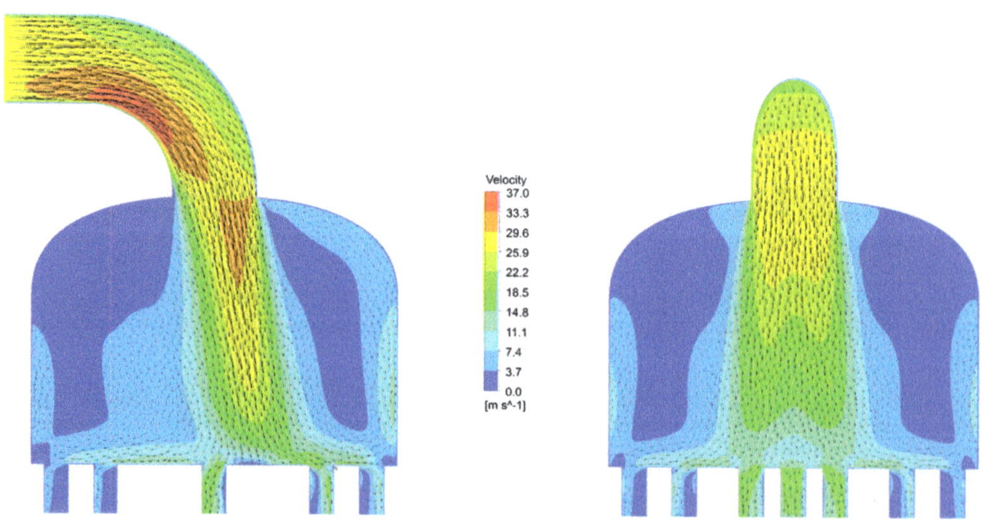

Figure 4. Syngas velocity distribution within the intake header for the clean heat exchanger.

In order to visualize the flow maldistribution in the tube bundle, Figure 5 depicts the velocity field of the syngas on a cross-section at the mid-length of the bundle, together with the numbering scheme adopted for the identification of the tubes (dashed lines clarify the orientation of the inlet nozzle). A noticeable flow non-uniformity could be detected, with syngas velocities up to 17 m/s in some of the preferentially fed tubes (tubes 2 and 7). However, several tubes facing the dead zone that formed below the 90° bend characterizing the inlet nozzle showed velocities in the order of 1 m/s. From a quantitative point of view, it is possible to express the flow maldistribution in the tube bundle in terms of a velocity deviation δ, defined as [23]:

$$\delta = \frac{v - V}{V} \tag{1}$$

where v is the local average fluid velocity in a tube and V is the average fluid velocity in the whole tube bundle, equal to 4.6 m/s (design value). The histogram in Figure 6 shows the distribution of the velocity deviation in the tubes; the quantity on the vertical axis is the ratio between the number n of tubes characterized by a velocity deviation that falls within a specific interval and the total number N of tubes in the bundle. As can be noticed, only about 3.5% of the tubes presented with a local average velocity almost equal to the design value. The local average fluid velocity was lower than 4.6 m/s in about 67% of the tubes. On the contrary, the velocity deviation exceeded 100% in about 16% of the tubes (preferentially fed tubes). These results confirm the highly questionable assumption of bundle uniform flow for shell-and-tube heat exchangers [23,24].

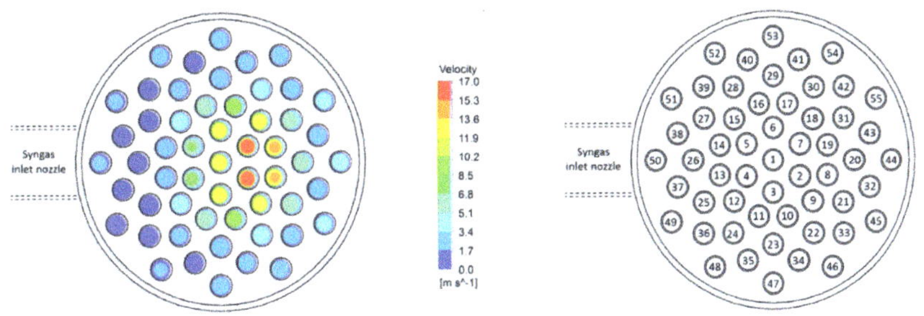

Figure 5. Syngas velocity distribution on a cross-section at the mid-length of the tube bundle for the clean heat exchanger (numbering scheme used to identify tubes reported on the right).

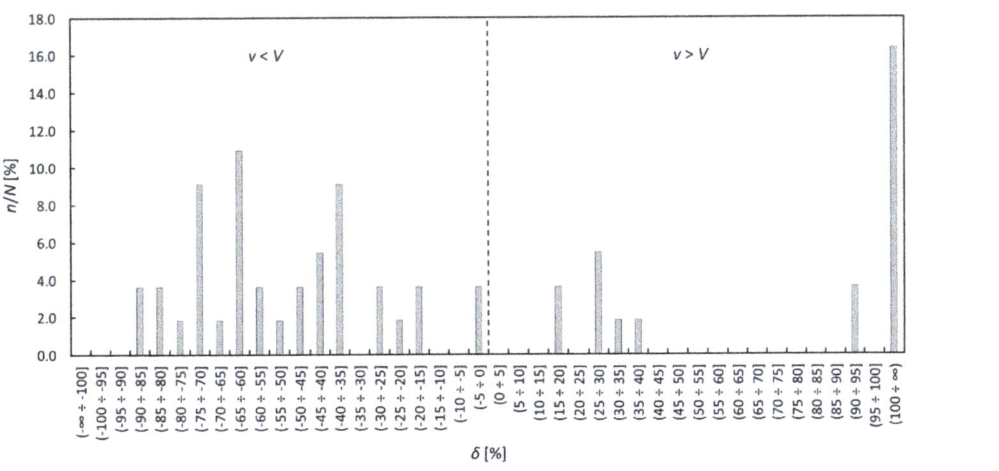

Figure 6. Velocity deviation distribution in the tube bundle for the clean heat exchanger.

3.2. Temperature Distribution Analysis

The temperature field of the syngas within the intake header for the clean heat exchanger is illustrated in Figure 7 on the same section planes considered in Figure 4. The syngas temperature distribution reflects the velocity field analyzed in the previous paragraph. The tubes aligned with the inlet nozzle were fed by a gas stream with a temperature slightly lower than 350 °C (syngas inlet temperature). At the same time, those facing the recirculating flow regions of the header were characterized by a lower entrance temperature. In fact, before entering these tubes, the gas flows over the upper tube plate (whose wall temperature ranged from 85 °C to 100 °C), separating the syngas from water.

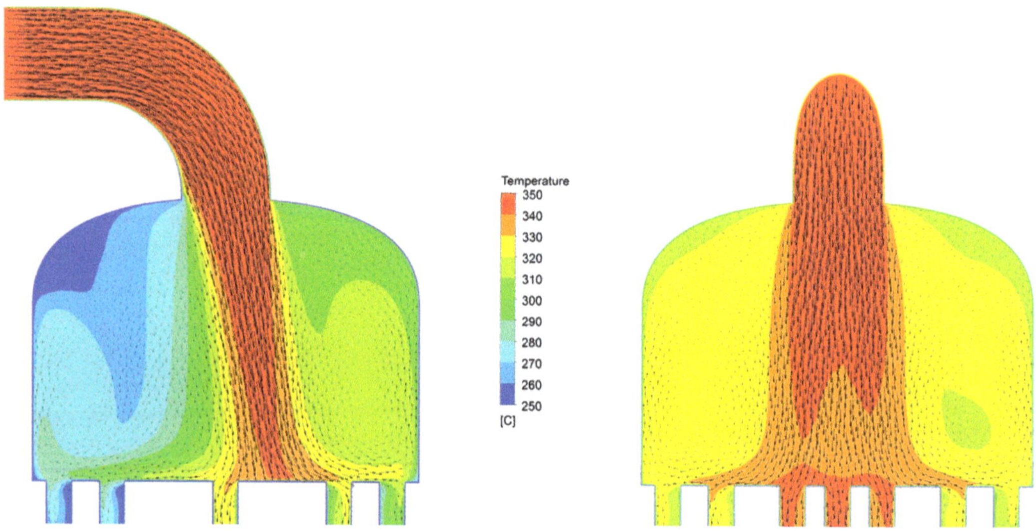

Figure 7. Syngas temperature distribution within the intake header for the clean heat exchanger.

In order to evaluate the gas cooling process within the tube bundle, the graph in Figure 8 outlines the evolution of the syngas temperature T along the tube axis (l refers to the tube length) for five characteristic tubes, whose location in the bundle is highlighted in red in the schematic (tube 1 is the central pipe, while tubes 44, 47, 50 and 53 are adjacent to the shell wall). The overall temperature drop ranged from 125 °C, for tube 1, to 140 °C, for tube 44. The maximum gas cooling was thus obtained in tube 44, in which the average fluid velocity was almost equal to the design value of 4.6 m/s (Figure 5). Therefore, the flow maldistribution imposed different heat transfer conditions for the tubes, generating a set of passages affected by flow conditions far different from those designed. This condition reflects a non-homogeneous condensation, reported in detail below. In contrast, the syngas temperature drop was lowest in tube 1, for which the velocity deviation exceeded 100%. The flow maldistribution finally resulted in a non-uniform gas exit temperature among the tubes.

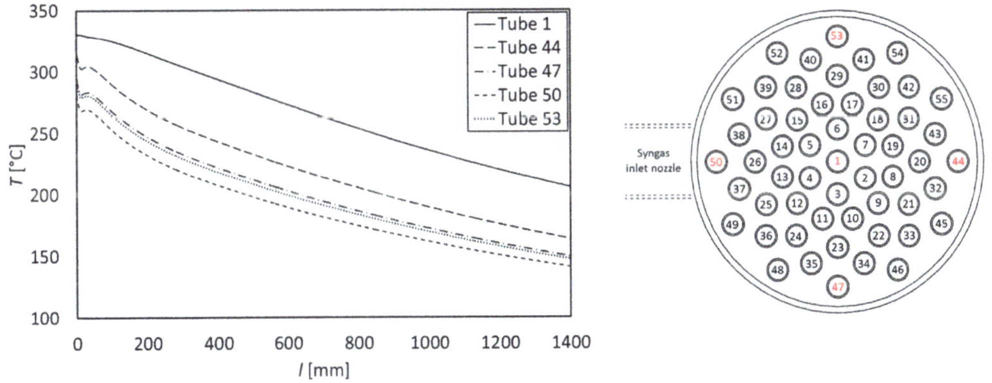

Figure 8. Evolution of syngas temperature along the tube axis for five characteristic tubes (location in the bundle highlighted in red in the schematic on the right).

4. Tar Condensation Assessment

For a qualitative assessment of tar droplet formation and deposition within the clean shell-and-tube heat exchanger, steady Eulerian–Lagrangian simulations were carried out. This analysis investigated if favorable conditions were established for tar condensation and predicted the potential deposition zones by tracking the droplet trajectories. Since the sampling of the tar produced in the biomass gasification plant of Parma was not available at the moment of writing, the composition and concentration of tar in the syngas were assumed based on literature data. In accordance with Casari et al. [20], who studied the deposition of tar in the piping system downstream of the heat exchanger under examination, the tar compound concentrations found by Dufour et al. [25] were employed. As can be observed from Table 4, tar components were grouped here according to the classification system proposed by Kiel et al. [14], and the dew point for the different tar classes was calculated as a function of tar-class concentration with the model developed in [14] (Figure 1).

Table 4. Tar compound classification, tar-class concentrations, and corresponding dew point.

Tar Class	Major Compounds	Concentration [mg/m^3]	Dew Point [°C]
Class 1	Undetectable	0	-
Class 2	Phenol, toluene	1000	98.4
Class 3	m-Xylene	1745	−23.3
Class 4	Phenanthrene, naphthalene	2250	108.4
Class 5	Pyrene	5	162.6

4.1. Simulation Strategy

As explained above, the condensation of tar occurs when the local temperature of the syngas drops until the tar dew point is reached, leading to aerosol formation [8]. Such droplets are likely to keep increasing in size as the condensation process continues. In the simplified simulation strategy adopted, based on the Eulerian Lagrangian approach, a distribution of fixed-diameter particles (discrete phase) was injected onto the previously solved gas flow field (continuous phase).

$$\frac{d\mathbf{u}_\mathrm{P}}{dt} = \mathbf{F}_\mathrm{D} + \frac{\rho_\mathrm{P} - \rho}{\rho_\mathrm{P}}\mathbf{g}, \tag{2}$$

Particles were released at the same local velocities as the gas flow from the inflow boundary of the inlet nozzle, with equally spaced, randomly positioned injection points. In accordance with Casari et al. [20], the particle diameter distribution was specified starting from the experimental results of Vasudevan et al. [26]. In particular, three characteristic diameters representing the reported particle size distribution were considered for each tar class: 0.1 μm, 5.6 μm, and 12.3 μm. These fixed-diameter particles represented 54.2%, 21.4%, and 24.4%, respectively, of the total volume of injected particles. Since the particle volume fraction was very low (<<10%), it was assumed that particles did not affect the fluid flow (one-way coupling approach). The trajectories of individual particles were computed by integrating a force balance equation on the particle, where the left-hand side represents the inertial force per unit mass acting on the particle and $\mathbf{u}_p$ is the particle velocity vector. The first and the second term on the right-hand side are the drag force and the buoyancy force per unit particle mass, respectively—where ρ_p is the particle density, ρ is the gas density, and $\mathbf{g}$ is the gravity acceleration vector. In the present analysis, the drag coefficient for spherical particles was expressed according to the empirical correlation of Schiller and Naumann [27]. Furthermore, the turbulent dispersion of particles in the fluid phase was predicted using the stochastic model of Gosman and Ioannides [28].

In order to assess if favorable conditions were established for tar condensation within the heat exchanger, the convective heat transfer between particles and fluid was accounted for with the model proposed by Ranz and Marshall [29,30]. If the particle temperature reaches the tar dew point, a droplet with a diameter equal to that of the particle is assumed to form instantly. Injected particles can thus be considered as probes, whose temperature is monitored to detect the potential condensation of tar in the syngas flow. When the particle impacts a surface, the computation of its trajectory is interrupted (null restitution coefficients), and the impact temperature is evaluated; if this temperature is lower than the tar dew point, tar droplet deposition is supposed to occur. The thermophysical properties of tar relevant to particle trajectory and heat transfer calculations were taken from the work of Euh et al. [31]. The density, specific heat, and thermal conductivity of tar were set equal to 1153 kg/m^3, 1470 J/(kg·K)] and 0.150 W/(m·K), respectively.

4.2. Potential Tar Deposition Zones

The analysis of particle impact temperature for the different tar classes showed that tar condensation/deposition was possible only for class 5. For the tar-class concentrations taken into consideration, the temperature of injected particles could only reach the dew point with the class 5 tar within the heat exchanger. A completely different scenario was found by Casari et al. [20] in the piping system downstream of the heat exchanger due to the lower temperatures of the syngas. Except for class 3, whose dew point falls below 0 °C, tar deposits inside the piping were predicted for each tar class.

Figure 9, which shows the class 5 tar, depicts the impact patterns for the considered particle diameters by means of colored particle plots. Each dot, representing a single particle hitting the surface, is colored red or blue depending on whether its impact temperature was higher or lower than the tar dew point. In the second case, tar droplet deposition is assumed to occur. For the sake of clarity, heat exchanger surfaces were grouped into three zones, which were analyzed separately: (i) the upper zone, comprising the inlet nozzle, and the upper header and plate, (ii) the tube bundle, and (iii) the bottom zone, which comprises the bottom header and plate, and the outlet nozzle. As can be noticed, favorable conditions were not established for tar condensation within the upper zone due to the high-temperature values of the gas flow (Figure 7). The deposition of tar droplets was first detected in the lower part of the tube bundle. In particular, tar deposition started earlier for the tubes adjacent to the shell wall, in which the syngas reached the lowest temperatures (Figure 8). However, fewer tar deposits formed in these tubes than in the preferentially fed ones, due to the flow maldistribution in the bundle (Figure 5). Therefore, tar droplet deposition also occurred on the bottom zone surfaces. Within the bottom header, the presence of low-temperature recirculating flow regions—analogous to those illustrated in Figure 7 for the intake header—promotes the condensation of tar, which also deposits on the outlet nozzle surface.

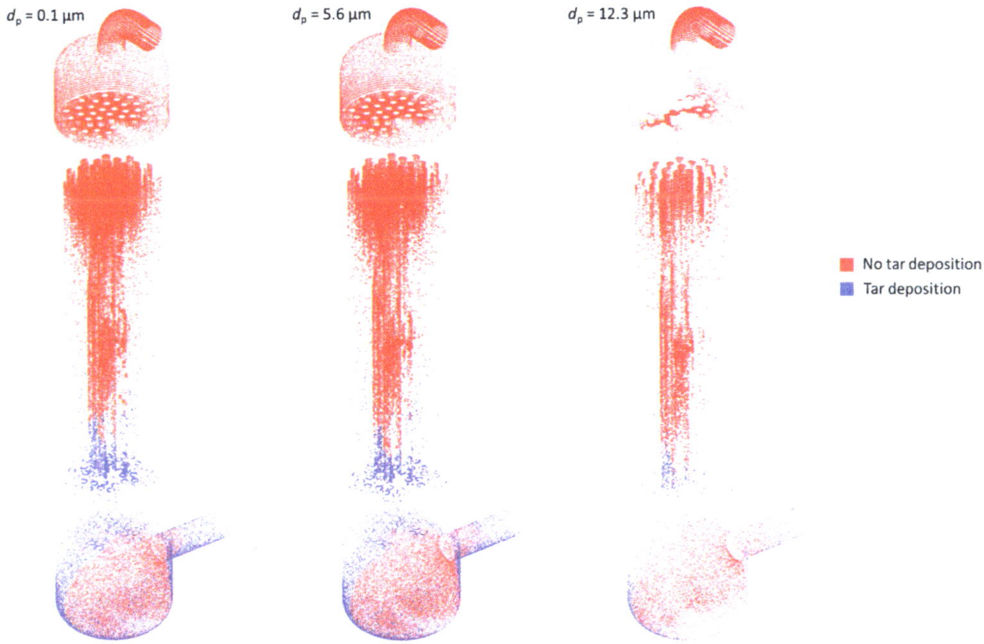

Figure 9. Tar class 5 deposition patterns according to droplet diameter.

The preceding considerations are further clarified by Figure 10, which still refers to the class 5 tar. The particle impacts along the heat exchanger height h are reported as a function of the impact temperature T, for particles with a diameter of 0.1 µm and 12.3 µm. Each dot, representing a single particle hitting the surface, is colored red, green, or blue depending on whether the impact took place over the upper zone, the tube bundle, or the bottom zone surfaces, respectively. For an immediate identification of impacts that, in accordance with the adopted strategy, resulted in tar droplet deposition, a dashed line is drawn in correspondence to the class 5 tar dew point (equal to 162.6 °C).

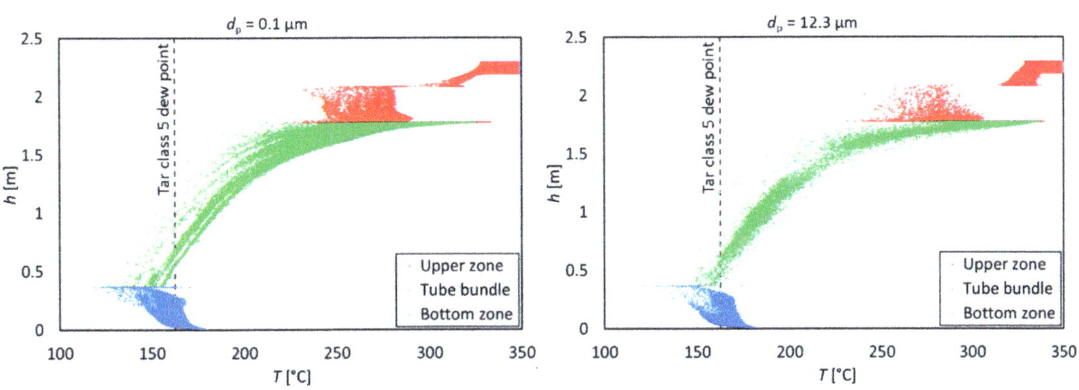

Figure 10. Particle impacts along with the heat exchanger height as a function of impact temperature, for particles with a diameter of 0.1 µm and 12.3 µm (tar class 5).

5. Fouled Heat Exchanger Analysis

5.1. On-Field Fouling Observations

The visual inspection of shell-and-tube heat exchanger surfaces during maintenance operations revealed the presence of solid deposits due to the condensation of tar. As can be noted from Figure 11, which shows the aspect of the upper tube plate before (left) and after (right) mechanical cleaning, the deposit build-up resulted in the clogging of several tubes. The presence of solid tar on the upper plate affected a significant portion of the surface, and no preferential regions can be detected. The condensation of tar during the heat exchanger operation was due to a combination of tar concentration, syngas velocity, and temperature conditions, together with flow recirculation and vortex phenomena responsible for carrying the tar droplets through the heat exchanger. Therefore, the visual inspection of the actual heat exchanger is fundamental to detect the fouling magnitude and the position of the fouled and clogged tubes. As remarked in the description of the numerical simulation strategy, the tar concentration was imposed according to literature data [20,26] and was not directly related to the operating conditions of the considered plant. Therefore, only a qualitative comparison was possible between the numerical condensation pattern and the actual one.

Figure 11. Upper tube plate before (**left**) and after (**right**) mechanical cleaning.

Starting from these on-field detections, a representative clogged configuration for the heat exchanger was determined, in which 21 tubes (38%) were considered wholly blocked by deposits. Moreover, to investigate the progressive performance degradation resulting from the accumulation of tar deposits on tube bundle heat transfer surfaces, a further fouled model for the heat exchanger was developed. In this model, which represents an intermediate fouled condition between the clean configuration and the clogged one, a uniform-thickness tar layer was assumed to reduce the cross-sectional area of each of the aforementioned 21 tubes by 50%. The numerical model setup for both the fouled and clogged configurations was analogous to that described for the heat exchanger with clean surfaces. The only difference to the previous setup concerned the definition of the computational domain, which involved an additional solid domain for tar deposits.

5.2. Performance Deterioration Assessment

Table 5 reports the performance of the shell-and-tube heat exchanger for each of the configurations taken into consideration. For the sake of completeness, the conditions reported in Table 5 refer to the clean (the heat exchanger is not affected by tar deposition), fouled (the tube bundle of the heat exchanger presents a reduced passage area due to a partial obstruction coming from tar deposition), and clogged (the worst operating condition for the heat exchanger, with several tubes completely obstructed by tar deposits) conditions. In particular, the heat transfer rate from syngas (Q_{syngas}) to water (Q_{water}) and ambient air (Q_{air}), the heat transfer effectiveness (ε), and the tube-side flow pressure drop (Δp_{syngas}) are outlined. The heat transfer effectiveness is defined as the ratio between the actual heat transfer rate from the syngas and the maximum possible heat transfer rate in the heat exchanger, equal to 30,828 W. The heat transfer rate from syngas to water (Q_{water}) progressively decreased as tar deposits accumulate on the tube bundle. For the fouled configuration, this was primarily due to the additional conductive resistance represented by the tar layer, which resulted in a reduction in the overall heat transfer coefficient. For the clogged configuration, the decrease in the rate of heat transfer was instead mainly due to the reduced total heat transfer surface. As can be seen, the heat transfer rate from syngas to ambient air (Q_{air}) showed an opposite behavior, since it increased due to solid deposit build-up. This phenomenon was due to a progressive reduction in the gas cooling within the tube bundle, which led to an increase in the heat transfer between the syngas and ambient air through the bottom header wall. Nevertheless, the increase in Q_{air} did not compensate for the decrease in Q_{water} and, due to this, the heat transfer effectiveness decreases, passing from the clean to the clogged configuration. Therefore, the deposition of tar caused an increase in the temperature of the gas leaving the heat exchanger, and the pressure drop of the tube-side flow (Table 5). These factors resulted in a reduction in the fuel gas density that fed the internal combustion engine of the combined heat and power plant and, hence, a decrease in its power. At the same time, the reduction in the syngas cooling within the heat exchanger decreased the capacity of the device in promoting the condensation of tar, which was able to reach the engine, causing fouling problems for its components.

Table 5. Heat exchanger performance for the considered configurations.

Configuration	Q_{syngas} [W]	Q_{water} [W]	Q_{air} [W]	ε	Δp_{syngas} [Pa]
Clean	21,404	18,193 (85.0%)	3211 (15.0%)	0.694	244
Fouled	21,053	17,806 (84.6%)	3247 (15.4%)	0.683	253
Clogged	20,796	17,526 (84.3%)	3270 (15.7%)	0.675	269

5.3. Flow Distribution Modification

The velocity and temperature distributions of syngas in the tube bundle for the considered configurations of the heat exchanger are depicted in Figure 12 by means of three-dimensional histograms. The velocity and temperature values refer to the cross-section at the mid-length of the bundle and were taken at the center of each tube. The position of the fouled/clogged tubes is also indicated according to the adopted numbering scheme. These velocity and temperature representations help in the interpretation of the data reported in Table 5. As stated above, fouling and clogging phenomena induced the modification of heat exchanger performance by reducing the total heat transfer surface. In fact, due to the flow maldistribution, even in the presence of clogged tubes, the distributions of the syngas flow rate (velocity) and temperature through the tubes appeared almost unchanged. However, the gas velocity values in the peripheral tubes slightly increased when passing from the clean to the clogged condition. Similarly, the gas temperature values in the peripheral tubes were not so different when comparing the clean and the clogged conditions. This means that the configuration adopted for the heat exchanger design affected (i) the performance of the clean device (design condition), and also (ii) its sensitivity to the presence of fouling or, more generally, the off-design operating conditions.

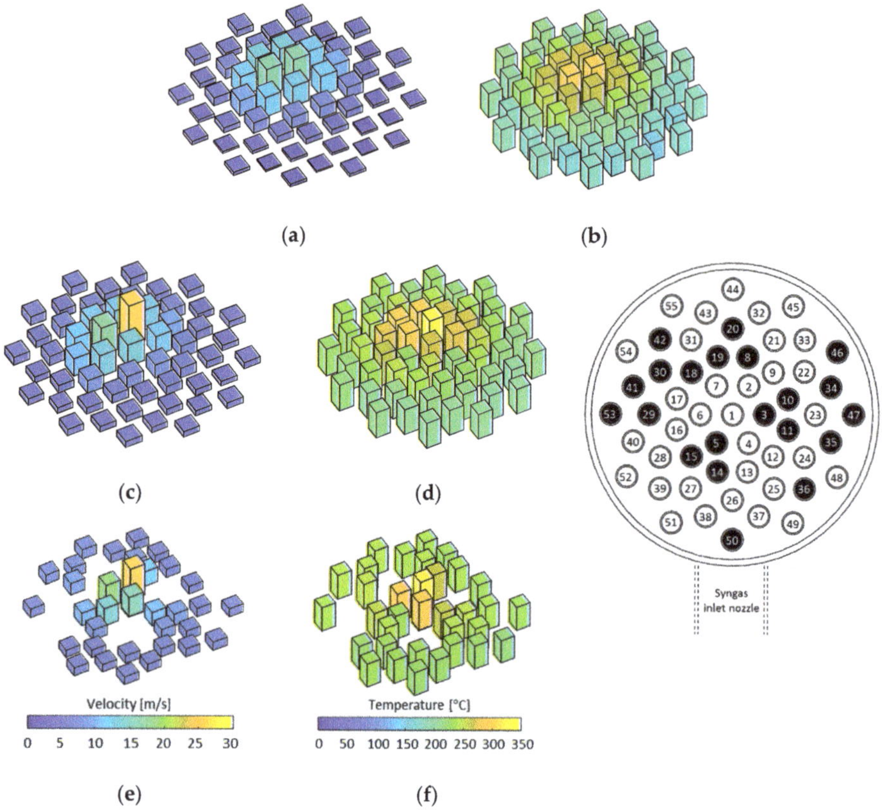

Figure 12. Syngas velocity and temperature distributions on a cross-section at the mid-length of the tube bundle for: (**a**,**b**) the clean heat exchanger; (**c**,**d**) the fouled heat exchanger; (**e**,**f**) the clogged heat exchanger (position of fouled/clogged tubes reported on the right by solid black color).

6. Conclusions

In the present paper, a numerical analysis of clean and degraded operating conditions of a shell-and-tube heat exchanger used in a syngas-fed power plant is proposed. The syngas was obtained through a gasification process based on lignocellulosic wood chips and pellets. Depending on its temperature, a condensing tar could be generated within the plant. The numerical model was developed based on the actual heat exchanger geometry, and the clean operating condition was studied according to on-field measurements. The analysis of the designed geometry showed a relevant flow maldistribution in relation to the intake header, which was responsible for a non-uniform flow distribution over the tubes. These flow and thermal fields were compared to those coming from the analyses of fouled and clogged conditions. The latter analyses were carried out by generating a degraded flow domain originating from on-field detections of the device after several operating hours. The condensing tar imposed a progressive obstruction of the flow passages and resulted in a modification of the heat exchanger performance during its operation. The present analysis shows how fouled and clogged conditions cause a decrease in heat exchanger effectiveness, imposing even more dangerous operating conditions for the internal combustion engine. The capability of the heat exchanger to condense, and thus, separate the tar from the syngas flow has to be kept at a sufficient level to preserve safe conditions for the combustion engine. With any reduced effectiveness, the capability of the heat exchanger to condense tar and separate it from the gas stream decreases, and the safety margin of the plant operation is reduced accordingly. Furthermore, maintenance intervals have to be scheduled according to the tar concentration and the design features of the heat exchanger. Fluid dynamic analysis is fundamental to discovering how the tubes and the intensity of the tar deposition are involved in flow passage.

Author Contributions: Conceptualization, N.A., N.C., M.P., A.S., A.V.; methodology, N.A., M.P., A.S.; software, N.A.; validation, N.A., M.P., A.S.; formal analysis, N.A., M.P., A.S.; investigation, N.A., N.C., M.P., A.S., A.V.; data curation, N.A.; writing—original draft preparation, N.A.; writing—review and editing, N.A., M.P., A.S.; project administration, M.P., A.S.; funding acquisition, M.P. All authors have read and agreed to the published version of the manuscript.

Funding: This work was carried out within the framework of the research project "SYNBIOSE—Gassificazione di biomasse lignocellulosiche in sistemi di cogenerazione di piccola taglia (<200 kW) per applicazioni nel settore terziario" (CUP G96G16000800003) funded by "Cassa per i servizi energetici e ambientali" within the call "Bando di gara per progetti di ricerca di cui all'art. 10, comma 2, lettera b) del decreto 26/1/2000, previsti dal Piano triennale 2012–2014 della ricerca di sistema elettrico nazionale e dal Piano operativo annuale 2013".

Conflicts of Interest: The authors declare no conflict of interest. The funders had no role in the design of the study, in the collection, analyses, or interpretation of data, in the writing of the manuscript, or in the decision to publish the results.

References

1. Srirangan, K.; Akawi, L.; Moo-Young, M.; Chou, C.P. Towards sustainable production of clean energy carriers from biomass resources. *Appl. Energy* **2012**, *100*, 172–186. [CrossRef]
2. Molino, A.; Chianese, S.; Musmarra, D. Biomass gasification technology: The state of the art overview. *J. Energy Chem.* **2016**, *25*, 10–25. [CrossRef]
3. Heidenreich, S.; Foscolo, P.U. New concepts in biomass gasification. *Prog. Energy Combust. Sci.* **2015**, *46*, 72–95. [CrossRef]
4. Ahrenfeldt, J.; Thomsen, T.P.; Henriksen, U.; Clausen, L.R. Biomass gasification cogeneration—A review of state of the art technology and near future perspectives. *Appl. Therm. Eng.* **2013**, *50*, 1407–1417. [CrossRef]
5. Belgiorno, V.; De Feo, G.; Della Rocca, C.; Napoli, R.M.A. Energy from gasification of solid wastes. *Waste Manag.* **2003**, *23*, 1–15. [CrossRef]
6. Asadullah, M. Biomass gasification gas cleaning for downstream applications: A comparative critical review. *Renew. Sustain. Energy Rev.* **2014**, *40*, 118–132. [CrossRef]
7. Han, J.; Kim, H. The reduction and control technology of tar during biomass gasification/pyrolysis: An overview. *Renew. Sustain. Energy Rev.* **2008**, *12*, 397–416. [CrossRef]
8. Li, C.; Suzuki, K. Tar property, analysis, reforming mechanism and model for biomass gasification—An overview. *Renew. Sustain. Energy Rev.* **2009**, *13*, 594–604. [CrossRef]
9. Morf, P.O. Secondary Reactions of Tar during Thermochemical Biomass Conversion. Ph.D. Thesis, Swiss Federal Institute of Technology, Zurich, Switzerland, 2001.
10. Milne, T.A.; Evans, R.J.; Abatzoglou, N. *Biomass Gasifier "Tars": Their Nature, Formation and Conversion*; National Renewable Energy Laboratory (NREL): Golden, CO, USA, 1998.
11. Evans, R.J.; Milne, T.A. Molecular Characterization of the Pyrolysis of Biomass. 1. Fundamentals. *Energy Fuels* **1987**, *1*, 123–137. [CrossRef]
12. Evans, R.J.; Milne, T.A. Molecular Characterization of the Pyrolysis of Biomass. 2. Applications. *Energy Fuels* **1987**, *1*, 311–319. [CrossRef]
13. Evans, R.J.; Milne, T.A. Chemistry of Tar Formation and Maturation in the Thermochemical Conversion of Biomass. In *Developments in the Thermochemical Biomass Conversion*; Bridgwater, A.V., Boocock, D.G.B., Eds.; Springer: Dordrecht, The Netherlands, 1997; Volume 2.
14. Kiel, J.H.A.; van Paasen, S.V.B.; Neeft, J.P.A.; Devi, L.; Ptasinski, K.J.; Janssen, F.J.J.G.; Meijer, R.; Berends, R.H.; Temmink, H.M.G.; Brem, G.; et al. *Primary Measures to Reduce Tar Formation in Fluidised-Bed Biomass Gasifiers*; Energy Research Centre of the Netherlands (ECN): Petten, The Netherlands, 2004.
15. Devi, L.; Ptasinski, K.J.; Janssen, F.J.J.G. A review of the primary measures for tar elimination in biomass gasification processes. *Biomass Bioenergy* **2003**, *24*, 125–140. [CrossRef]
16. Anis, S.; Zainal, Z.A. Tar reduction in biomass producer gas via mechanical, catalytic and thermal methods: A review. *Renew. Sustain. Energy Rev.* **2011**, *15*, 2355–2377. [CrossRef]
17. Thapa, S.; Bhoi, P.R.; Kumar, A.; Huhnke, R.L. Effects of Syngas Cooling and Biomass Filter Medium on Tar Removal. *Energies* **2017**, *10*, 349. [CrossRef]

18. Thapa, S.; Indrawan, N.; Bhoi, P.R.; Kumar, A.; Huhnke, R.L. Tar reduction in biomass syngas using heat exchanger and vegetable oil bubbler. *Energy* **2019**, *175*, 402–409. [CrossRef]
19. Gambarotta, A.; Manganelli, M.; Morini, M. A model for filter diagnostics in a syngas-fed CHP plant. *Energy Procedia* **2018**, *148*, 400–407. [CrossRef]
20. Casari, N.; Pinelli, M.; Suman, A.; Candido, A.; Morini, M. Deposition of syngas tar in fuel supplying duct of a biomass gasifier: A numerical study. *Fuel* **2020**, *273*, 117579. [CrossRef]
21. Bergman, T.L.; Lavine, A.S.; Incropera, F.P.; DeWitt, D.P. *Fundamentals of Heat and Mass Transfer*, 8th ed.; John Wiley & Sons Inc.: Hoboken, NJ, USA, 2017.
22. *ANSYS CFX User Manual—V 20.2.*; ANSYS, Inc.: Canonsburg, PA, USA, 2020.
23. Mohammadi, K.; Malayeri, M.R. Parametric study of gross flow maldistribution in a single-pass shell and tube heat exchanger in turbulent regime. *Int. J. Heat Fluid Flow* **2013**, *44*, 14–27. [CrossRef]
24. Kim, M.I.; Lee, Y.; Kim, B.W.; Lee, D.H.; Song, W.S. CFD modeling of shell-and-tube heat exchanger header for uniform distribution among tubes. *Korean J. Chem. Eng.* **2009**, *26*, 359–363. [CrossRef]
25. Dufour, A.; Masson, E.; Girods, P.; Rogaume, Y.; Zoulalian, A. Evolution of Aromatic Tar Composition in Relation to Methane and Ethylene from Biomass Pyrolysis-Gasification. *Energy Fuels* **2011**, *25*, 4182–4189. [CrossRef]
26. Vasudevan, T.V.; Gokhale, A.J.; Mahalingam, R. Phoretic Phenomena in Tar Vapor-Particulate Mixture Separation from Fuel Gas Streams. *Can. J. Chem. Eng.* **1985**, *63*, 903–910. [CrossRef]
27. Schiller, L.; Naumann, A. A drag coefficient correlation. *Z. VDI* **1933**, *77*, 318–320.
28. Gosman, A.D.; Ioannides, E. Aspects of computer simulation of liquid-fueled combustors. *J. Energy* **1983**, *7*, 482–490. [CrossRef]
29. Ranz, W.E.; Marshall, W.R. Evaporation from drops: Parts I. *Chem. Eng. Prog.* **1952**, *48*, 141–146.
30. Ranz, W.E.; Marshall, W.R. Evaporation from drops: Parts II. *Chem. Eng. Prog.* **1952**, *48*, 173–180.
31. Euh, S.H.; Kafle, S.; Lee, S.Y.; Lee, C.G.; Jo, L.; Choi, Y.S.; Oh, J.H.; Kim, D.H. Establishment and validation of tar fouling mechanism in wood pellet boiler using kinetic models. *Appl. Therm. Eng.* **2017**, *127*, 165–175. [CrossRef]

energies

Article

Research on Capacity Allocation Optimization of Commercial Virtual Power Plant (CVPP)

Songkai Wang [1,2,*], Rong Jia [2,3], Xiaoyu Shi [3,*], Chang Luo [4], Yuan An [3], Qiang Huang [1], Pengcheng Guo [1], Xueyan Wang [3] and Xuewen Lei [1]

1 School of Water Resources and Hydropower, Xi'an University of Technology, Xi'an 710048, China; sy-sj@xaut.edu.cn (Q.H.); guoyicheng@xaut.edu.cn (P.G.); leixuewen456@163.com (X.L.)
2 Key Laboratory of Smart Energy in Xi'an, Xi'an University of Technology, Xi'an 710048, China; jiarong@xaut.edu.cn
3 School of Electrical Engineering, Xi'an University of Technology, Xi'an 710048, China; anyuan@xaut.edu.cn (Y.A.); wxy000113@163.com (X.W.)
4 Hanjiang-to-Weihe River Valley Water Diversion Project Construction Co., Ltd., Xi'an 710048, China; luochang@hwrvwd.cn
* Correspondence: 1170413040@stu.xaut.edu.cn (S.W.); b21016@xaut.edu.cn (X.S.)

Citation: Wang, S.; Jia, R.; Shi, X.; Luo, C.; An, Y.; Huang, Q.; Guo, P.; Wang, X.; Lei, X. Research on Capacity Allocation Optimization of Commercial Virtual Power Plant (CVPP). *Energies* **2022**, *15*, 1303. https://doi.org/10.3390/en15041303

Academic Editors: Michele Pinelli, Alessio Suman and Nicola Casari

Received: 6 January 2022
Accepted: 22 January 2022
Published: 11 February 2022

Abstract: Commercial virtual power plants (CVPP) connect the form of renewable energy resource portfolio to the power market and reduce the risk of the unstable operation of a single renewable energy. Combining different kinds of large-scale renewable energy in CVPP to provide capacity services like base load, peak shaving, and valley-filling, etc., for the system loads is an urgent problem to be solved. Therefore, it is valuable to analyze the capacity allocation ratio of the CVPP to maximize the utilization of all kinds of energy, especially for the large-scale multi-energy base. This paper proposed a multi-energy coordinated operation framework by considering various load demands, including base load and peak shaving for the capacity allocation of CVPP based on the world's largest renewable energy resource base on the upstream area of the Yellow River. The main procedures of this framework are as follows: (1) A paratactic model satisfying base load and peak shaving is proposed to determine the ability of the CVPP operation model's capacity services to meet the different demands of the power system load. (2) A hybrid dimension reduction algorithm with a better convergence rate and optimization effect solves the proposed paratactic model based on the ReliefF and the Adaptive Particle Swarm Optimization (APSO). The results show that the large-scale CVPP with different compositions can achieve both of the goals of a stable base load output and stable residual load under different weather conditions. Compared with the operation on sunny days, the base load fluctuation and residual load fluctuation of CVPP on rainy days are reduced by 14.5% and 21.9%, respectively, proving that CVPP can alleviate renewable energy's dependence on weather and improve energy utilization.

Keywords: commercial virtual power plants; capacity allocation; base load; peak shaving; hybrid dimension reduction algorithm

1. Introduction

With the continuous increasing proportion of the renewable energy (RER) in the power grid, scholars around the world have proposed virtual power plant (VPP) technology in recent years to realize the integration and control of this RER [1]. Through advanced communication technology and software management systems, VPPs can be considered as the aggregation and optimization of RER, energy storage facilities, controllable loads, and other types of power resources in the power grid [2,3]. VPPs can participate in the power grid operators to coordinate the contradiction between the power grid and RER and realize the optimal allocation and efficient utilization of resources. According to the different functions of VPPs, the EU Fenix project divides VPPs into two categories [4]. One is technical virtual power plants (TVPP), which provides the system operation, management, balance, and auxiliary services for distribution and transmission system operators on the technical management level [5]. The other is commercial virtual power plants (CVPP), which refers to the virtual power plant from commercial revenue and connects renewable energy resources as a portfolio to the power market [6]. The investment portfolio between renewable energy resources provides diversified energy for the power market and reduces the risk of unbalanced operation of single renewable energy of the power market. Arranging the capacity of various renewable energies in the most appropriate proportion to meet the different needs of the market has become a new problem. Allocating the capacity of renewable energy in CVPP to meet different load requirements and giving full play to the benefits of renewable energy will be the main problem of VPP dispatching operation. Therefore, it is valuable to analyze the capacity allocation ratio of the CVPP to maximize the utilization of all kinds of energy, especially for the large-scale multi-energy base.

Countries worldwide have launched research on the capacity allocation of CVPP. Reference [7] discussed the best capacity of the VPP with wind power and pumped-hydro by considering both the maximum benefit and maximum renewable energy ratio. However, the type of energy is too singular. Reference [8] investigated a unit optimal commitment method by considering the electricity price for a VPP to participate in the electricity market. Reference [9] took the maximum economic benefit of the dispatching model as the objective function. A VPP including thermal power, wind power, CHPS, and pumped storage is constructed. However, photovoltaic was not discussed. Reference [10] proposed a new multi-objective genetic algorithm to evaluate the VPP's capacity proportioning by aiming at the lowest cost. However, the load side demand was not considered. Reference [11] considered the spare capacity and carbon emission cost to propose an improved heuristic method for the capacity planning of the VPP. Reference [12] proposed a VPP project framework by considering the societal benefits and the market proceeds, aiming at cost, revenue, and environmental emission. Reference [13] established a microgrid-type VPP structure and proposed a capacity proportioning model with a game-theory algorithm. However, the scale of the VPP was small. Reference [14] developed a new microgrid two-layer game model by a collaborative subsequence game of different energies to obtain the optimal capacity of a VPP. Reference [15] proposed random scene creation and robust majorization to reduce uncertainty and optimize the VPP capacity proportioning program. Reference [16] configured the capacities of wind/solar/gas microgrids, aiming at providing demand-side services, dealing with emergencies, and supporting system toughness. However, hydropower was not discussed. Reference [17] proposed an island-type VPP capacity proportioning optimization model in view of measuring the cost of renewable power generation. However, this lacked application to large-scale energy scenes. Reference [18] established an optimal model of the electricity and gas combination VPP to obtain optimal operation time and capacity of different energy sources. Reference [19] proposed a new intelligent algorithm that combined the GA and the water drop algorithm to obtain the position and capacity of distributed generation in the microgrid-type VPP. However, the proposed algorithm is too complex. Reference [20] established a mathematical model by using the AHP to analyze the multi-energy capacity ratio and economy. Reference [21] emulated a typical combined cycle power generation unit to analyze the susceptibility of carbon emission and electricity price. Reference [22] considered energy cooperation among different regions to build a multi-energy VPP capacity ratio model, and minimizing the system cost, by proposing a random scene creation. However, the impact of the load side was not discussed. Reference [23] proposed a two-layer model for the full life cycle of the muti-energy system to obtain the minimum economic cost and the optimal control strategy.

The past studies have many common deficiencies on the capacity allocation of CVPP: (1) Capacity allocation research is usually from the perspective of power generation. The built model aims to minimize power generation cost or maximize economic benefit [24], which does not consider the load demand, and the operating mode is too singular. (2) The capacity allocation studies of VPPs are mainly for the small capacity systems, microgrids, and Combined Cooling Heating and Power (CCHP) [25,26]. Only a few studies have been conducted on sizeable virtual power plants with large-scale renewable energy. (3) The previous research always chose typical GA, PSO, DE, and other algorithms to solve the model [27,28], which could easily fall into a local convergence and gain inaccurate results.

In order to supplement and innovate the previous research, this paper provides a new study framework for the capacity allocation of CVPP by considering different load demands. The main contributions are as follows: (1) A paratactic model combining the base load type (BLT), which is responsible for the stable part of the power system load, and the peak shaving type (PST), which is responsible for the peak part of the power system load, is proposed for determining the operation model of CVPP. (2) A hybrid dimension reduction algorithm on ReliefF and APSO is a new avenue for resolving the proposed paratactic model. The built framing is tested by the world's largest renewable energy resource base on the upstream area of the Yellow River.

The research structure of this paper is as follows: the CVPP's capacity allocation modeling and the hybrid dimension reduction algorithm based on ReliefF and APSO are described in Section 2. Section 3 probes into a case study, and Section 4 examines its results. Finally, the conclusions of the study are drawn in Section 5.

2. Method

By uniting the CVPP generation power curve and the load curve, this paper divides the CVPP capacity allocation model into the base load type (BLT), which is responsible for the stable part of power system load, and the peak shaving type (PST), which is responsible for the peak part of the power system load. A hybrid dimension reduction algorithm on ReliefF and APSO is a new avenue for resolving the paratactic model. The flowchart of the proposed the CVPP capacity allocation framework is shown in Figure 1.

2.1. *CVPP Capacity Allocation Modeling*

2.1.1. Base Load Type

The base load type is responsible for the stable part of the power system load to operate under high-efficiency conditions as far as possible. The model is used to discuss the practicability of CVPP and compute the most steady capacity allocation.

The base load type CVPP analyzed in this section includes wind power, photovoltaic, and hydropower. Aiming at the minimum fluctuation of CVPP output, the model calculates various energy capacity allocations as follows

$$\min F_1 = \sqrt{\frac{1}{T-1}\sum_{t=1}^{T}(P_t - P(av))} \tag{1}$$

$$P_t = \alpha P_p(t) + (1-\alpha)\beta P_w(t) + (1-\alpha)(1-\beta)P_h(t) \tag{2}$$

where F_1 denotes the standard deviation of P_t; P_t denotes the total out power of CVPP, the F_1 value and the total output fluctuation are in the direct ratio; $P_p(t)$, $P_w(t)$, and $P_h(t)$ denote the photovoltaic power, wind power, and hydropower generation, respectively, the calculation formulas will be introduced in Section 2.2; α and β denote the weight—α is the proportion of photovoltaic power in total CVPP output, $(1-\alpha)\beta$ is the wind power proportion in total CVPP output, $(1-\alpha)(1-\beta)$ is the hydropower proportion in total CVPP output. Then, when α = 1, it is combined into 100% photovoltaic; when α = 0 and β = 1, means 100% wind power generation; when α = 0 and β = 0, means 100% hydropower generation.

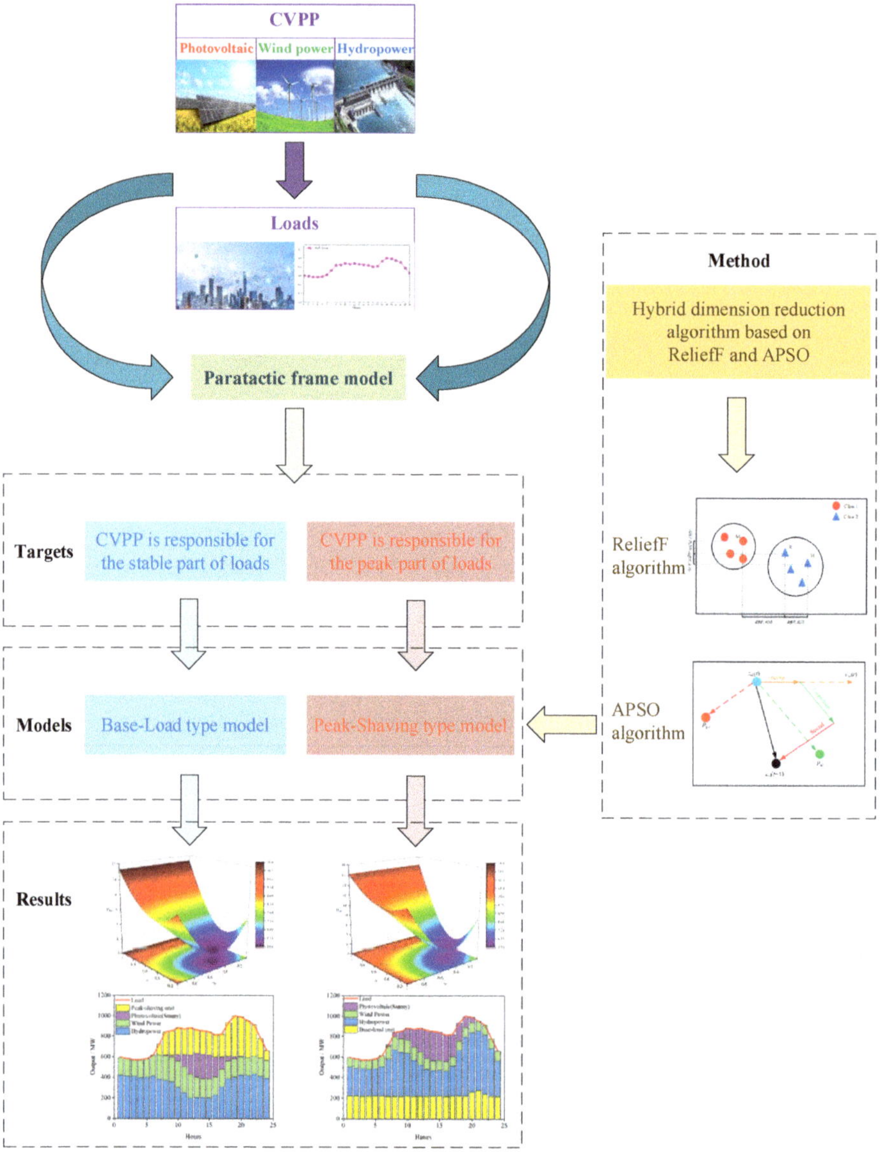

Figure 1. The flowchart of the proposed CVPP capacity allocation framework.

2.1.2. Peak Shaving Type

The peak shaving type is responsible for the peak part of the power system load, making the residual load allocated to power sources with poor regulation capacity such as thermal power more stable. The peak shaving type will reduce the number of startups and shutdowns of thermal power units, save startup consumption, improve the overall operational efficiency of the power station.

The peak shaving type aims at minimizing the mean square deviation of residual load after deducting wind power, photovoltaic, and hydropower in CVPP from grid load [29]. The formula is as follows

$$\min F_2 = \sqrt{\frac{1}{T}\sum_{t=1}^{T}\left[R_t - \frac{1}{T}\sum_{t=1}^{T} R_t\right]^2} \tag{3}$$

$$R_t = L_t - \alpha P_P(t) - (1-\alpha)\beta P_W(t) - (1-\alpha)(1-\beta)P_H(t) \tag{4}$$

where F_2 denotes the mean square deviation of R_t; R_t denotes the residual load after deducting wind power, photovoltaic, and hydropower, the larger the F_2 value, the greater the residual load fluctuation, and the smaller the F_2 value, the smaller the residual load fluctuation; $P_p(t)$, $P_w(t)$, and $P_h(t)$ denote the photovoltaic power, wind power, and hydropower generation, respectively; the remaining variables are the same as shown in Section 2.1.1.

2.2. Power Modeling

The CVPP in this paper only includes photovoltaic, wind, and hydropower.

(1) Wind power model

The wind power output P_w is the power generation of wind turbines under different wind speed conditions, and will be calculated as follows [30]

$$P_w = \begin{cases} 0 & v \le v_i \quad or \quad v \ge v_o \\ P_r \frac{v - v_i}{v_r - v_i} & v_i \le v \le v_r \\ P_r & v_r \le v \le v_o \end{cases} \tag{5}$$

where P_r denotes the rated wind output under the rated conditions; v denotes the real-time wind speed; v_i denotes the cut-in wind speed; v_o denotes the cut-out wind speed; v_r denotes the rated wind speed.

(2) Photovoltaic model

The photovoltaic output P_p is linearly related to the solar light intensity, and is calculated as follows [31]

$$P_p = P_r \frac{G}{G_r}[1 + \alpha_T(T - T_r)] \tag{6}$$

where P_r denotes the rated photovoltaic output under the rated conditions; G denotes the actual solar irradiance (W/m^2); G_r denotes the rated solar irradiance (1000 W/m^2); a_T denotes the temperature coefficient; T denotes the actual surface temperature of the photovoltaic cells (°C); T_r denotes the rated surface temperature of the photovoltaic cells (25 °C) [32].

(3) Hydropower model

This study only deals with reservoir power plants. The high water period, low water period, environmental flows (such as meltwater, rainfall, evaporated water, etc.), and tributary flows are all not considered for simplifying the calculation in this study. The output power P_h will be calculated as follows [33]

$$P_h = \eta_h g h \rho Q \tag{7}$$

where g denotes the gravitational acceleration (9.81 m/s^2); η_h denotes the efficiency of the generator; ρ denotes the density of the water (1000 kg/m^3); Q denotes the water flow (m^3/s); h denotes the height of the water drop (m).

The constraints of wind power, photovoltaic, and hydropower are as follows [34]

$$0 \le P_W \le P_{W,\max} \tag{8}$$

$P_{W,max}$ denotes the wind power rated output (MW);

$$0 \leq P_P \leq P_{P,\max} \tag{9}$$

$P_{P,max}$ denotes the photovoltaic rated output (MW);

$$V_{t+1} = V_t + (I_t - Q_t)\Delta t \tag{10}$$

V_{t+1} and V_t denotes the reservoir storage (m^3) at the end and the beginning, respectively; I_t denotes the reservoir inflow (m^3/s);

$$V_t^l \leq V_t \leq V_t^u \tag{11}$$

$V_t{}^l$ and $V_t{}^u$ denote the lower and upper limits for reservoir storage (m^3), respectively;

$$Q_{t\min} \leq Q_t \leq Q_{t\max} \tag{12}$$

$Q_t{}^l$ and $Q_t{}^u$ denote the lower and upper limits for river discharge flow (m^3/s), respectively;

$$N_t^l \leq N_t \leq N_t^u \tag{13}$$

$N_t{}^l$ denotes the lower limits for hydropower output (MW) (The output of the reservoir must be discharged for meeting the downstream irrigation, water supply, navigation, etc.); $N_t{}^u$ denotes the upper limits for hydropower output (MW);

$$N_s = N_t - N_t^l \tag{14}$$

N_s denotes the schedulable output (MW), which refers to the output that the hydropower can offer in CVPP on the satisfying water dispatching premise.

2.3. *Hybrid Algorithm Based on ReliefF and APSO*

2.3.1. APSO Algorithm

Virtual power plant optimal scheduling is a typical power system optimization problem with high dimension, nonlinearity, and multi-constrained. Compared with genetic algorithms and other algorithms, the PSO algorithm is a better algorithm for optimizing and solving the problem.

Each particle in the PSO algorithm is described by position and velocity vectors [35]. Assuming the total number of particles is M, the position and velocity of the nth particle in dimension d are as follows

$$\begin{cases} X'_n = (x'_{n1}, x'_{n2}, \ldots, x'_{nd})^T, n = 1, 2, \ldots, M \\ V'_n = (v'_{n1}, v'_{n2}, \ldots, v'_{nd})^T, n = 1, 2, \ldots, M \end{cases} \tag{15}$$

Each particle adjusts its speed and position by tracing the last best individual position and best group position, which are expressed as follows

$$\begin{cases} P'_n = (P'_{n,1}, P'_{n,2}, \ldots, P'_{n,d}) \\ P'_g = (P'_{g,1}, P'_{g,2}, \ldots, P'_{g,d}) \end{cases} \tag{16}$$

where P_n denotes the individual optimal position of the nth particle; P_g denotes the best position of the group obtained from all particles in the previous iteration.

The speed and position update formula of the PSO algorithm is presented as follows

$$\begin{cases} X_{nd}^{k+1} = X_{nd}^k + V_{nd}^{k+1} \\ V_{nd}^{k+1} = \omega' V_{nd}^k + c'_1 r'_1 (P_{nd}^k - X_{nd}^k) + c'_2 r'_2 (P_{gd}^k - X_{nd}^k) \end{cases} \tag{17}$$

where ω denotes the inertia weight factor. c'_1 and c'_2 denote the learning factor; r'_1 and r'_2 denote the uniformly distributed random numbers (0, 1).

The PSO algorithm often has premature convergence and other problems in dealing with multi-extreme function problems. The Adaptive Particle Swarm Optimization (APSO) algorithm is adopted by adding inertial frames during speed update to help particles quit of local extreme values [36]. The formula is as follows

$$\omega' = (\omega'_{max} - \omega'_{min}) \times \exp(-(\tau \times \frac{k}{K_{max}})^2) + \omega'_{min} \tag{18}$$

where ω'_{max} and ω'_{min} denote the maximum and minimum inertia coefficients; K_{max} denotes the maximum Iterations; τ denotes the empirical value, generally within (20, 55). Larger ω' makes APSO have strong whole region searchability, while smaller ω' tends to local search. The APSO can achieve different search results by changing the inertia coefficient. With the gradual reduction in inertia coefficient, the algorithm also has the initial global search to the later local search.

2.3.2. ReliefF Algorithm

Although the APSO algorithm can solve the traditional PSO algorithm's problems, it is easy to fall into local optimization and premature convergence to a certain extent. Sometimes, a single algorithm cannot obtain the optimization results because of its defects. For example, the APSO algorithm has strong global searchability and weak local searchability in the initial stage, which will result in low computational efficiency in the initial step of the CVPP optimal scheduling model. Therefore, selecting an appropriate algorithm to supplement this shortcoming is the focus of this paper.

Feature selection is a prevalent dimensionality reduction method with solid local searchability. It refers to selecting the feature subset that makes a certain evaluation standard optimal from the original feature set. Its purpose is to select some of the most valuable features to reduce the data feature dimension and make the chosen optimal feature subset approximate or even better prediction results before feature selection. It improves the generalization ability and the calculation efficiency of the model, the data's actual utility, and reduces the frequency of dimensional disasters.

The core content of the ReliefF algorithm is the correlation between features and dataset class marks [37]. ReliefF algorithm randomly selects a sample R from the training set. Find k samples that belong to the same class as R and are closest to it, which is called *NearHit*, and then find k samples that belong to a different class from R and are closest to it, which is called *NearMiss*. The weight is obtained according to the following algorithm:

(1) Calculate ${S_A}^{Hit}$, the distance between R and *NearHit* on each feature A.
(2) Calculate ${S_A}^{Miss}$, the distance between R and *NearMiss* on each feature A.
(3) Compare the two distances ${S_A}^{Hit}$ and ${S_A}^{Miss}$. If ${S_A}^{Hit}$ is greater than ${S_A}^{Miss}$, each feature of the A is helpful to distinguish the same kind and different kinds. If ${S_A}^{Hit}$ is less than ${S_A}^{Miss}$, each feature of the A hurts, distinguishing the same class and different classes and reduces the weight of the feature.
(4) Repeat the above process m times to gain the average weight of each feature, as is shown in Formula (19)

$$W(A) = -\sum_{j=1}^{k} diff(A, R, H_j)/mk + \sum_{C \notin class(R)} [\frac{p(C)}{1 - p(class(R))} \sum_{j=1}^{k} diff(A, R, M_j(C))]/mk \tag{19}$$

where $diff(A,R,H_j)$ denotes the difference between sample R and H_j in feature A, as is shown in Formula (20); $M_j(C)$ denotes the *j-th* nearest sample in category C; $p(C)$ denotes the target probability of class C, given by Formula (21). When the number of samples is approximately the same, there are $p(C) = 1/C$.

$$diff(A, R_1, R_2) = \frac{|R_1[A] - R_2[A]|}{\max(A) - \min(A)} \quad (20)$$

$$p(C) = \frac{N_c}{\sum_{c=1}^{C} N_c} \quad (21)$$

After m repetitions, each feature gets an average weight. The larger the average weight, the better the feature is at distinguishing different categories. The smaller the average weight, the worse the feature is at distinguishing different categories.

2.3.3. ReliefF–APSO Hybrid Algorithm

This paper combines the ReliefF with the APSO to obtain a better hybrid algorithm by using the advantages of the ReliefF algorithm's strong local searchability and the APSO algorithm's strong global optimization ability. The ReliefF–APSO hybrid algorithm can make the particle positions in a sub-population relatively concentrated and learning relatively easy, and improve the search efficiency, spending limited time on the most effective search. The structure diagram of the ReliefF–APSO hybrid algorithm is shown in Figure 2.

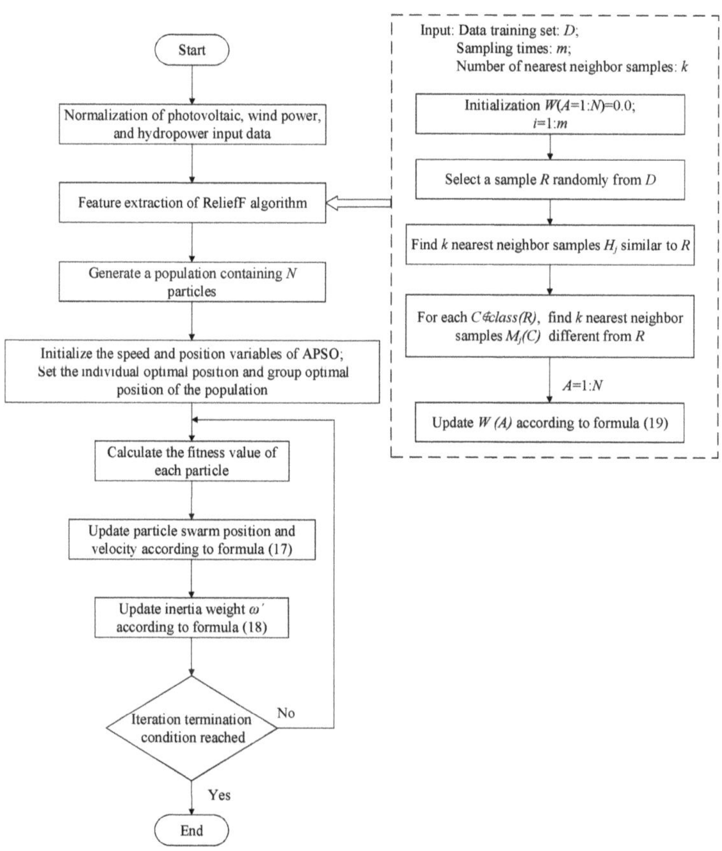

Figure 2. Flowchart for the ReliefF–APSO hybrid algorithm.

The steps of the hybrid algorithm are as follows:

(1) The given data set is normalized by Z-score before training;
(2) Using the Formula (19) to extract the features of the ReliefF algorithm and take the first d features with relatively large weight as the training set and test set of APSO, $d = 100$;
(3) Generate population, set the number of particles N, set each particle as a random number vector within $(-1, 1)$, and set the number of neurons and hidden layer nodes. In the experiment, N takes 20;
(4) Initialize the speed and position variables of APSO, and set the individual optimal position and group optimal position of the population;
(5) Calculate the fitness value of each particle;
(6) Update the position and velocity of the adaptive particle swarm according to Formulas (17) and (18);
(7) Judge whether the maximum number of iterations is reached. If so, stop the iteration. Otherwise, turn to step (5) and continue the iteration.

3. Case Study

The built framing is tested using the world's largest renewable energy resource base on the upstream area of the Yellow River in China. This area is in the southeast of Qinghai Province, China, and is rich in photovoltaic, wind power, hydropower, and other renewable energy resources [38]. The location of the upstream area of the Yellow River is shown in Figure 3.

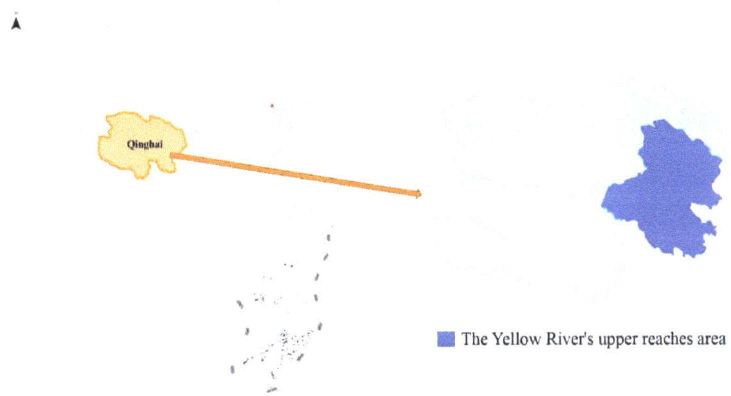

Figure 3. Location of the upstream area of the Yellow River.

The data of wind speed, wind force, temperature, and solar radiation of Gonghe observation point and Xinghai observation point in the upstream area of the Yellow River are provided by the China Integrated Meteorological Information Sharing System (CIMISS). The mean daily wind speed and daily maximum total irradiance in 2019 are shown in Figure 4. The Yellow River's historical hydrological data provides hydropower output.

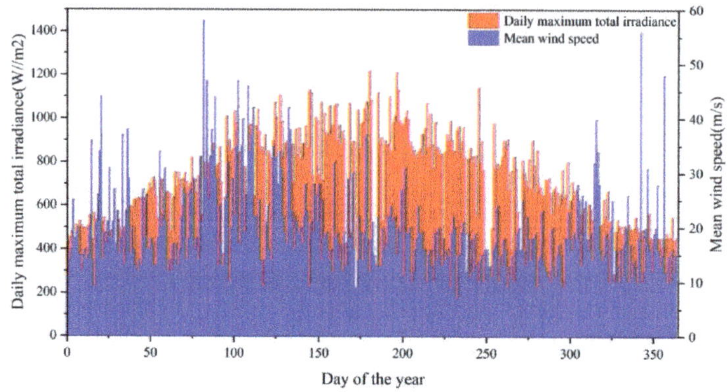

Figure 4. Data map for wind and solar resources of the upstream area of the Yellow River.

4. Results and Discussion

Due to the instability of renewable energy output, it can only participate in the medium- or long-term power market, which cannot be a member of the spot market that needs flexible adjustment [39]. The integration of renewable energy through CVPP can improve the stability and flexibility of the system and enable renewable energy to participate more in the day-ahead and real-time market. This paper selects the daily scale as the research scope, and the typical daily load curve data and typical daily renewable energy output data including sunny and rainy days, are selected to avoid the impact of renewable energy and load uncertainty. The typical daily hourly output of renewable energies and the daily hourly load curve of the target area in the study area are shown in Figure 5.

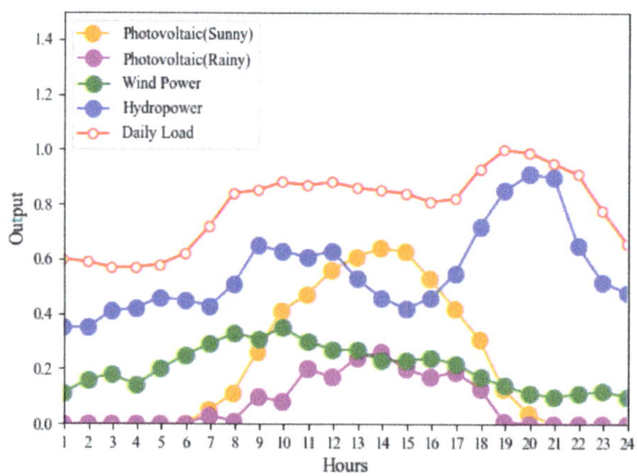

Figure 5. The hourly output and daily load curve.

Using Python3.7 to simulate the proposed paratactic model. This section sets the total installed capacity of CVPP to be 1000 MW. This chapter assumes that at the beginning of the calculation, each kind of energy has the same capacity proportion and contribution to the CVPP. The value range of weight α and β from Equations (2) and (4) is (0, 1), and the change step is 0.01.

4.1. Comparative Analysis of ReliefF–APSO Hybrid Algorithm, APSO, and PSO

The ReliefF–APSO hybrid algorithm, APSO, and PSO are respectively used for solving the base load type operation model on sunny days. In the PSO algorithm, $w = 0.6$, learning factor $c_1 = c_2 = 2$. In the ReliefF–APSO hybrid algorithm and APSO algorithm, $w_{max} = 0.9$, $w_{min} = 0.4$, and learning factor $c_{1a} = c_{2a} = 2$. The number of particles is 20 and the maximum number of iterations is 100. The results of the algorithm comparison are shown in Figure 6.

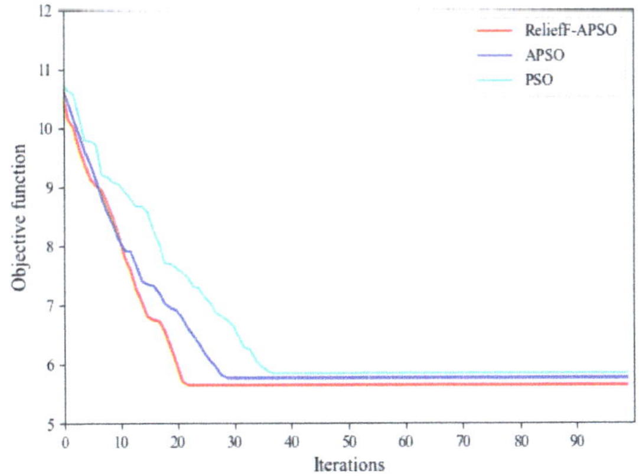

Figure 6. Algorithm comparison between the ReliefF–APSO, APSO, and PSO.

The ReliefF–APSO hybrid algorithm gains the lowest objective function and finds the optimal solution in the 23rd iteration. In contrast, the APSO algorithm finds the optimal solution in the 30th iteration, and the PSO algorithm finds the optimal solution in the 38th iteration. The convergence rate of the ReliefF–APSO hybrid algorithm is 23.33% quicker than the APSO and 39.47% quicker than the PSO. Moreover, the smoothness of the curve shows that the optimization effect of the ReliefF–APSO hybrid algorithm is better than that of the APSO and PSO.

4.2. Base load Type Operation Model

The base load type is responsible for the relatively stable "basic" part of the load, which takes the minimum fluctuation of the total output power of CVPP as the optimization goal. Peak shaving is distributed to thermal power and other energy. The results of the energy proportion of the base load type operation on sunny and rainy days are shown in Figure 7.

When the output of CVPP is all composed of photovoltaic output, the total output fluctuates the most because compared with wind power and hydropower, photovoltaic fluctuates the most in a day. When the output composition of CVPP is 25% photovoltaic, 27% wind power, and 48% hydropower, the fluctuation of total output on sunny days is the smallest. The small proportion of photovoltaic is because photovoltaic can only provide daytime output, but its fluctuation is too strong to maintain the stability of total output. Wind power has output all day, so the proportion is slightly higher than that of photovoltaic.

When the output composition of CVPP is 4% photovoltaic, 18% wind power, and 78% hydropower, the total output fluctuation on rainy days is the smallest. On rainy days, the system's total output needs to be mainly borne by hydropower with stronger regulation capacity because the PV output is too small, and the wind power output fluctuates strongly.

The smallest F_1 is 5.65 on sunny days and 4.83 on rainy days, which means the CVPP output on rainy days is more stable. That is because on rainy days, the proportion of photovoltaic is small, and the proportion of hydropower is high, which is more conducive to maintaining the stability of the overall output. The largest F_1 is 10.63 on sunny days and 13.09 on rainy days, which means compared to the single energy mode, the CVPP could reduce the base load fluctuation by 46.8% on sunny days and 63.1% on rainy days. The operation results of the base load type in different weather are shown in Figure 8.

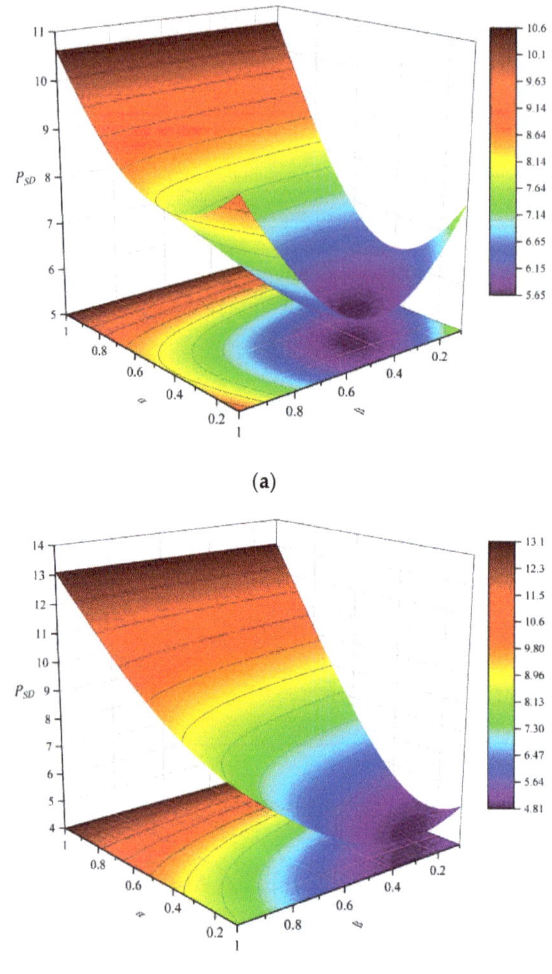

(a)

(b)

Figure 7. The energy proportion of the base load type operation. (a) Sunny days; (b) rainy days.

Although the output composition of the base load type CVPP is different on sunny and rainy days, both operation results show that the stability of the base load is well maintained. On sunny days, the output of photovoltaic is remarkable between 9:00 and 19:00, and meets the crest value between 12:00 and 16:00. To ensure photovoltaic consumption and maintain the stability of total output, the hydropower output needs to be reduced during this period. On rainy days, the photovoltaic output is unstable and undulates, which leads to the proportion of photovoltaic being very small. The stability of the base load is mainly guaranteed by hydropower and wind power. As is shown in Figure 9, the results show that the CVPP output fluctuation on rainy days is 14.5% more than on sunny days.

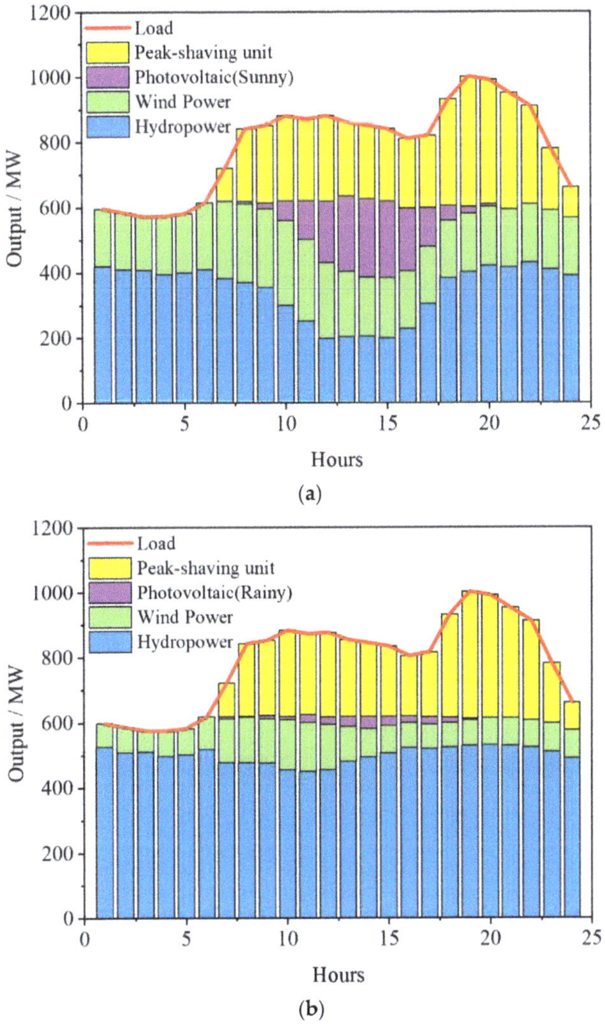

Figure 8. The operation results of the base load type. (**a**) Sunny days; (**b**) rainy days.

4.3. Peak shaving Type Operation Model

The peak shaving type takes the minimum mean square deviation of the residual load after deducting wind power, photovoltaic, and hydropower from the grid load as the optimization goal. It makes the residual load distributed to the power supply with poor regulation capacity more stable, such as thermal power. The results of the energy proportion of the peak shaving type operation on sunny and rainy days are presented in Figure 10.

When the output composition of CVPP is 29% photovoltaic, 13% wind power, and 58% hydropower, the residual load fluctuation is the smallest. The higher proportion of PV is due to the high photovoltaic output on sunny days, which can participate more in peak shaving in the early peak hours to reduce the demand for residual load at peak hours and stabilize thermal power output.

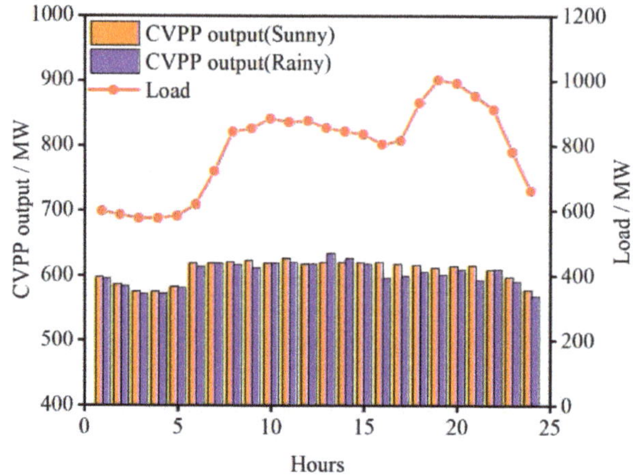

Figure 9. The CVPPV output of the base load type.

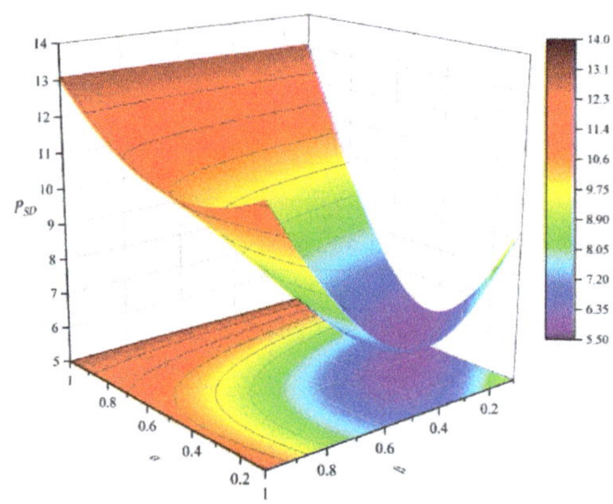

(**a**)

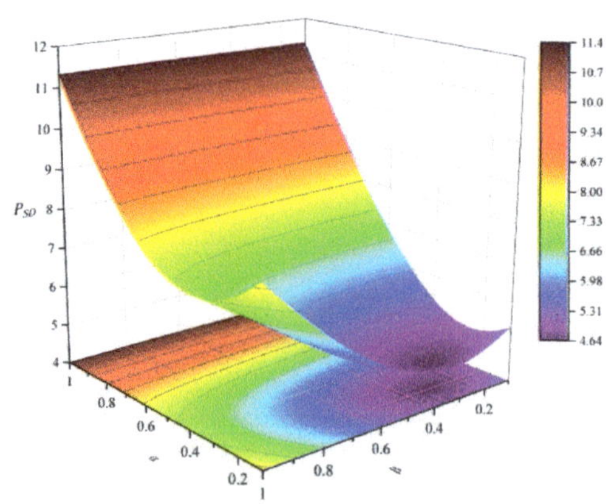

(**b**)

Figure 10. The energy proportion of the peak shaving type operation. (**a**) Sunny days; (**b**) rainy days.

On rainy days, the hydropower station needs to consider flood control factors and cannot fully participate in peak shaving. Therefore, the proportion of hydropower should not be too high, and wind power and some photovoltaic power are needed to assist peak shaving. When the output composition of the virtual power plant is 12% photovoltaic, 22% wind power, and 66% hydropower, the residual load fluctuation is the smallest, and the system peak shaving is mainly completed by hydropower and wind power. The operation results of the peak shaving type in different weather are shown in Figure 10.

The smallest F_2 on sunny days is larger than on rainy days, which means the residual load on rainy days is more stable than on sunny days. That is because, on sunny days, the proportion of photovoltaic is large. However, it cannot provide the peak shaving power at the evening peak, so more thermal power and other energy are needed to help hydropower and wind power for peak shaving. The largest F_2 is 13.09 on sunny days and 11.34 on rainy days, which means compared to the single energy mode, the CVPP could reduce the residual load fluctuation by 54.3% on sunny days and 58.8% on rainy days. The operation results of the peak shaving type in different weather are shown in Figure 11.

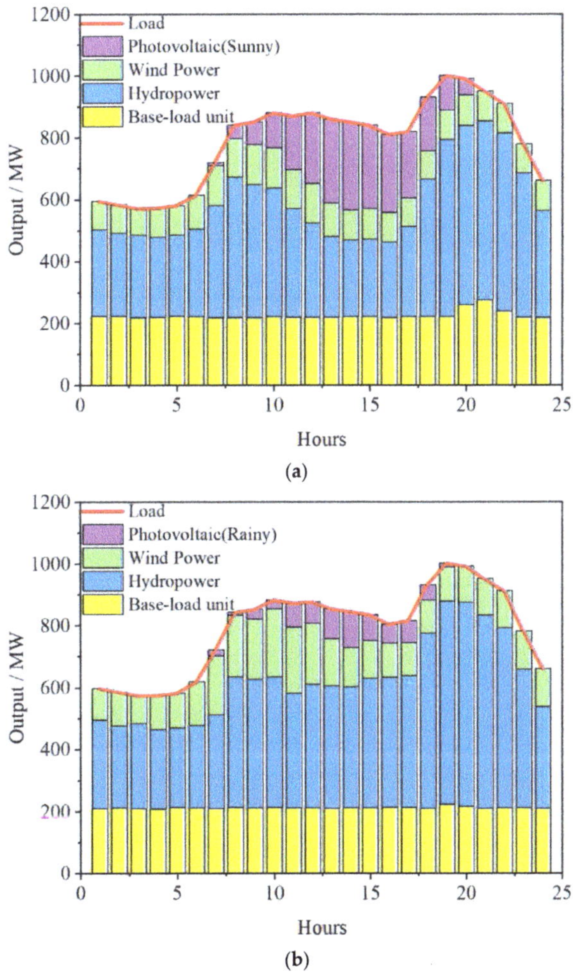

Figure 11. The operation results of the peak shaving type. (**a**) Sunny days; (**b**) rainy days.

Both operation results of sunny days and rainy days show that the residual load of the system has good stability through the CVPP peak shaving. Since the photovoltaic has no output in the evening, the peak shaving assignment is completed by hydropower and wind power during the evening peak period from 19:00 to 22:00. As is shown in Figure 12, on sunny days, photovoltaic has almost no output from 20:00, thus hydropower and wind power cannot satisfy the system peak shaving capacity. Then, the residual load of the system increases, which means that more thermal power and other units partake in the peak shaving. On rainy days, due to the small proportion of photovoltaic and large proportion of hydropower, the peak shaving capacity can be met by the CVPP without additional units. It can be seen that the residual load fluctuation on rainy days is 21.9% more than on sunny days.

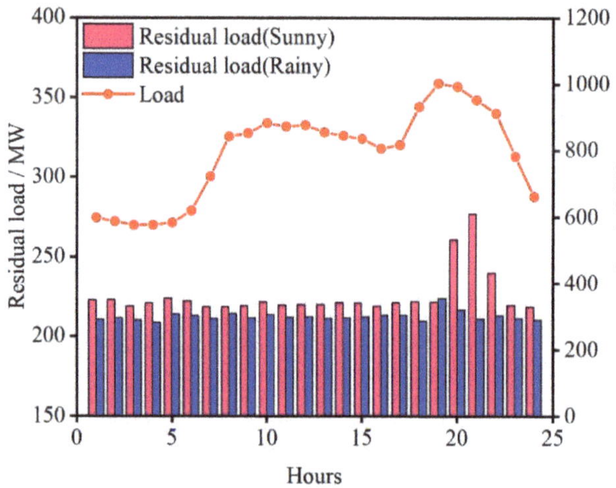

Figure 12. The residual load of the peak shaving type.

The results of this section confirm that the integration of renewable energy through CVPP can provide more stable and flexible power generation resources for system load. Compared to the single energy mode, the CVPP could reduce the base load fluctuation and the residual load fluctuation on both sunny and rainy days, while the base load fluctuation is reduced by 46.8% on sunny days and 63.1% on rainy days, and the residual load fluctuation is reduced by 54.3% on sunny days and 58.8% on rainy days. The base load fluctuation and residual load fluctuation of CVPP on rainy days are reduced by 14.5% and 21.9% more than on sunny days, respectively, proving that CVPP can alleviate renewable energy dependence on weather and improve energy utilization.

5. Conclusions

This paper's research shows that the percentage of CVPP composition is related to the region and available resources. At the same time, large-capacity CVPP can be responsible for different functions of base load and peak shaving in the power market, which can select different renewable energy ratios according to different weather. The convergence rate of the proposed ReliefF–APSO hybrid algorithm is 23.33% quicker than the APSO and 39.47% quicker than the PSO, and the optimization effect of the ReliefF–APSO hybrid algorithm is also better than that of the APSO and PSO. The output compositions of the CVPP on sunny days are 25% photovoltaic, 27% wind power, and 48% hydropower for the base load type, and 39% photovoltaic, 23.2% wind power, and 37.8% hydropower for the peak shaving type. The output compositions of the CVPP on rainy days are 4% photovoltaic, 18% wind power, and 78% hydropower for the base load type, and 12% photovoltaic, 22% wind power, and 66% hydropower for the peak shaving type. The research of this paper will provide the basis for the flexible capacity allocation construction of CVPP and provide diversity for the selection of VPP services on the user side of the power market.

In practical projects, the services to be provided by CVPP are not limited to base load and peak shaving, and there will be more and more complex working conditions. At the same time, there are many changes in climate and seasons, and the operation of CVPP needs more accurate research.

Author Contributions: For research articles, conceptualization, S.W. and X.S.; methodology, S.W. and R.J.; software, S.W.; validation, C.L., and P.G.; formal analysis, S.W. and X.W.; investigation, S.W. and Y.A.; resources, Q.H. and X.S.; data curation, S.W. and X.L.; writing—original draft preparation, S.W.; writing—review and editing, R.J. and P.G.; visualization, S.W. and C.L.; supervision, R.J.; project administration, R.J. and Y.A.; funding acquisition, R.J. and Q.H. All authors have read and agreed to the published version of the manuscript.

Funding: This work was supported by the National Natural Science Foundation of China (no. 51779206), the Key Projects of Science and Technology Department of Shaanxi Province (2018ZDXM-GY-169), and the Key Industry Innovation Chain Project of Science and Technology Department of Shaanxi Province (2019ZDLGY18-03).

Institutional Review Board Statement: Not applicable.

Informed Consent Statement: Not applicable.

Data Availability Statement: Not applicable.

Conflicts of Interest: The authors declare no conflict of interest.

References

1. Shafiekhani, M.; Ahmadi, A.; Homaee, O.; Shafie-khah, M.; Catalão, J.P. Optimal bidding strategy of a renewable-based virtual power plant including wind and solar units and dispatchable loads. *Energy* **2022**, *239*, 122379. [CrossRef]
2. Rouzbahani, H.M.; Karimipour, H.; Lei, L. A review on virtual power plant for energy management. *Sustain. Energy Technol. Assess.* **2021**, *47*, 101370. [CrossRef]
3. Naval, N.; Yusta, J.M. Virtual power plant models and electricity markets—A review. *Renew. Sustain. Energy Rev.* **2021**, *149*, 111393. [CrossRef]
4. Kieny, C.; Berseneff, B.; Hadjsaid, N.; Besanger, Y.; Maire, J. On the concept and the interest of virtual power plant: Some results from the European project FENIX. In Proceedings of the 2009 IEEE Power & Energy Society General Meeting, Calgary, AB, Canada, 26–30 July 2009; pp. 1–6. [CrossRef]
5. Foroughi, M.; Pasban, A.; Moeini-Aghtaie, M.; Fayaz-Heidari, A. A bi-level model for optimal bidding of a multi-carrier technical virtual power plant in energy markets. *Electr. Power Energy Syst.* **2021**, *125*, 106397. [CrossRef]
6. Elgamal, A.H.; Kocher-Oberlehner, G.; Robu, V.; Andoni, M. Optimization of a multiple-scale renewable energy-based virtual power plant in the UK. *Appl. Energy* **2019**, *256*, 113973. [CrossRef]
7. Papaefthymiou, S.V.; Papathanassiou, S.A. Optimum sizing of wind-pumped-storage hybrid power stations in island systems. *Renew. Energy* **2014**, *64*, 187–196. [CrossRef]
8. Peik-Herfeh, M.; Seifi, H.; Sheikh-El-Eslami, M.K. Decision making of a virtual power plant under uncertainties for bidding in a day-ahead market using point estimate method. *Int. J. Electr. Power Energy Syst.* **2013**, *44*, 88–98. [CrossRef]
9. Pandžić, H.; Kuzle, I.; Capuder, T. Virtual power plant mid-term dispatch optimization. *Appl. Energy* **2013**, *10*, 134–141. [CrossRef]
10. Dufo-López, R.; Bernal-Agustín, J.L.; Yusta-Loyo, J.M.; Domínguez-Navarro, J.A.; Ramírez-Rosado, I.J.; Lujano, J.; Aso, I. Multi-objective optimization minimizing cost and life cycle emissions of stand-alone PV–wind–diesel systems with batteries storage. *Appl. Energy* **2011**, *88*, 4033–4041. [CrossRef]
11. Olamaei, J.; Nazari, M.E.; Bahravar, S. Economic environmental unit commitment for integrated CCHP-thermal-heat only system with considerations for valve-point effect based on a heuristic optimization algorithm. *Energy* **2018**, *159*, 737–750. [CrossRef]
12. Li, J.; Sang, C. Discussion on optimal planning and operation framework for integrated energy system. *Electr. Power Constr.* **2015**, *36*, 41–48. [CrossRef]
13. Shunping, J.; Shoupeng, W.; Fang, F. Game theoretical analysis on capacity configuration for microgrid based on multi-agent system. *Electr. Power Energy Syst.* **2021**, *125*, 106485. [CrossRef]
14. Fang, F.; Zhongyan, Z.; Shunping, J.; Shiyan, H. Two-Layer Game Theoretic Microgrid Capacity Optimization Considering Uncertainty of Renewable Energy. *IEEE Syst. J.* **2021**, *15*, 4260–4271. [CrossRef]
15. Xue, M.; Zhao, B.; Zhang, X.; Jiang, Q. Integrated plan andevaluation of grid connected microgrid. *Autom. Electr. Power Syst.* **2015**, *39*, 6–13. [CrossRef]
16. Yuan, C.; Liu, G.; Wang, Z.; Chen, X.; Illindala, M.S. Economic power capacity design of distributed energy resources for reliable community microgrids. *Energy Procedia* **2017**, *142*, 2561–2567. [CrossRef]

17. Jing, Z.; Luo, Z. An IGDT model for capacity configuration optimization of island microgrid. *Energy Procedia* **2019**, *158*, 2774–2779. [CrossRef]
18. Zhang, X.; Shahidehpour, M.; Alabdulwahab, A.S.; Abusorrah, A. Security-constrained co-optimization planning of electricity and natural gas transportation infrastructures. *IEEE Trans. Power Syst.* **2015**, *30*, 2984–2993. [CrossRef]
19. Moradi, M.H.; Abedini, M. A novel method for optimal DG units capacity and location in microgrids. *Int. J. Electr. Power Energy Syst.* **2016**, *75*, 36–44. [CrossRef]
20. Zhang, H.; Xie, Z.; Lin, H.C.; Li, S. Power Capacity Optimization in a Photovoltaics-Based Microgrid Using the Improved Artificial Bee Colony Algorithm. *Applies Sci.* **2020**, *10*, 2990. [CrossRef]
21. Javadi, M.A.; Hoseinzadeh, S.; Ghasemiasl, R.; Heyns, P.S.; Chamkha, A.J. Sensitivity analysis of combined cycle parameters on exergy, economic, and environmental of a power plant. *J. Therm. Anal. Calorim.* **2019**, *139*, 519–525. [CrossRef]
22. Maraver, D.; Sin, A.; Sebastián, F.; Royo, J. Environmental assessment of CCHP (combined cooling heating and power) systems based on biomass combustion in comparison to conventional generation. *Energy* **2013**, *57*, 17–23. [CrossRef]
23. Xiao, H.; Pei, W.; Dong, Z.; Kong, L. Bi-level planning for integrated energy systems incorporating demand response and energy storage under uncertain environments using novel meta model. *CSEE J. Power Energy Syst.* **2018**, *4*, 155–167. [CrossRef]
24. Ju, L.; Tan, Q.; Lu, Y.; Tan, Z.; Zhang, Y.; Tan, Q. A CVaR-robust-based multi-objective optimization model and three-stage solution algorithm for a virtual power plant considering uncertainties and carbon emission allowances. *Int. J. Electr. Power Energy Syst.* **2019**, *107*, 628–643. [CrossRef]
25. Nosratabadi, S.M.; Hooshmand, R.A.; Gholipour, E. A comprehensive review on microgrid and virtual power plant concepts employed for distributed energy resources scheduling in power systems. *Renew. Sustain. Energy Rev.* **2017**, *67*, 341–363. [CrossRef]
26. Zhang, T.; Wang, M.; Wang, P.; Gu, J.; Zheng, W.; Dong, Y. Bi-stage stochastic model for optimal capacity and electric cooling ratio of CCHPs—A case study for a hotel. *Energy Build.* **2019**, *194*, 113–122. [CrossRef]
27. Sharafi, M.; ELMekkawy, T.Y. Multi-objective optimal design of hybrid renewable energy systems using PSO-simulation based approach. *Renew. Energy* **2014**, *68*, 67–79. [CrossRef]
28. Das, P.; Das, B.K.; Rahman, M.; Hassan, R. Evaluating the prospect of utilizing excess energy and creating employments from a hybrid energy system meeting electricity and freshwater demands using multi-objective evolutionary algorithms. *Energy* **2022**, *238*, 121860. [CrossRef]
29. Wang, X.; Chen, L.; Chen, Q.; Mei, Y.; Wang, H. Model and Analysis of Integrating Wind and PV Power in Remote and Core Areas with Small Hydropower and Pumped Hydropower Storage. *Energies* **2018**, *11*, 3459. [CrossRef]
30. Heide, D.; Von Bremen, L.; Greiner, M.; Hoffmann, C.; Speckmann, M.; Bofinger, S. Seasonal optimal mix of wind and solar power in a future, highly renewable Europe. *Renew. Energy* **2010**, *35*, 2483–2489. [CrossRef]
31. Sasikumar, C.; Manokar, A.M.; Vimala, M.; Winston, D.P.; Kabeel, A.E.; Sathyamurthy, R.; Chamkha, A.J. Experimental studies on passive inclined solar panel absorber solar still. *J. Therm. Anal. Calorim.* **2019**, *139*, 3649–3660. [CrossRef]
32. Manokar, A.M.; Vimala, M.; Sathyamurthy, R.; Kabeel, A.E.; Winston, D.P.; Chamkha, A.J. Enhancement of potable water production from an inclined photovoltaic panel absorber solar still by integrating with flat-plate collector. *Environ. Dev. Sustain.* **2020**, *22*, 4145–4167. [CrossRef]
33. Ming, B.; Liu, P.; Guo, S.; Zhang, X.; Feng, M.; Wang, X. Optimizing utility-scale photovoltaic power generation for integration into a hydropower reservoir by incorporating long- and short-term operational decisions. *Appl. Energy* **2017**, *204*, 432–445. [CrossRef]
34. Wang, S.; Jia, R.; Shi, X.; An, Y.; Huang, Q.; Guo, P.; Luo, C. Hybrid time-scale optimal scheduling considering multi-energy complementary characteristic. *IEEE Access* **2021**, *9*, 94087–94098. [CrossRef]
35. Hossain, M.A.; Pota, H.R.; Squartini, S.; Zaman, F.; Guerrero, J.M. Energy scheduling of community microgrid with battery cost using particle swarm optimisation. *Appl. Energy* **2019**, *254*, 113723. [CrossRef]
36. Anand, H.; Ramasubbu, R. A real time pricing strategy for remote micro-grid with economic emission dispatch and stochastic renewable energy sources. *Renew. Energy* **2018**, *127*, 779–789. [CrossRef]
37. Wen, X.; Xu, Z. Wind turbine fault diagnosis based on ReliefF-PCA and DNN. *Expert Syst. Appl.* **2021**, *178*, 115016. [CrossRef]
38. Fang, W.; Huang, Q.; Huang, S.; Yang, J.; Meng, E.; Li, Y. Optimal sizing of utility-scale photovoltaic power generation complementarily operating with hydropower: A case study of the world's largest hydro-photovoltaic plant. *Energy Convers. Manag.* **2017**, *136*, 161–172. [CrossRef]
39. Du, C.; Wang, X.; Wang, X.; Shao, C. A Block-Based Medium-Long Term Energy Transaction Method. *IEEE Trans. Power Syst.* **2016**, *31*, 4155–4156. [CrossRef]

energies

Review

Numerical Modeling of Energy Systems Based on Micro Gas Turbine: A Review

Fabrizio Reale [1,*] and **Raniero Sannino** [2]

1 Institute of Sciences and Technologies for Sustainable Energy and Mobility (STEMS), National Research Council (CNR), 80125 Naples, Italy
2 Italian National Agency for New Technologies (ENEA), Energy and Sustainable Economic Development, 00123 Rome, Italy; raniero.sannino@enea.it
* Correspondence: fabrizio.reale@cnr.it

Abstract: In the context of the great research pulse on clean energy transition, distributed energy systems have a key role, especially in the case of integration of both renewable and traditional energy sources. The stable interest in small-scale gas turbines can further increase owing to their flexibility in both operation and fuel supply. Since their not-excellent electrical efficiency, research activities on micro gas turbine (MGT) are focused on the performance improvements that are achievable in several ways, like modifying the Brayton cycle, integrating two or more plants, using cleaner fuels. Hence, during the last decades, the growing interest in MGT-based energy systems encouraged the development of many numerical approaches aimed to provide a reliable and effective prediction of the energy systems' behavior. Indeed, numerical modeling can help to individuate potentialities and issues of each enhanced layout or hybrid energy system, and this review aims to discuss the various layout solutions proposed by researchers, with particular attention to recent publications, highlighting the adopted modeling approaches and methods.

Keywords: micro gas turbine; distributed energy system; humid air turbine; numerical modeling; hybrid energy systems; MGT-ORC; SOFC-MGT; EFMGT

Citation: Reale, F.; Sannino, R. Numerical Modeling of Energy Systems Based on Micro Gas Turbine: A Review. *Energies* **2022**, *15*, 900. https://doi.org/10.3390/en15030900

Academic Editors: Michele Pinelli, Alessio Suman and Nicola Casari

Received: 7 December 2021
Accepted: 24 January 2022
Published: 26 January 2022

1. Introduction

The reduction of greenhouse gas emissions and the energy transition are fundamental topics within the actual EU recovery plan NextGenerationEU [1]. A transition between the present and a scenario in which energy should be obtained mainly or exclusively by renewable and/or no-carbon sources will spend some decades and one of the researchers' goals is to reduce this time as minimum as possible. In the meantime, the research is also focused on low-medium term solutions. In this context, distributed energy systems and small-scale smart energy grids represent a viable solution to contribute to greenhouse and pollutant emissions reduction, mainly owing to the benefits of combined heat and power operation [2,3]. Micro gas turbines (MGTs) in the target of 10–100 kW are suitable for distributed generation since their flexibility and versatility in terms of fuel type, variable load, and combined heat and power (CHP) applications [4,5]. During the last two decades of the twentieth century and the first years of the new millennium, industrial and academic research efforts were focused on the development of small-scale energy systems based on micro gas turbines. In more recent times, the research focus was addressed to alternative fuels (i.e., biogas, syngas, methane-hydrogen blends, ammonia), strategies to increase performance levels, and the integration in more complex hybrid energy systems including the presence of renewable sources. The recent frontiers in MGTs research have been encouraged by some attractive features like wide flexibility and versatility, which overcome the limited electrical efficiency. Although the significance of experimental data is universally recognized, numerical modeling is capable to provide a fundamental contribution to the research even in the field of micro gas turbines [6,7]. The variety of modeling approaches is almost boundless, and the researchers set the constraints and the numerical features according to the goal of the investigation and the related accuracy level. In some cases, MGTs are simply modeled as a black box [8], while in other studies mono-dimensional models can be adopted [9]; however, the mainly adopted approach allows to build a zero-dimensional (0D) model of micro gas turbines, starting from the well-known gas turbine theory [10]. In general, a simple thermodynamic analysis appears to be useful in several cases, like steady operation and thermo-economic evaluation, especially if the focus is on the design operating point: these studies usually involve fixed characteristic parameters for each component and the neglecting of some peculiar losses. The introduction of characteristic maps of rotating components and also of the regenerative heat exchanger can lead to more accurate predictions of MGT plant performance. In these cases, a zero-dimensional analysis can be coupled with experimental information of single components, to increase the quality of numerical results and to extend the analysis to off-design conditions with an adequate capability of simulating the actual behavior of the energy system. Research studies are mainly addressed the steady-state behavior since the focus is on single operating points or the time intervals are taken into account allowing to consider part-load operation as a quasi-steady-state one. Nevertheless, in some cases, the introduction of time-based analysis like dynamic or mono-dimensional models is considered for specific purposes.

This review is focused on the recent progress in micro gas turbine numerical modeling, including a detailed investigation on the plant layouts that stand out about optimal energetic performance. The work is organized as follows: the simple and recuperated cycle have been summarized in Section 2, while Section 3 deals with the enhanced layouts (humidified layouts or externally fired MGTs); hybrid energy systems (coupling with Organic Rankine Cycles, Solid Oxide Fuel Cells or solar plants and the integration in smart energy grids) have been reviewed in Section 4. Section 5 provides an outline of the CFD investigation applied to a micro gas turbine, and finally, the conclusions of this work are presented in Section 6.

2. Simple and Recuperated Cycles

Excluding layout modifications and any system integration, which will be reviewed in the next sections, micro gas turbines follow two schemes: a simple cycle and a regenerative cycle. Figure 1 shows the base MGT scheme in the case of the simple Brayton cycle. The main components are a compressor, a turbine, and a combustion chamber; also, auxiliaries (e.g., fuel compressor and stack fan) and electrical generator should be considered. This scheme ensures the simplest configuration, but the electrical efficiency is not satisfactory, hence it is mainly adopted in the case of micro-turboshaft or aero-derivative MGTs. Hoseinimaab et al. [11] proposed an approach to off-design performance analysis without the introduction of characteristic maps for a simple turboshaft or a double turbo-shaft engine. Altharazi et al. [12] implemented a model of a micro turboshaft, a Kingtech K180, with the commercial software Gasturb and focused their studies on the adoption of liquid biofuels such as palm oil methyl ester (POME) or TPOME in blend with Jet-A.

Adding a recuperator to the base layout allows to realize a regenerative Brayton cycle (Figure 2) and to increase the combustor inlet temperature and the system efficiency, consequently. In this way, it is possible to recover part of the thermal energy available in the exhaust gas that heats the compressed air and increase its temperature. Researchers' attention is mainly addressed on this configuration, owing to the higher efficiency levels, close to 30% in various commercial MGTs. The approaches are varied and can be better described by thorough literature analysis.

Gopisetty and Treffinger [8] studied a trigeneration system based on a Capstone C65 MGT which is described as a black-box model with the adoption of empirical coefficients, in order to consider part-load operations and the power correction related to the variations of ambient air temperatures; transient operations were neglected since the time response of MGT is higher with respect to other components (heat exchangers). Malinowski and Lewandowska [13] discussed an analytical model for a part-load operation that involves exergy analysis. Model validation was done starting from Capstone C30 experimental data available in the literature. Nipkey et al. [14] adopted a commercial software (IPSEpro), with libraries obtained in their previous papers; the subject of the study was a Turbec T100 modeled by compressor and turbine characteristic maps. Other parameters such as loss coefficients or recuperator effectiveness were obtained through the design case calculations. The authors validated the model in off-design conditions, in the case of adoption of natural gas as fuel, and analyzed the effects on MGT related to the adoption of biogas as fuel. Cáceres et al. [15] defined a biogas-fueled microturbine system in Matlab/Simulink. They proposed to simulate a recuperated MGT with its main controls (speed governor, fuel injection control, and exhaust gases temperature), focusing the analysis mainly on the thermodynamic model of the biogas production process. Caresana et al. [16] analyzed the effect of ambient conditions on MGT performance: the study was performed on a Turbec T100 and, also in this case, compressor and turbine maps were implemented in order to simulate the behavior of the rotating components. The numerical model was validated through a comparison against experimental data. Their analysis highlighted that T100 electrical power decreases with the temperature at a rate of about 1.22 %/°C.

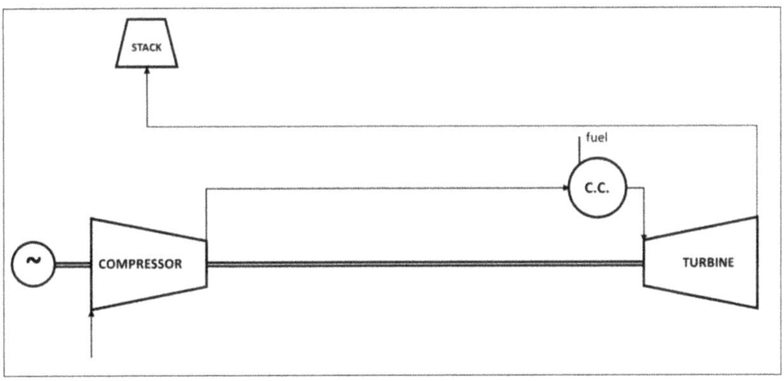

Figure 1. Simple Brayton cycle MGT layout.

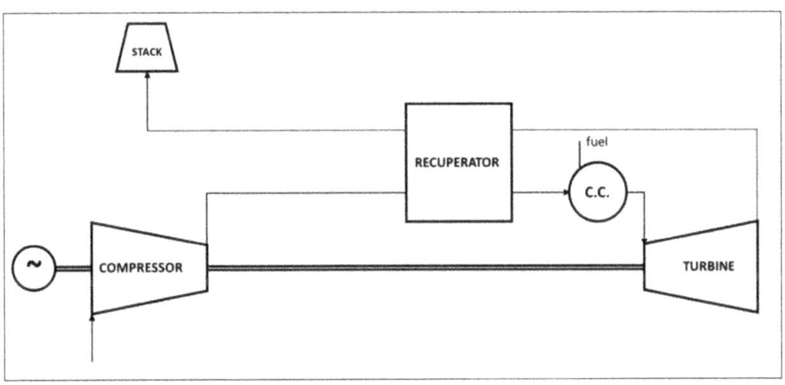

Figure 2. Recuperated MGT.

Gimelli and Sannino [17] implemented a multi-variable multi-objective methodology aimed to calibrate and validate a thermodynamic model of energy systems and applied it to Capstone C30 and Turbec T100 plants; the two MGT models have been investigated more in detail in [18] and [19], respectively. The same authors [9] also proposed a one-dimensional model of the Capstone C30 in which rotating components are modeled through the adoption of characteristic operating maps in the GT-SUITE environment. Duan et al. [20] proposed a nonlinear model of a regenerative cycle in which the dynamic response is considered but several parameters were fixed (i.e., isentropic exponents, specific heat capacity, pressure losses) while the volume effect of the pipeline is ignored. The model was established in Matlab and the subject of study was a Capstone C30. Zornek et al. [21] adopted a numerical simulation program developed and validated at DLR [22–24] to simulate Turbec T100 behavior in case of low calorific fuels supply. Despite this work being mainly experimental, the authors used a validated model of the MGT plant to investigate the system behavior with and without the embedded control and to optimize the operating range of the turbine. Visser et al. [25] proposed a 0D/1D approach to simulate the performance of turboshaft micro turbines below 30 kW. They started with a 3 kW MTT MGT, modeled with a Gas turbine Simulation Program (GSP), and focused their study on a 1D model of the recuperator [26]. di Gaeta et al. [27] proposed a dynamic model of a controlled commercial MGT (a Turbec T100). A first-order differential equation (ODE) described the dominant dynamics of the MGT imposed by its control system, while the differential equation is coupled to a set of nonlinear maps numerically derived from the matching of a detailed 0D thermodynamic model of the MGT evaluated over a wide range of operating conditions.

In Table 1, a summary of the reviewed literature has been listed.

Table 1. Recuperated micro gas turbine reviewed.

MGT Model	Software	Analysis Feature	Ref.
Capstone C65	Modelica	Black-box	[8]
Capstone C30	Mathcad	Part-load correlation	[13]
Turbec T100	IPSEpro	Steady-state thermodynamic with compressor and turbine characteristic maps	[14]
20 kW MGT	Matlab/Simulink	Dynamic	[15]
Turbec T100	Matlab/Simulink	quasi-stationary components and static maps	[16]
Capstone C30	-	Numerical methodology validation	[17,18]
Turbec T100	-	Numerical methodology validation	[17,19]
Capstone C30	GT-SUITE	1D	[9]
Capstone C30	Matlab/Simulink	Dynamic non-linear	[20]
Turbec T100	Matlab/Simulink	Low LHV combustion	[21]
Turbec T100	Matlab/Simulink	Component maps, simulation of the T100 control algorithms	[22–24]
3 kW Turbocharger	GSP	Cycle modeling for different component efficiencies and losses	[25]
3 kW Turbocharger	GSP	1D recuperator	[26]
Turbec T100	In-house code/Matlab	Steady-state—0D model with compressor and turbine characteristic maps/ Dynamic first-order ODE	[27]

3. Enhanced Layouts

3.1. Humidified Gas Turbines

Steam or water injection in a micro gas turbine plant is standing out in recent years as a solution to improve Brayton cycle efficiency and net power, also mitigating greenhouse and pollutant gas emissions. Water/steam addition leads to a reduction of temperature peaks into the combustion chamber and, consequently, a reduction of nitric oxide (NOx) formation. Research activities are addressed on the determining of several layouts, which can be summarized into three main categories, i.e., the evaporation of the water injected, the injection of steam, and the presence of a humidification tower with a recirculation water loop [28].

Parente et al. [29,30] found that the application of the evaporative cycle to recuperated micro gas turbines could increase both power output and electrical efficiency. They analyzed the effects on two different MGTs (with an electrical power of 100 and 500 kW) and they showed through a thermo-economic analysis that evaporative MGT cycles had lower specific investment costs with respect to traditional MGTs as regard CHP applications. De Paepe et al. [31] studied the influence of steam injection on a Turbec T100 performance. The proposed layout, shown in Figure 3, has been modeled with Aspen Plus and the simulations predicted a rise in electrical efficiency of 2.2% for 5% of steam instead of air.

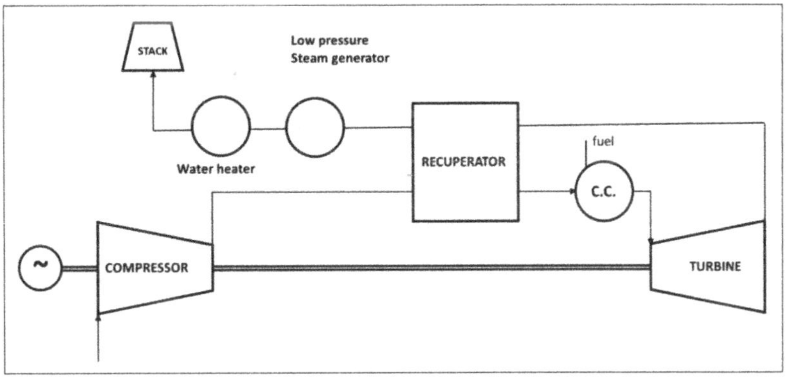

Figure 3. Steam Injection micro gas turbine (with injection before the recuperator) [31].

Montero et al. [32] examined the application of humidified cycle on a Turbec T100 by adding a saturator between the compressor and the recuperator and introducing an economizer just after the recuperator. Results of their simulations highlighted that the mHAT layout (Figure 4) leads to an improvement of the overall efficiency. Ali et al. [33] presented a numerical comparison between three different layouts: an EGR cycle, a STIG, and an mHAT. Results of simulations showed that electrical efficiency increases by 6% for humid and steam cycles and decreases by 9% for the EGR layout, with respect to the base case of a recuperated 100 kW MGT. HAT cycle had the best performance also as regard NOx emissions reduction, despite the combustor should be modified.

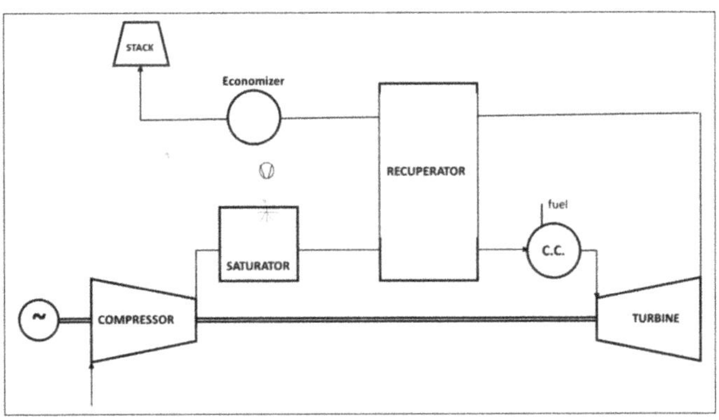

Figure 4. Micro Humid Air gas Turbine (mHAT) [32].

Reale and Sannino [34] proposed an H-STIG layout in which both water and steam are injected into the micro gas turbine. Water is injected just after the compressor, while a low amount of steam is injected directly into the fuel feeding system. The comparison between a validated model of Turbec T100 and results of simulations of the modified energy system highlighted the benefit of a small quantity of water (almost 30 g/s with respect to an air mass flow of 800 g/s), namely an improvement of net power and overall efficiency up to

20% and 8.9%, respectively. H-STIG layout (Figure 5) allows also to increase the admissible percentage of H_2 blended with methane up to 30%$_{vol}$, tripling the experimental limit of 10%$_{vol}$ for the dry cycle.

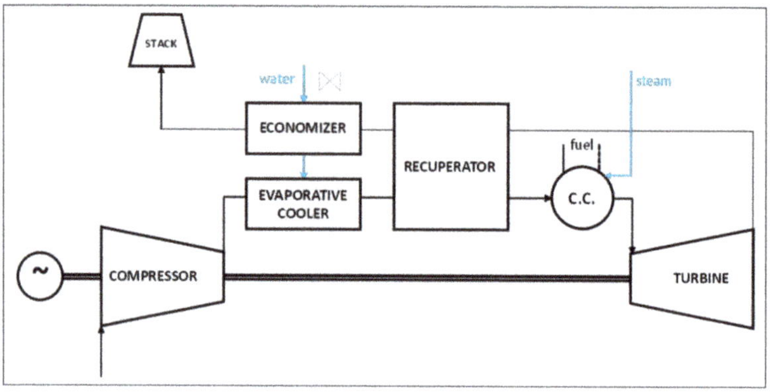

Figure 5. H-STIG layout [34].

De Paepe et al. [35] compared the impact of different humidified micro gas turbine concepts on the plant performance: micro Humid Air Turbine Plus (mHAT+), Advanced Humid Air Turbine (AHAT), and the REgenerative EVAPoration (REVAP) cycle concept. Results of simulations carried out with Aspen Plus highlighted that the optimal heat recovery can be accomplished using the REVAP cycle concept (Figure 6), with an increase of 4.3% regarding electrical efficiency.

De Paepe et al. [36] also analyzed the potentialities of the M-power cycle concept. This cycle is characterized by a specific saturator module that combines the actions of the aftercooler, evaporator, and recuperator. Numerical simulations allowed us to analyze the performance of the MGT by varying the main parameters of the saturator. Wan et al. [37] proposed an mHAT cycle in which the exhaust is discharged with pressure below ambient (BAHAT). Numerical models were implemented in gPROMS. Results of numerical simulations highlighted that the specific work increases by about 20%, while the electrical efficiency increases by about 2 percentage points with respect to the mHAT cycle.

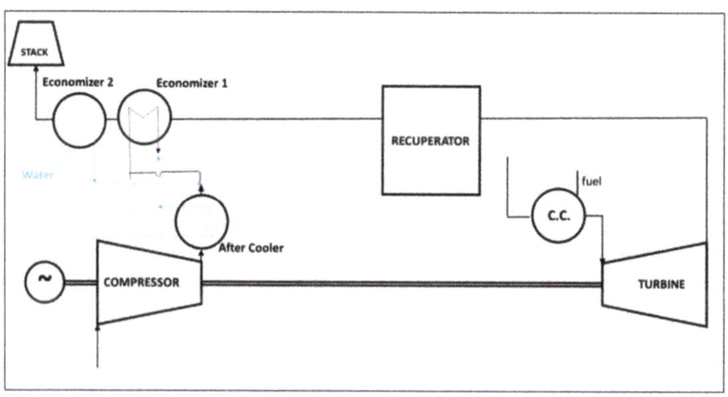

Figure 6. REVAP layout [35].

In Table 2, a summary of the reviewed literature on humidified gas turbine plants has been listed.

Table 2. Summary of humidified gas turbine plants.

MGT Model	Software	Layout Feature	Ref.
100 and 500 kW	In-house (TEMP, SAT, TRANSEO)	Turbomachinery performance maps	[29,30]
Turbec T100	Aspen Plus	Compressor and turbine efficiency assumed constant	[31]
Turbec T100	Aspen Plus	Performance map for Compressor, chocked turbine	[32]
Turbec T100	Aspen Hysys	Thermodynamic model with chemical equilibrium in the combustor estimated by minimization of the total Gibbs free energy	[33]
Turbec T100	Thermoflex/Ansys CFX	Turbomachinery performance maps, RANS RSM	[34]
Turbec T100	Aspen Plus	Performance map for Compressor, chocked turbine	[35]
Turbec T100	Aspen Plus	Performance map for Compressor, chocked turbine	[36]
Theoretical	gPROMS/GateCycle	Multiflash database, Advanced Redlich-Kwong-Soave equation of state	[37]

3.2. Externally Fired Micro Gas Turbines

The research on externally fired micro gas turbines (EFMGTs) deals with the combustion of both solid and liquid fuels, like coal and biomass, as well as the heat supplied by renewable sources. As suggested by the layout label, the main feature of an EFMGT is related to the working fluid that not directly participates in the combustion process, but is heated by an external source through a heat exchanger. The adoption of this layout allows the reduction of the requirements in terms of cleaning and composition of fuels and, consequently, of exhaust gases. In other words, in an externally fired micro gas turbine is possible to burn non-standard and critical liquid or gaseous fuels besides solid fuels. While this solution is, of course, mandatory for solid fuels, it can be considered also in the case of liquid or gaseous biofuels, since their use in traditional gas turbines involves several problems, especially without any physical or chemical treatment. EFMGTs "move" this problem to the heat exchanger, since the heat is provided by placing the combustor in the hot exhaust-air stream at the turbine outlet. The externally fired micro gas turbines are usually derived by commercial MGTs, with the main difference in the plant layout represented by a heat exchanger that replaces the internal combustion chamber. The heat exchanger type varies with the type of input thermal power: in the case of fuels derived by coal or biomass, it should operate at high-temperature levels and the layout can present also a combustion chamber; in the case of renewable source, it should operate also for medium-low temperatures, and the burner, if any, is usually located just after the turbine.

As mentioned in [38], one of the early experimental layouts of EFGT dates back to 1992: it was the BM/367/92-Be project, based on a metallic high-temperature heat exchanger (HTHE) with a turbine inlet temperature of 1073 K and HAT with 500 kW. The adoption of EFMGT layouts in case of combustion of solid biomass was analyzed by Traverso et al. [39], proposing a layout in which a heat exchanger and an external combustion chamber replace the internal combustor of a regenerative micro gas turbine (Figure 7); this scheme also involves an air bypass on the recuperator cold side. The EFMGT was modeled by TRANSEO software, once validated with experimental data, both for steady and unsteady behavior [40–42].

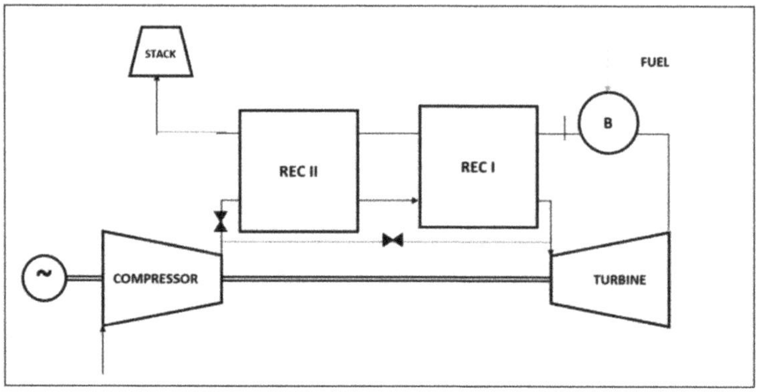

Figure 7. Externally fired micro gas turbine with two recuperators [39].

Pantaleo et al. [43] proposed a comparison between an IFMGT and an EFMGT. Thermodynamic analysis was carried out through the adoption of the software Gate-cycle, investigating several blends of natural gas and biomass. The authors focused their attention mainly on a thermo-economic comparison. As expected, a trade-off between efficiency and profitability has been noticed along with the significance of a dual fuel MGT for the integration of renewable and fossil fuels. Riccio and Chiaramonti [44] proposed a dual fuel layout, in which both internal combustor and external furnace coexist in the same layout (Figure 8). Fresh air is first heated by the exhausts coming from a furnace, in which biomass is burned, and then comes into a combustion chamber fueled by natural gas. This solution can lead to the flexible use of solid and gaseous fuels. The study was based on a steady-state thermodynamic matching analysis of a Turbec T100 carried out by AMOS, an in-house software.

Datta et al. [45] performed a parametric thermodynamic analysis of 100 kW EFMGT, in order to identify the ideal operating conditions for an EFGT decentralized plant by analyzing the energetic and exergetic performance variations with thermodynamic operating parameters. Once the optimal pressure ratio of the turbine has been fixed, the most desirable boundary conditions arise from a trade-off between higher thermal and exergetic efficiency versus smaller heat exchanger size.

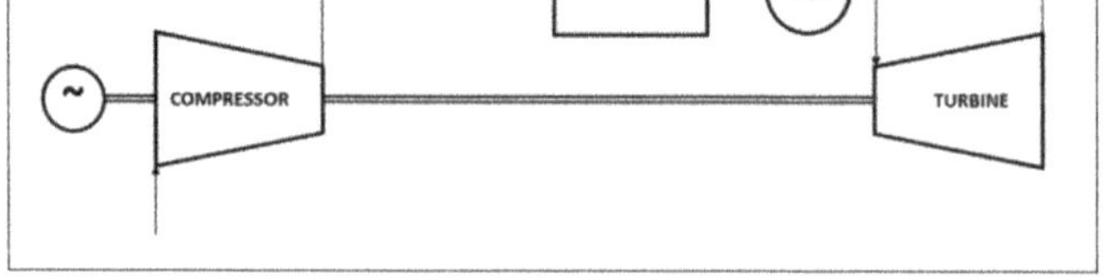

Figure 8. Dual Fuel micro Gas Turbine with both internal combustor and external furnace [44].

In Table 3, a summary of the reviewed papers dealing with externally fired MGT has been listed.

Table 3. Externally fired micro gas turbine reviewed.

Starting MGT Model	Software	Layout Specifications	Ref.
Elliott TA-80R	In-house (TEMP, TRANSEO)	Two recuperators and main valves for cycle control	[39–42]
Turbec T100	Gate-Cycle	Dual fuel NG and biomass	[43]
Turbec T100	In-house (AMOS)	Dual fuel	[44]
100 kW MGT	In-house	Single recuperator	[45]

3.3. Further Layouts

Several researchers have proposed original and noteworthy plant schemes that are hard to categorize in a single group, due to the uncommon conceptualization of the components' layout.

Cameretti et al. [46] proposed a recuperated MGT layout with both recuperator bypass and partial exhaust gas recirculation (EGR) aimed to find out operating conditions that satisfactorily meet NOx reduction and cycle efficiency. Numerical simulations have been carried out by an in-house code, while CFD analysis was done using a Fluent solver. Majoumerd et al. [47] extended the adoption of a Turbec T100 thermodynamic model already validated in [14] to a more complex system with a CO_2 capture unit; the authors compared an EGR (Figure 9) and a HAT cycle with a focus on improved carbon capture efficiency. The base MGT plant showed a higher cost for CO_2 capture compared to EGR and HAT layouts and the authors suppose that this feature will support the deployment of carbon capture in distributed power generation; the detailed analysis of EGR and HAT cycle modeling with IPSEpro software has been disclosed in [48]. Ali et al. [49] modeled in Aspen HYSYS a micro gas turbine with CO_2 injection and EGR, coupled with an amine-based CO_2 capture plant; base MGT, CO_2 injection, and pilot-scale carbon capture plant models were validated against experimental data. Hampel et al. [50] investigated a promising plant layout made by pure hydrogen-fueled T100 MGT coupled with a dehydrogenation unit that processes liquid organic hydrogen carriers (LOHC). The advantage of such an energy system is that the hydrogen needed to feed the MGT is obtained by the dehydrogenation endothermal process which is supplied by the gas turbine exhaust gas; the numerical analysis showed the thermodynamic feasibility for three of the four investigated cases.

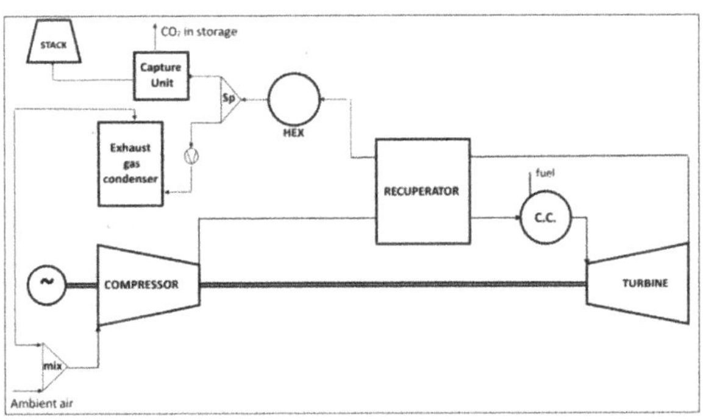

Figure 9. An EGR micro gas turbine with CO_2 capture [47].

4. Hybrid Energy Systems Based on MGT

The assumption of a hybrid energy system is the multiplicity of energy conversion that can be achieved by the integration of two or more devices or by the combination of two or more fuels for the same device, as exhaustively described by Manwell [51]. The hybrid energy systems are usually made by a renewable energy plant coupled with thermal prime movers like internal combustion engines or gas turbines; alternative hybrid layouts involve the waste heat recovery approach, in which two or more energy conversion systems are coupled based on the temperature level of the heat source, like a gas turbine and organic Rankine cycle (ORC) and solid oxide fuel cell (SOFC) with a gas turbine, that will be discussed in the following subsections. According to this summary of hybrid energy systems, it is clear the convenience of MGT, whose use is favorable due to the flexibility in the operating domain and to the high-quality recoverable heat at the turbine exhaust. The main advantage of numerical modeling applied to novel hybrid plant configuration is the capability to individuate potentialities and issues of each layout, although the need for experimental data in order to validate the models. The hybridization can be obtained by substituting MGT components with more convenient elements or by introducing new components, such as heat exchangers, which can turn into a thermal connection with other energy systems. The hybrid energy systems based on micro gas turbines reviewed in this work are four: a first one in which the Brayton cycle is the topping one, a second in which the MGT is the bottoming one, a third in which MGTs are connected with solar power plants and a fourth in which the MGT is a node of a small-scaled smart energy grid. In the latter case, the hybridization is achieved by the integration of different energy conversion plants in a single smart grid, whose control system manages the matching between supply and demand.

4.1. MGT—ORC Systems

Organic Rankine Cycle (ORC) represents a technology of great interest as waste heat recovery (WHR) system in the medium and low-grade energy recovery, also in the typical range of MGT exhaust gases temperature [52,53]. The ORCs are closed-cycle plants whose simple layout includes a pump, a heat exchanger for the organic fluid temperature increase, an expander that generates electrical power, and a condenser; the hybridization is performed by the MGT exhaust gas which flows into the ORC heat exchanger with the aim of evaporating the compressed organic fluid. The thermal recovery performed by ORC allows to increase the combined cycle efficiency up to 15% more than base MGT [54]: by this way, MGTs can become advantageous for small-scale power plants, thanks to the environmental and fuel diversification [55].

Chacartegui et al. [56] presented an integrated energy system based on an mHAT cycle and two bottomed ORC plants. The topping cycle is combined with a medium-temperature ORC and a low-temperature ORC. The proposed scheme is shown in Figure 10 and its global efficiency is above 50%. The thermal heat which is available at the exhausts of the recuperated mHAT is used both as thermal input of the ORCs and to preheat the liquid water needed on the saturator.

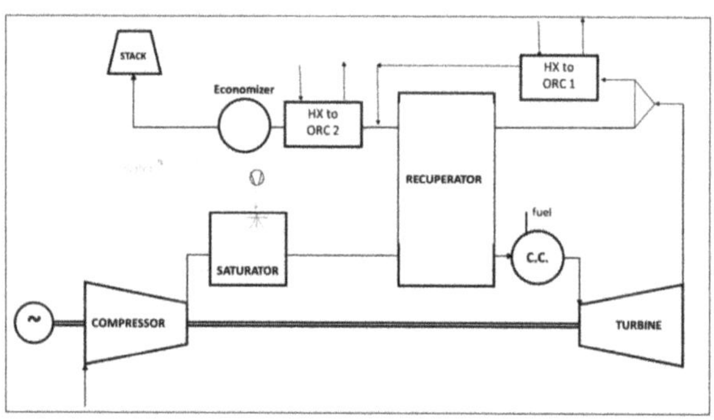

Figure 10. Integrated layout with mHAT and two ORCs [56].

Moradi et al. [57] investigated a micro-CHP system consisting of a fluidized-bed gasifier, a 100 kW steam injected micro gas turbine, and a bottomed ORC cycle, along with a heat recovery steam generator (HRSG) that produces steam both for the gasification process and the STIG. Numerical simulations have been performed with Aspen Plus and included syngas purification systems, an improved gasifier model, and the operating limits of MGT. Despite the results of syngas fueled cycle with steam injection showing an increased electrical power but a slight reduction of electrical efficiency, the integration of different energy systems has been proved to be capable of providing energy and environmental benefits. Nazari et al. [58] defined an integrated energy system composed of an EFMGT, a transcritical ORC, and a Li-Br/H_2O absorption refrigeration cycle (EFGT-ORC-ARC). In this way, the two renewable sources (solar and biomass) allow the combined production of electricity, hot water, and cooling load. A steady-state numerical study was carried out with several constraints (i.e., ideal gas mixture, compressor, turbines, and pumps with constant isentropic efficiencies) in order to obtain an exergo-economic assessment through a thermodynamic code developed using Equation Engineering Solver (EES) software. Compared to the base case, the optimized integrated plant can achieve up to 17% higher electrical power, up to 23% higher cooling power, and about 200% higher heating power, with a first law efficiency of 54%. Nazari and Porkhial [59] proposed a multi-generation system composed of a solar field, a biomass-fueled externally fired gas turbine, a steam Rankine Cycle, an ORC, an absorption chiller, and a desalination plant. The whole system has been modeled by EES software and the results showed the good potential of potable water production, a higher cooling capacity compared to the available literature, along with an achievable exergy efficiency of 21.5%. Xiao et al. [60] presented a novel configuration combining a solar (a heliostat field) micro gas turbine with steam injection and a bottomed ORC. A numerical model was developed and validated by experimental or referenced data and results of simulations in representative days (i.e., spring equinox, summer solstice, autumn equinox, and winter solstice) highlighted that the proposed system increases the system flexibility in dealing with complex loading conditions and the solar energy has a significant impact on fuel saving. Cameretti et al. [61] proposed the integration of an MGT with an ORC system and a solar field, whose aim is to increase the MGT recuperator inlet temperature. The authors carried out both thermodynamic numerical simulations of the entire system and CFD analysis of the MGT combustor in the case of natural gas and biogas as fuel. Such plant layout maximizes the heat from solar source with a reasonable collectors' surface achieving satisfactory performance in CHP mode as well as at part-load operation. Baccioli et al. [62] investigated the improvement of the anaerobic digestion plant installed in Viareggio (Italy) by adding to the biogas fueled 600 kW MGT an ORC system. Transient and steady-state simulations have been performed in the AmeSIM environment and the authors considered three different ORC sizes (30–40-50 kW) and individuated that the most valuable solution from a thermo-economic point of view is the 40 kW case. Reale et al. [63] studied an integrated gasifier-MGT-ORC energy system; the convenience of including a syngas fuel from gasification of organic municipal waste has been also examined. Thermodynamic simulations have been carried out with Thermoflex, while the issues related to the combustion process of biogas-syngas blends are analyzed by CFD simulations in Ansys CFX.

Table 4 lists the summary of the reviewed literature about the MGT-ORC plant.

Table 4. Summary of MGT-ORC plants.

MGT Model	Software	MGT Model Feature	Ref.
Theoretical	EES	Genetic algorithm optimization	[56]
Turbec T100	Aspen Plus	Performance maps	[57]
Theoretical	EES-based	Mass and energy balance	[58]
Theoretical	EES-based	Mass and energy balance	[59]
Turbec T100	MATLAB	Mass and energy balance, performance maps	[60]
Turbec T100, Capstone C30	Thermoflex/Ansys Fluent	Performance maps/3-step oxidation mechanism within a finite rate—eddy dissipation model	[61]
Capstone C600s	AmeSIM	Performance maps	[62]
Turbec T100	Thermoflex/Ansys CFX	Performance maps/RANS RSM	[63]

4.2. SOFC—MGT Systems

Solid Oxide Fuel Cells (SOFCs) convert the chemical energy of a fuel into electrical energy through high-efficiency electrochemical processes; the typical operating temperature of a SOFC makes convenient the coupling with MGTs. In particular, the heat rejected from the fuel cell is transferred to the compressed air with the purpose of substituting the MGT combustion chamber.

Costamagna et al. [64] analyzed the design and off-design behavior of a hybrid system based on a recuperated MGT and a high-temperature SOFC reactor, considering also the MGT variable speed control system. Results of simulations showed a potential of achieving an efficiency of more than 60% at the design point and greater than 50% at part load, considering a SOFC reactor of about 250 kW and a 50 kW MGT. Ferrari et al. [65] implemented the model of a hybrid MGT-SOFC system using the in-house TRANSEO tool developed in MATLAB/Simulink environment, which includes an ejector model based on 1D-CFD equations able to predict the dynamic behavior of the component. The off-design analysis of the hybrid system showed that FC temperatures too high or too low should be avoided because they could drive the compressor operation too close to the surge line or the MGT overspeed. Kaneko et al. [66] proposed the dynamic model of a 35 kW hybrid system based on a biomass-fueled SOFC coupled with an MGT; the main features of the model are two controllers of power and SOFC temperature that provide an actual estimation of the time response for each component under different boundary conditions. Bang-Møller et al. [67,68] studied a hybrid CHP plant combining a two-stage gasifier, a SOFC stack, and an MGT: numerical simulations allowed to compare three different combinations (gasifier-SOFC, gasifier-MGT, and gasifier-SOFC-MGT) and to individuate that the gasifier-SOFT-MGT layout presents the best energetic- and exergetic-based electrical efficiency. Bakalis and Stamatis [69] studied the integration of a Siemens-Westinghouse SOFC with four commercially available recuperated MGT plants (Capstone C30 and C60, Parallon 75 kW, and Turbec T100); the model has been carried out with Aspen Plus and considers the off-design behavior of rotating components through user-defined Fortran subroutines. The referring layout is shown in Figure 11: the SOFC model consists of a pre-reformer (PR), an indirect internal reformer (IIR), an afterburner (AB), a preheater, and, obviously, a cathode-anode stack. The results showed the increasing trend of system efficiency with the decrease of MGT power for a fixed SOFC generator.

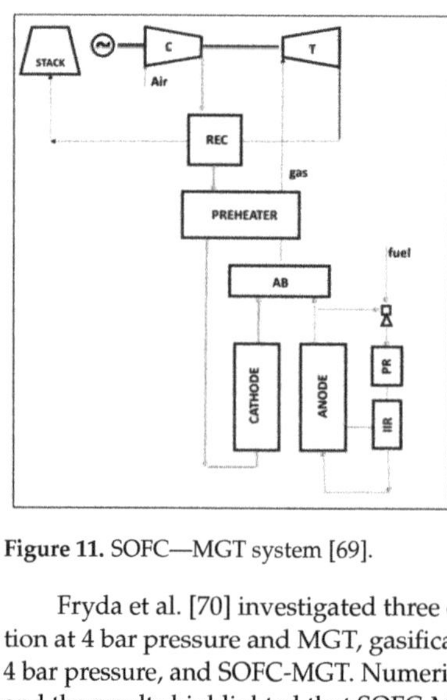

Figure 11. SOFC—MGT system [69].

Fryda et al. [70] investigated three different small-scale CHP configurations: gasification at 4 bar pressure and MGT, gasification at 1.4 bar pressure and SOFC, gasification at 4 bar pressure, and SOFC-MGT. Numerical simulations were carried out using Aspen Plus, and the results highlighted that SOFC-MGT was more efficient with respect to both simple

MGT and SOFC. Zhao et al. [71] proposed a hybrid SOFC-HAT-CCHP (combined cooling, heating, and power) system based on methanol-reforming. The model is based on Aspen Plus and the main outcome shows that the introduction of humid air turbines increased the power generation reducing at the same time the coal consumption rate. A SOFC-MGT-ORC hybrid plant for power generation is proposed by Ebrahimi et al. [72]: the authors highlighted through a parametric analysis that this combination can be useful to convert as much as possible fuel to power, since MGT can use high-quality SOFC stack gases and ORC uses medium quality MGT exhaust gases. Results of numerical simulations showed that the electrical efficiency of the energy system can reach about 61%. Wongchanapai et al. [73] performed a sensitivity analysis on a direct internal reforming biogas SOFC-MGT hybrid CHP plant, focusing on energy and exergy balances. The four parameters considered were air-steam mixtures as reforming agents, fuel utilization factor, turbine inlet temperature, and compression ratio, and the several sets of information resulting from the analysis represent valuable knowledge for direct-biogas SOFC operational performance. Krummrein et al. [74] investigated the performance related to the use of biogas fuel for their hybrid power plant SOFC-MGT demonstrator. Results of steady state simulations showed that the thermodynamic parameters are only slightly influenced by the substitution of methane with biogas. Di Carlo et al. [75] presented a SOFC-MGT system integrated with a biomass gasifier whose model has been validated against experimental data. Results highlighted that wood-gas rich in H_2 and CO can be used in SOFCs and that the total efficiency of the entire system is between 36% and 44%, also considering the presence of the gasifier. Perna et al. [76] investigated the performance of a SOFC-MGT with a biomass downdraft gasifier (BG). The comprehensive system model has been validated against experimental data, and the numerical simulations were carried out considering electrical, thermal, and combined heat and power production: such integrated energy conversion system significantly improves the electric efficiency and accordingly enhances the exploitation of biomass as a primary energy source. Isfahani et al. [77] proposed a hybrid system based on SOFC and MGT with capturing CO_2 and a three-reactors chemical looping for hydrogen generation from natural gas; their model allows the MGT combustor to burn the non-reacted hydrogen. A sensitivity analysis has been performed by varying several SOFC parameters, and the direct correlation between plant efficiency and operating pressures has been demonstrated. Zhen et al. [78] proposed an integrated distributed energy system consisting of SOFC and MGT with solar thermochemical hydrogen production. The system can be divided into three parts: the power plant (SOFC-MGT), the solar thermal methanol decomposition aimed at hydrogen-enriched fuel production, and a waste heat recovery system. Exhaust gases from MGT heat first the compressed air and hydrogen-rich syngas and afterward pass through the Li-Br absorption chiller. The DES numerical model is validated by comparing the results with published literature. Karimi et al. [79] proposed a SOFC-MGT-ORC energy system combined with a rice straw biomass gasifier. The authors carried out a parametric analysis aimed to individuate through a multi-objective optimization the best compromise between exergy efficiency and cost rate. Pantaleo et al. [80] analyzed a particular case of study of a major Italian coffee roasting and processing plant, with an operating cycle of 6 h per day. They proposed and numerically compared three different solutions: an intermittent waste heat recovery via thermal energy storage (TES) and ORC, a recuperated MGT coupled to the existing modulating gas burner, and a non-regenerative topping MGT with the direct recovery of the turbine outlet air for the roasting process. The authors concluded that the investigated solutions are not economically profitable for an operating cycle of 6 h/day.

In Table 5, a summary of the literature reviewed in this subsection has been listed.

Table 5. MGT-SOFC plants reviewed.

MGT Model	Software	Feature Y	Ref.
50 kW	MATLAB	Performance maps	[64]
50 kW	In-house (TRANSEO, MATLAB/Simulink based)	Transient	[65]
5 kW	MATLAB/Simulink	Dynamic	[66]
Theoretical	In-house (Dynamic Network Analysis)	Thermodynamic model	[67,68]
Capstone C30 and C60, Parallon 75 kW, Turbec T100	Aspen Plus	Subroutines for turbomachinery	[69]
Theoretical	Aspen Plus	Thermodynamic model with fixed efficiencies	[70]
Theoretical	Aspen Plus Engineering Equation Solver N.A.	Mass and energy balance 0D and 1D	[71] [72] [73]
3 kW	In-house (MGTS^3—Micro Gas Turbine Steady State Simulator, MATLAB/Simulink based)	Performance maps	[74]
Turbec T100	ChemCAD	TIT variability with SOFC (Subroutines for SOFC and gasifier?)	[75]
Turbec T100	Aspen Plus		[76]

4.3. Solar—MGT

Kautz and Hansen [81] proposed an EFMGT based on a Turbec T100 in which the recuperator became the heat exchanger linked to the external burner. In their work, a further heat exchanger could be added in order to consider an additional rate of input thermal power by solar panels (Figure 12); the numerical computation has been carried out with Aspen Plus software. Despite an externally fired gas turbine allowing the burning of "dirty" fuels along with the decrease of fuel consumption, using the solar source as the main heat gain and the biomass as backup fuel can lead to a fully renewable energy system.

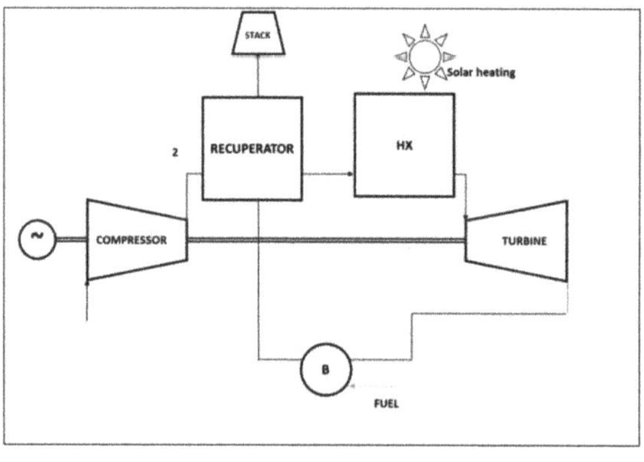

Figure 12. Solar—EFMGT layout [81].

Aichmayer et al. [82] compared two different configurations of a solar receiver integrated into a small-scale dish-mounted hybrid solar micro gas-turbine system. In the first layout, the solar receiver is pressurized and located just before the turbine. The second configuration is an atmospheric one, in which the receiver is after the turbine. Steady-state thermodynamic models were realized in Matlab, using also the REFPROP database, while the receiver design was carried out by a validated coupled CFD-FEM analysis. The pressurized receiver stood out as the only suitable component for a micro gas turbine-based solar dish system, due to the assumed pressure drop constraint. Yang et al. [83] proposed a thermodynamic model and a real-time control strategy of a solar MGT coupled with a thermochemical energy storage (TCES); the mathematical model involved the actual direct normal irradiance (DNI) and the disturbance for both short and long periods. The simulations showed a low risk of surging or choking and highlighted the better performance of constant turbine outlet temperature operation control compared to constant shaft speed control. Cameretti et al. [84] studied a hybrid plant consisting of a micro gas turbine with a solar field with a tower-placed receiver (Figure 13). Simulations on the proposed layout with solar heating and internal combustion chamber were carried out on both the integrated energy system (with Thermoflex software) and the combustor (CFD analysis).

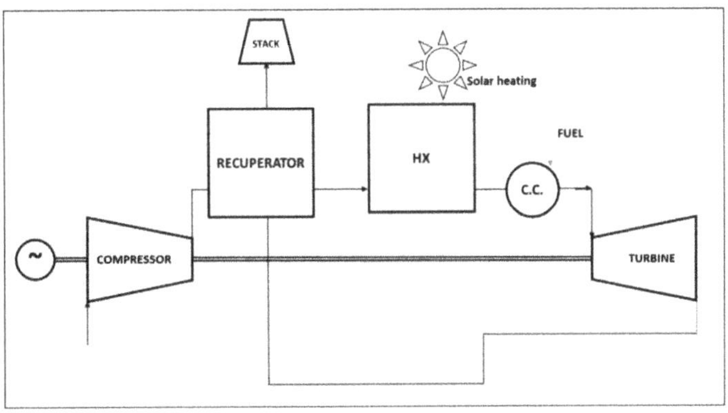

Figure 13. Solar-MGT scheme [84].

Wang et al. [85] also proposed a solar receiver MGT scheme in which the receiver is located between the recuperator and the combustor. Giostri et al. [86] described a solarized micro-gas turbine model coded in Matlab. In particular, a raytracing model allows to estimate the optical performance of the solar tower system, while the recuperated MGT was modeled starting from previously validated models of the same authors [87]; the thermo-economic indexes have been also analyzed. Solar-MGT, without any storage system, appears currently not competitive if compared with PVs, as regards the levelized cost of electricity (LCOE) and the authors proposed hybridization with gas turbine compatible fuels as a solution to both increase performance levels and reduce LCOE. Aichmayer et al. [88] proposed a thermo-economic analysis of a solar dish MGT combined cycle power plant consisting of an array of hybrid solar dish micro gas turbines and a single large steam cycle. Each MGT size is 80 kW while the combined power plant nominal power is 150 MW. The performance of this power plant is analyzed through an in-house thermo-economic tool, DYESOPT. Results of simulations highlighted that the proposed optimal design presents a cost 46.1% higher than a contemporary molten-salt tower CSP plant and that at this moment it is an unattractive investment. In the case of solar-MGT plants, numerical simulation is also useful to investigate components that establish the link between MGT and the solar plant. For example, Giovannelli et al. [89] studied the effect of cavity geometry on solar receiver effectiveness. They designed a high-temperature cavity solar receiver for a 5 kW Dish-MGT system and analyzed it through a CFD analysis. The adopted software is Ansys Fluent, and the results of the configurations comparison highlighted their differences and contributed to individuate the better one.

In Table 6, a summary of the reviewed literature has been listed.

Table 6. Summary of Solar-MGT layouts.

Starting MGT Model	Software	Layout Specifications	Ref.
Turbec T100	Aspen Plus	Single recuperator/ solar + biomass	[81]
Compower	MATLAB	Solar (pressurized and ambient configurations)	[82]
10 kW MGT	MATLAB/Simulink	Solar-MGT-TCES	[83]
Capstone C30	Thermoflex/Ansys Fluent	Solar heating + internal combustor	[84]
25 kW MGT	MATLAB + In-house (Axtur)	Solar, recuperated MGT with internal combustor	[85]
100–200 kW	MATLAB	Solar- recuperated MGT + auxiliary heater	[86,87]
80 kW	In-house (DYESOPT)	Array of dish-MGT combined with steam plant	[88]
5 kW	Ansys Fluent	Dish-MGT	[89]

4.4. MGTs in Distributed Energy Networks and Smart Energy Grids

MGTs represent a valuable solution as nodes of smart energy grids in which both renewable and traditional sources are considered. With respect to the layouts discussed in previous sections, the key feature of these hybrid energy plants is the energy management system: most of the research activity on this topic is focused on modeling and simulation of management and control algorithms, paying the attention to time response and dynamic behavior of each component of the smart grid.

Xu et al. [90] proposed a simplified dynamic micro gas turbine model for long-term analysis of distribution networks based on single and multiple MGTs. The model of a Capstone C30 has been validated by experimental data and it is capable of accurately predicting the dynamic behaviors of the turbine plants, although a lower order of mathematical equations representing the system. Rist et al. [91] proposed an MGT model focused on solving the economic dispatch problem, optimizing a CHP unit in a smart-grid environment. The authors compared results of simulations in case of four different scenarios in which an MGT can be driven: electricity, heat, maintenance cost, revenue; actual demand profiles (i.e., a full-service restaurant, a large and a small hotel, a residential building) and electricity tariffs are also included. The authors concluded that the integration of an MGT in a smart grid presents economic benefits in all the considered cases. Kanchev et al. [92] presented a microgrid integration of a prosumer (PV, batteries, ultracapacitors, inverter, and local energy management) and an MGT. Photovoltaic panels are associated with a storage system (batteries and ultracapacitors as long-term storage and fast dynamic power regulator, respectively). The authors consider local energy management and microgrid energy management, the latter including long-term energy management and a short-term power balancing. Moreover, the microgrid has been studied and modeled in order to define the specifications of the control system: a determinist operational power planning has been proposed along with strategies and experimental results of the individuated solution. The same authors [93] studied an urban microgrid including residential load, two 30 kW MGT, one 60 kW MGT and 12 homes with 3.6 kW photovoltaic panels with energy storage: a microgrid central energy management has been designed and implemented into supervisory control and data acquisition. Rivarolo et al. [94] studied the impact of not controllable renewable energy systems (wind turbines or photovoltaic panels) on the performance of a poly-generation smart grid, focusing the attention also on the economic aspect. The grid is composed of PV panels, a wind generator, a 100 kW MGT, a 20 kW reciprocating internal combustion engine, a 100 kW adsorption chiller. Thermal storage tanks are also present in the grid, since both MGT and ICE operate as CHP. Simulations are carried out by in-house software (EPoMP) and highlighted that the result for 17 kW of PV power is optimum compared to a 3 kW wind generator. Oh et al. [95] developed and tested hardware and software of a distributed intelligent management system. The microgrid is composed of an MGT, battery energy storage system, and smart loads. Yan et al. [96] developed an EMS of an urban microgrid in Matlab. The case study regarded two 30 kW and one 60 kW MGTs, 55 kW PV panels, and 110 kW of rated load, into two different scenarios (a sunny and a cloudy day, respectively). For each scenario, the comparison of the results highlighted that system cost and pollution are lower with optimization with respect to the cases without any optimization.

Di Gaeta et al. [27] proposed a hybrid energy grid in which the energy storage is made adopting hydrogen as an energy vector and both programmable and unprogrammable sources are present. The smart grid is composed of solar photovoltaic panels, wind generators, a micro gas turbine, water electrolyzers, and a hydrogen tank. The authors modeled the dynamic behavior of a controlled MGT (a Turbec T100) and embedded it in a hybrid energy grid model in order to simulate advanced scenarios based on a mix of renewable and non-renewable energy sources. Results of simulations highlighted that the combined adoption of MGT and water electrolyzers with renewable sources allows to compensate the power fluctuations related to renewable energy sources and to reduce fossil fuels consumption (37.5% in case of partial hybridization, up to 41.5% in case of full hybridization), carbon dioxide and pollutant emissions. Mohammadi et al. [97] proposed a stochastic method and an adaptive modified firefly algorithm to simulate and predict the non-linear and complex optimization problem related to the economic energy dispatch. The subject of study is a microgrid based on wind turbines, PV, MGT, fuel cell, and a NiMH battery as energy storage.

Table 7 lists the summary of the reviewed literature concerning MGT-based energy systems involved in distributed energy networks and smart grids.

Table 7. Summary of MGT in DES and smart-grid.

Smart Grid Components	MGT Size/Model	Ref.
Multiple MGTs	Capstone C30	[90]
MGT in CHP applications	100 kW	[91]
PV, batteries, ultracapacitors, MGT	-	[92]
MGTs, PV, energy storage	30 kW (2), 60 kW (1)	[93]
PV, wind gen., MGT, ICE, chiller, thermal storage	100 kW	[94]
MGT, energy storage	-	[95]
PV, MGTs	30 kW (2), 60 kW (1)	[96]
PV, water electrolyzer, hydrogen tank, MGT	Turbec T100	[27]
PV, wind turbine, MGT, Fuel cell, NiMH-Battery	30 kW	[97]

5. The Role of CFD Analysis of the Combustion Process in Case of Alternative Fuels

In several studies, 0D analysis of the entire system is coupled with a CFD analysis of the combustion process. This choice allows us to overcome the great simplification of the combustion models considered for thermodynamic analysis. The computational fluid dynamic approach represents a valuable solution to provide more realistic information of the combustion process, especially in the case of alternative or critical fuels. In fact, the potential issues related to supplying the combustor chamber with unconventional fuels, cannot be detected through a 0D analysis, and a deeper study is mandatory: hence, the results of thermodynamic analysis become both boundary and initial conditions of CFD analysis. As stated above, CFD is useful to highlight critical issues in case of adoption of critical fuels and to contribute to overcoming these limits. In recent years, the focus was mainly addressed on hydrogen and biomass-derived fuels. As regards the hydrogen, the research activity follows two directions: new designs of combustion chambers, to allow the burning of 100% hydrogen, and the individuation of commercial combustors operating limits, including strategies to fix the limits, in case of hydrogen-enriched blends based on natural gas or other fuels.

Cappelletti et al. [98] proposed a redesign of the Turbec T100 combustor chamber for 100% hydrogen fueling. The study was carried out through the CFD approach using the commercial software Ansys Fluent. Simulations highlighted that the flame ignites into the premixing conduits, that are not designed to withstand high temperatures. The proposed modifications to the fuel injection system and to the main flow allowed to move the flame position far from the walls but it does not ensure an adequately premixed flame. Devriese et al. [99] showed a CFD-based design of the combustion chamber, compressor, and turbine of a 100 kW hydrogen-fueled MGT focusing, as regards the combustion chamber, on the micromix design; steady Reynolds averaged Navier Stokes (RANS) simulations were carried out by Ansys Fluent. Tuccillo et al. [100] compared the behavior of two different commercial combustor geometries in the case of methane-hydrogen blends supply: a lean-premixed combustor (Turbec T100) and an RQL combustion chamber (Capstone C30). The response of the two combustors for increasing hydrogen/methane ratios was analyzed in order to establish combustor limits in terms of both flashback occurrence and pollutant formation. Results showed that the typical fuel addition of lean premixed combustors, from both main and pilot lines, presents the risk of flashback phenomena for hydrogen concentration higher than 10%, while the pseudo-RQL combustor can operate also with 20% hydrogen blends. Reale et al. [101] compared experimental data and results of CFD simulations of a Turbec T100 MGT fueled by three different hydrogen/methane blends (H_2 up to 10% in volume). The authors highlighted, as regard carbon monoxide, the significant deviation due to the limited prediction capability of the adopted simple two-step kinetic mechanism. Meziane et al. [102] presented a 3D CFD simulation of the combustion process of an RQL type can combustor. They compared temperature and pollutants emission levels with different hydrogen/methane blend compositions up to 90% of H_2 in molar fraction, finding out that CO and NO emissions decrease by 60% and 14% respectively with only 10% of H_2 in the fuel blend. Furthermore, the authors highlighted that the use of a detailed scheme involves two opposite effects: a good one, namely a smoother profile of temperature distribution in the combustor, and an unfavorable one, i.e., an increased deviation in pollutants prediction. Shih et al. [103] studied the effects of hydrogen content in methane/hydrogen blends on the combustion process occurring in a can-type combustor. Results indicated good combustion performance at moderate hydrogen fractions up to 50%. Calabria et al. [104] studied the combustion process of a Turbec T100 at part load for both methane and methane/hydrogen blends. Boundary and inlet conditions are taken from experimental campaigns and results of 0D numerical analysis. As regards the hydrogen-methane blends, the authors highlighted that early combustion occurs into the main swirling vanes already with a percentage of hydrogen of 15% in volume, while for lower hydrogen content the combustion is regular yet at part load; finally, a strategy to slightly increase the hydrogen percentage limit based on the fuel distribution between main and pilot has been proposed.

The discussion on biomass-derived fuels regards mainly gaseous fuels such as biogas or syngas, and the CFD analysis allows us to investigate the effects of the adoption of these low calorific fuels on the combustion process. Bazzoyar et al. [105] studied the behavior of a new biogas MGT combustor, starting from a 12 kW Blandon MT, coupling a RANS k-ω model with a steady diffusion flamelet approach. The numerical results were compared with experiments for several testing points: the presence of CO_2 requires fuel-rich mixtures at startup, while can help to reduce the NOx formation in the primary zone. A proper design of the combustor allows us to control CO emissions. Ilbas et al. [106] investigated the effects of hydrogen addition also to biogas and turbulator angle variations on combustion process behavior. They highlighted an average increment in the flame temperature of the biogas of about 15% in case of hydrogen addition. Potentialities and issues related to the adoption of alternative, critical, or low calorific fuels into a Turbec T100 combustion chamber were studied in various papers through CFD analysis as regard synthesis gas (Cadorin et al. [107], Calabria et al. [108]) or biogas (Calabria et al. [109]). Results of simulations described that the presence of hydrogen in synthesis gas can lead to combustion anomalies within the main swirling ducts [107,108] and the issues related to the biogas adoption can be overcome through a dual-fuel approach and modifications to the fuel feeding system, in order to ensure stable combustion [109].

In Table 8, a summary of the reviewed literature has been listed.

Table 8. CFD analysis on MGT supplied by unconventional fuel.

MGT/Combustor Type	Software	Turbulence Model	Fuel	Ref.
Turbec T100 modified	Ansys Fluent	RANS k-ω SST	H_2	[98]
Turbec T100 modified	Ansys Fluent	RANS realizable k-ε	H_2	[99]
Turbec T100, Capstone C30	Ansys Fluent	RANS realizable k-ε /DES	H_2/CH_4	[100]
Turbec T100	Ansys CFX	RANS k-ε	H_2/CH_4	[101]
RQL can	Ansys Fluent	RANS k-ω	H_2/CH_4	[102]
Can	-	RANS k-ε	H_2/CH_4	[103]
Turbec T100	Ansys CFX	RANS RSM	H_2/CH_4	[104]
Derived by Bladon MT/	-	RANS k-ω SST	biogas	[105]
tubular	Ansys Fluent	RANS k-ε	H_2/biogas	[106]
Turbec T100	Ansys CFX	RANS RSM	NG, syngas	[107]
Turbec T100	Ansys CFX	RANS RSM	NG, H_2/CH_4, biogas	[108]
Turbec T100	Ansys CFX	RANS RSM	biogas	[109]

6. Conclusions

Despite their commercial market having never really exploded, the research interest in MGT-based energy systems in recent years grew up, thanks to their flexibility and versatility along with the suitability of this energy system for hybrid small-scale plants. Studies and research activities are mainly addressed to the individuation of enhanced layouts able to overcome those performance limits typical of small-scale gas turbines and to the integration of MGTs with other systems in hybrid and complex energy systems able to guarantee both high-performance levels and the diversification of the energy sources. In this context, numerical modeling can represent the best way to investigate innovative solutions capable of extending the operating range of micro gas turbines.

In this review, the authors focused their work on the wide and various fields of a modeling approach for several layouts of MGT-based energy systems, with the aim of highlighting the main model features for all the analyzed cases. It is clear that the choice of numerical approaches and mathematical tools depends on the focus and aim of each study; moreover, a factor that should not be neglected is the preeminent expertise of each author/group, along with the main purpose of a single work, which can be focused on chemical, thermodynamic, thermo-economic or further aspects. Furthermore, the choice between commercial or in-house tools depends also on the authors' background. Nonetheless, some remarks can be specified, and this review aims to provide a broad overview of numerical tools, their application, and their possible combination. In particular, a simple thermodynamic analysis is a good choice to simulate the MGT behavior in design conditions, even if in some cases MGT can be described as a black box. These approaches are useful, for example, in thermo-economic analysis or when only power and overall performance levels are required. More complex models need to be implemented when off-design behavior assumes a prominent role, especially in the case of hybrid systems like MGT coupled with SOFC or ORC. The introduction of characteristic performance maps of compressor and turbine can be a solution to have more reliable results since the variations of rotating component behavior with load, rotational speed, and ambient conditions are considered. Off design maps or corrective factors can be assumed also for combustor and heat exchangers, although experimental data should be available to perform an accurate validation of specific components models. Dynamic models as well as energy management system modeling are needed in the case of variable loads and integration in distributed smart energy grids. In some cases, CFD analysis of the combustion process can permit to individuate issues related to the adoption of alternative feeding fuels such as biomass-derived or high hydrogen content fuels.

Author Contributions: Conceptualization, F.R.; methodology, F.R. and R.S.; writing—original draft preparation, F.R. and R.S.; writing—review and editing, F.R. and R.S.; visualization, F.R.; project administration, F.R.; funding acquisition, F.R. All authors have read and agreed to the published version of the manuscript.

Funding: This review was partly carried out as a part of the Italian national research project "Biofeedstock—PON ARS01_00985", funded by the Italian Ministry of University and Research (MUR).

Institutional Review Board Statement: Not applicable.

Informed Consent Statement: Not applicable.

Data Availability Statement: Not applicable.

Conflicts of Interest: The authors declare no conflict of interest.

Abbreviations

B	Burner
C.C.	Combustion chamber
CFD	Computational fluid dynamics
CHP	Combined Heat and Power
CCHP	Combined Cooling Heat and Power
EFMGT	Externally Fired micro Gas Turbine
EGR	Exhaust gas recirculation
EMS	Energy management system
HAT	Humid Air Turbine
HGT	Humidified Gas Turbine
H-STIG	Humidified—steam injected gas turbine
HX	Heat exchanger
LHV	Lower heating value [kJ/kgK]
MGT	Micro Gas Turbine
mHAT	Micro Humid Air Turbine
ORC	Organic Rankine Cycle
PV	Photovoltaic
RANS	Reynolds averaged Navier Stokes
RQL	Rich-quench-lean
STIG	Steam Injection Gas Turbine
SOFC	Solid Oxide Fuel Cell

References

1. European Commission, Recovery Plan for Europe. Available online: https://ec.europa.eu/info/strategy/recovery-plan-europe_en (accessed on 15 November 2021).
2. Pratt, R.G.; Balducci, P.J.; Gerkensmeyer, C.; Katipamula, S. *The Smart Grid: An Estimation of the Energy and CO_2 Benefits*; Technical Report; U.S. Department of Energy: Washington, DC, USA, 2010. Available online: https://www.pnnl.gov/main/publications/external/technical_reports/PNNL-19112.pdf (accessed on 21 January 2022). [CrossRef]
3. Yi, J.H.; Ko, W.; Park, J.K.; Park, H. Impact of carbon emission constraint on design of small scale multi-energy system. *Energy* **2018**, *161*, 792–808. [CrossRef]
4. Pilavachi, P.A. Mini- and micro-gas turbines for combined heat and power. *Appl. Therm. Eng.* **2002**, *22*, 2003–2014. [CrossRef]
5. Akorede, M.F.; Pouresmaeil, H.H.E. Distributed energy resources and benefits to the environment. *Renew. Sust. Energy Rev.* **2010**, *14*, 724–734. [CrossRef]
6. Al-Hamdan, Q.Z.; Ebaid, M.S.Y. Modeling and Simulation of a Gas Turbine Engine for Power Generation. *ASME J. Eng. Gas Turbines Power* **2006**, *128*, 302–311. [CrossRef]
7. Wendt, J.F. (Ed.) *Computational Fluid Dynamics: An Introduction*; Springer Science & Business Media: Berlin/Heidelberg, Germany, 1992; ISBN 978-3-540-53460-0.
8. Gopisetty, S.; Treffinger, P. Generic Combined Heat and Power (CHP) Model for the Concept Phase of Energy Planning Process. *Energies* **2017**, *10*, 11. [CrossRef]
9. Gimelli, A.; Sannino, R. A micro gas turbine one-dimensional model: Approach description, calibration with a vector optimization methodology and validation. *Appl. Therm. Eng.* **2021**, *188*, 116644. [CrossRef]
10. Saravanamuttoo, H.I.H.; Rogers, G.F.C.; Cohen, H.; Straznicky, P.V. *Gas Turbine Theory*, 6th ed.; Pearson: London, UK, 2009; ISBN 978-0-12-222437-6.

11. Hosseinimaab, S.M.; Tousi, A.M. A new approach to off-design performance analysis of gas turbine engines and its application. *Energy Convers. Manag.* **2021**, *243*, 114411. [CrossRef]
12. Altarazi, Y.S.M.; Abu Talib, A.R.; Gires, E.; Yu, J.; Lucas, J.; Yusaf, T. Performance and exhaust emissions rate of small-scale turbojet engine running on dual biodiesel blends using Gasturb. *Energy* **2021**, *232*, 120971. [CrossRef]
13. Malinowski, L.; Lewandowska, M. Analytical model-based energy and exergy analysis of a gas microturbine at part load operation. *Appl. Therm. Eng.* **2013**, *57*, 125–132. [CrossRef]
14. Nikpey Somehsaraei, J.; Mansouri, M.; Majoumerd, M.M.; Breuhaus, P.; Assadi, M. Performance analysis of a biogas-fueled micro gas turbine using a validated thermodynamic model. *Appl. Therm. Eng.* **2014**, *66*, 181–190. [CrossRef]
15. Cáceres, C.X.; Cáceres, R.E.; Hein, D.; Molina, M.G.; Pia, J.M. Biogas production from grape pomace: Thermodynamic model of the process and dynamic model of the power generation system. *Int. J. Hydrog. Energy* **2012**, *37*, 10111–10117. [CrossRef]
16. Caresana, F.; Pelagalli, L.; Comodi, G.; Renzi, M. Microturbogas cogeneration systems for distributed generation: Effects of ambient temperature on global performance and component's behavior. *Appl. Energy* **2014**, *124*, 17–27. [CrossRef]
17. Gimelli, A.; Sannino, R. A multi-variable multi-objective methodology for experimental data and thermodynamic analysis validation: An application to micro gas turbines. *Appl. Therm. Eng.* **2018**, *134*, 501–512. [CrossRef]
18. Gimelli, A.; Sannino, R. Thermodynamic model validation of Capstone C30 micro gas turbine. *Energy Procedia* **2017**, *126*, 955–962. [CrossRef]
19. Gimelli, A.; Sannino, R. A vector optimization methodology applied to thermodynamic model calibration of a micro gas turbine CHP plant. *Energy Procedia* **2018**, *148*, 2–9. [CrossRef]
20. Duan, J.; Sun, L.; Wang, G.; Wu, F. Nonlinear modeling of regenerative cycle micro gas turbine. *Energy* **2015**, *91*, 168–175. [CrossRef]
21. Zornek, T.; Monz, T.; Aigner, M. Performance analysis of the micro gas turbine Turbec T100 with a new FLOX-combustion system for low calorific fuels. *Appl. Energy* **2015**, *159*, 276–284. [CrossRef]
22. Panne, T.; Widenhorn, A.; Boyde, J.; Matha, D.; Abel, V.; Aigner, M. Thermodynamic Process Analysis of SOFC/GT Hybrid Cycles. In Proceedings of the 5th International Energy Conversion Engineering Conference and Exhibit (IECEC), St. Louis, MO, USA, 25–27 June 2007. Paper No. 2007-4833.
23. Henke, M.; Monz, T.; Aigner, M. Inverted Brayton cycle with exhaust gas recirculation—A numerical investigation. *J. Eng. Gas Turb. Power* **2013**, *135*, 091203. [CrossRef]
24. Henke, M.; Klempp, N.; Hohloch, M.; Monz, T.; Aigner, M. Validation of a T100 micro gas turbine steady-state simulation tool. In Proceedings of the ASME Turbo Expo 2015, Montreal, QC, Canada, 15–19 June 2015. Paper No. GT2015-42090.
25. Visser, W.P.J.; Shakariyants, S.A.; Oostveen, M. Development of a 3 kW Micro Turbine for CHP Applications. *J. Eng. Gas Turbine Power* **2011**, *133*, 042301. [CrossRef]
26. Visser, W.P.J.; Dountchev, I.D. Modeling thermal effects on performance of small gas turbines. In Proceedings of the ASME Turbo Expo 2015, Montreal, QC, Canada, 15–19 June 2015. Paper No. GT2015-42744.
27. di Gaeta, A.; Reale, F.; Chiariello, F.; Massoli, P. A dynamic model of a 100 kW micro gas turbine fuelled with natural gas and hydrogen blends and its application in a hybrid energy grid. *Energy* **2017**, *129*, 299–320. [CrossRef]
28. Jonsson, M.; Yan, J. Humidified gas turbines—A review of proposed and implemented cycles. *Energy* **2005**, *30*, 1013–1078. [CrossRef]
29. Parente, J.; Traverso, A.; Massardo, A.F. Micro Humid air cycle part A: Thermodynamic and technical aspects. In Proceedings of the ASME Turbo Expo 2003, Atlanta, GA, USA, 16–19 June 2003; pp. 221–229, Paper No. GT2003-38326.
30. Parente, J.; Traverso, A.; Massardo, A.F. Micro humid air cycle part B: Thermoeconomic analysis. In Proceedings of the ASME Turbo Expo 2003, Atlanta, GA, USA, 16–19 June 2003. Paper No. GT2003-38328.
31. De Paepe, W.; Delattin, F.; Bram, S.; De Ruyk, J. Steam injection experiments in a microturbine—A thermodynamic performance analysis. *Appl. Energy* **2012**, *97*, 569–576. [CrossRef]
32. Montero Carrero, M.; De Paepe, W.; Bram, S.; Parente, A.; Contino, F. Does humidification improve the micro Gas Turbine cycle? Thermodynamic assessment based on Sankey and Grassmann diagrams. *Appl. Energy* **2017**, *204*, 1163–1171. [CrossRef]
33. Ali, U.; Font-Palma, C.; Hughes, K.J.; Ingham, D.B.; Ma, L.; Pourkashanian, M. Thermodynamic analysis and process system comparison of the exhaust gas recirculated, steam injected and humidified micro gas turbine. In Proceedings of the ASME Turbo Expo 2015, Montreal, QC, Canada, 15–19 June 2015. Paper No. GT2015-42688.
34. Reale, F.; Sannino, R. Water and steam injection in micro gas turbine supplied by hydrogen enriched fuels: Numerical investigation and performance analysis. *Int. J. Hydrog. Energy* **2021**, *46*, 24366–24381. [CrossRef]
35. De Paepe, W.; Montero Carrero, M.; Bram, S.; Contino, F.; Parente, A. Waste heat recovery optimization in micro gas turbine applications using advanced humidified gas turbine cycle concepts. *Appl. Energy* **2017**, *207*, 218–229. [CrossRef]
36. De Paepe, W.; Pappa, A.; Montero Carrero, M.; Bricteux, L.; Contino, F. Reducing waste heat to the minimum: Thermodynamic assessment of the M-power cycle concept applied to micro Gas Turbines. *Appl. Energy* **2020**, *279*, 115898. [CrossRef]
37. Wan, K.; Zhang, S.; Wang, J.; Xiao, Y. Performance of humid air turbine with exhaust gas expanded to below ambient pressure based on microturbine. *Energy Convers. Manag.* **2010**, *51*, 2127–2133. [CrossRef]
38. Al-attab, K.A.; Zainal, Z.A. Externally fired gas turbine technology: A review. *Appl. Energy* **2015**, *138*, 474–487. [CrossRef]
39. Traverso, A.; Massardo, A.F.; Scarpellini, R. Externally Fired micro-Gas Turbine: Modelling and experimental performance. *Appl. Therm. Eng.* **2006**, *26*, 1935–1941. [CrossRef]

40. Traverso, A.; Magistri, L.; Scarpellini, R.; Massardo, A.F. Demonstration plant and expected performance of an externally fired microgas turbine for distributed power generation. In Proceedings of the ASME Turbo Expo 2003, Atlanta, GA, USA, 16–19 June 2003. Paper No. GT2003-38268.
41. Traverso, A.; Calzolari, F.; Massardo, A.F. Transient behaviour and control system for micro gas turbine advanced cycles. *J. Eng. Gas Turbines Power* **2005**, *127*, 340–347. [CrossRef]
42. Magistri, L. Hybrid systems for distributed generation. Ph.D. Thesis, University of Genoa, Genoa, Italy, 2003.
43. Pantaleo, A.M.; Camporeale, S.M.; Shah, N. Thermo-economic assessment of externally fired micro-gas turbine fired by natural gas and biomass: Applications in Italy. *Energy Convers. Manag.* **2013**, *75*, 202–213. [CrossRef]
44. Riccio, G.; Chiaramonti, D. Design and simulation of a small polygeneration plant cofiring biomass and natural gas in a dual combustion micro gas turbine (BIO_MGT). *Biomass Bioenergy* **2009**, *33*, 1520–1531. [CrossRef]
45. Datta, A.; Ganguly, R.; Sarkar, L. Energy and exergy analyses of an externally fired gas turbine (EFGT) cycle integrated with biomass gasifier for distributed power generation. *Energy* **2010**, *35*, 341–350. [CrossRef]
46. Cameretti, M.C.; Piazzesi, R.; Reale, F.; Tuccillo, R. Combustion Simulation of an Exhaust Gas Recirculation Operated Micro-gas Turbine. *J. Eng. Gas Turbines Power* **2009**, *131*, 051701. [CrossRef]
47. Majoumerd, M.M.; Somehsaraei, H.N.; Assadi, M.; Breuhaus, P. Micro gas turbine configurations with carbon capture—Performance assessment using a validated thermodynamic model. *Appl. Therm. Eng.* **2014**, *73*, 172–184. [CrossRef]
48. Nikpei, H.; Majoumerd, M.M.; Assadi, M.; Brreuhaus, P. Thermodynamic analysis of innovative micro gas turbine cycles. In Proceedings of the ASME Turbo Expo 2014, Düsseldorf, Germany, 16–20 June 2014. Paper No. GT2014-26917.
49. Ali, U.; Hughes, K.J.; Ingham, D.B.; Ma, L.; Pourkashanian, M. Effect of the CO_2 enhancement on the performance of a micro gas turbine with a pilot-scale CO_2 capture plant. *Chem. Eng. Res. Des.* **2017**, *117*, 11–23. [CrossRef]
50. Hampel, B.; Bauer, S.; Heublein, N.; Hirsch, C.; Sattelmayer, T. Feasibility Study on Dehydrogenation of LOHC Using Excess Exhaust Heat from a Hydrogen Fueled Micro Gas Turbine. In Proceedings of the ASME Turbo Expo 2015, Montreal, QC, Canada, 15–19 June 2015. Paper No. GT2015-43168.
51. Manwell, J.F. Hybrid Energy Systems. In *Encyclopedia of Energy*; Cleveland, C.J., Ed.; Elsevier: Amsterdam, The Netherlands, 2004; pp. 215–229. ISBN 978-0-12-176480-7. [CrossRef]
52. Invernizzi, C.; Iora, P.; Silva, P. Bottoming micro-Rankine cycles for micro-gas turbines. *Appl. Therm. Eng.* **2007**, *27*, 100–110. [CrossRef]
53. Mago, P.J.; Chamra, L.M.; Srinivasan, K.; Somayaji, C. An examination of regenerative organic Rankine cycles using dry fluids. *Appl. Therm. Eng.* **2008**, *28*, 998–1007. [CrossRef]
54. Javanshir, A.; Sarunac, N.; Razzaghpanah, Z. Thermodynamic Analysis of ORC and Its Application for Waste Heat Recovery. *Sustainability* **2017**, *9*, 1974. [CrossRef]
55. Bonolo de Campos, G.; Bringhenti, C.; Traverso, A.; Tomita, J.T. A Review on Combining Micro Gas Turbines with Organic Rankine Cycles. In Proceedings of the E3S Web of Conference, Savona, Italy, 4–6 September 2019; Volume 113.
56. Chacartegui, R.; Becerra, J.A.; Blanco, M.J.; Muñoz-Escalona, J.M. A Humid Air Turbine–Organic Rankine Cycle combined cycle for distributed microgeneration. *Energy Convers. Manag.* **2015**, *104*, 115–126. [CrossRef]
57. Moradi, R.; Marcantonio, V.; Cioccolanti, L.; Bocci, E. Integrating biomass gasification with a steam-injected micro gas turbine and an Organic Rankine Cycle unit for combined heat and power production. *Energy Convers. Manag.* **2020**, *205*, 112464. [CrossRef]
58. Nazari, N.; Mousavi, S.; Mirjalili, S. Exergo-economic analysis and multi-objective multi-verse optimization of a solar/biomass-based trigeneration system using externally-fired gas turbine, organic Rankine cycle and absorption refrigeration cycle. *Appl. Therm. Eng.* **2021**, *191*, 116889. [CrossRef]
59. Nazari, N.; Porkhial, S. Multi-objective optimization and exergo-economic assessment of a solar-biomass multi-generation system based on externally-fired gas turbine, steam and organic Rankine cycle, absorption chiller and multi-effect desalination. *Appl. Therm. Eng.* **2020**, *179*, 115521. [CrossRef]
60. Xiao, G.; Chen, J.; Ni, M.; Cen, K. A solar micro gas turbine system combined with steam injection and ORC bottoming cycle. *Energy Convers. Manag.* **2021**, *243*, 114032. [CrossRef]
61. Cameretti, M.C.; Ferrara, F.; Gimelli, A.; Tuccillo, R. Employing Micro-Turbine Components in Integrated Solar-mGT-ORC Power Plants. In Proceedings of the ASME Turbo Expo 2015, Montreal, QC, Canada, 15–19 June 2015. Paper No. GT2015-42572.
62. Baccioli, A.; Ferrari, L.; Vizza, F.; Desideri, U. Potential energy recovery by integrating an ORC in a biogas plant. *Appl. Energy* **2019**, *256*, 113960. [CrossRef]
63. Reale, F.; Sannino, R.; Calabria, R.; Massoli, P. Numerical study of a small-scale micro gas turbine-ORC power plant integrated with biomass gasifier. In Proceedings of the ASME Turbo Expo 2020, Online. 21–25 September 2020. Paper No: GT2020-15401.
64. Costamagna, P.; Magistri, L.; Massardo, A.F. Design and part-load performance of a hybrid system based on a solid oxide fuel cell reactor and a micro gas turbine. *J. Power Sources* **2001**, *96*, 352–368. [CrossRef]
65. Ferrari, M.L.; Traverso, A.; Magistri, L.; Massardo, A.F. Influence of the anodic recirculation transient behaviour on the SOFC hybrid system performance. *J. Power Sources* **2005**, *149*, 22–32. [CrossRef]
66. Kaneko, T.; Brouwer, J.; Samuelsen, G.S. Power and temperature control of fluctuating biomass gas fueled solid oxide fuel cell and micro gas turbine hybrid system. *J. Power Sources* **2006**, *160*, 316–325. [CrossRef]
67. Bang-Møller, C.; Rokni, M. Thermodynamic performance study of biomass gasification, solid oxide fuel cell and micro gas turbine hybrid systems. *Energy Convers. Manag.* **2010**, *51*, 2330–2339. [CrossRef]

68. Bang-Møller, C.; Rokni, M.; Elmegaard, B. Exergy analysis and optimization of a biomass gasification, solid oxide fuel cell and micro gas turbine hybrid system. *Energy* **2011**, *36*, 4740–4752. [CrossRef]
69. Bakalis, D.P.; Stamatis, A.G. Incorporating available micro gas turbines and fuel cell: Matching considerations and performance evaluation. *Appl. Energy* **2013**, *103*, 607–617. [CrossRef]
70. Fryda, L.; Panopoulos, K.D.; Kakaras, E. Integrated CHP with autothermal biomass gasification and SOFC–MGT. *Energy Convers. Manag.* **2008**, *49*, 281–290. [CrossRef]
71. Zhao, H.; Hou, Q. Thermodynamic performance study of the MR SOFC-HAT-CCHP system. *Int. J. Hydrog. Energy* **2019**, *44*, 4332–4349. [CrossRef]
72. Ebrahimi, M.; Moradpoor, I. Combined solid oxide fuel cell, micro-gas turbine and organic Rankine cycle for power generation (SOFC-MGT-ORC). *Energy Convers. Manag.* **2016**, *116*, 120–133. [CrossRef]
73. Wongchanapai, S.; Iwai, H.; Saito, M.; Yoshida, H. Performance evaluation of a direct-biogas solid oxide fuel cell-micro gas turbine (SOFC-MGT) hybrid combined heat and power (CHP) system. *J. Power Sources* **2013**, *223*, 9–17. [CrossRef]
74. Krummrein, T.; Henke, M.; Kutne, P.; Aigner, M. Numerical analysis of operating range and SOFC-off-gas combustor requirements of a biogas powered SOFC-MGT hybrid power plant. *Appl. Energy* **2018**, *232*, 598–606. [CrossRef]
75. Di Carlo, A.; Borello, D.; Bocci, E. Process simulation of a hybrid SOFC/mGT and enriched air/steam fluidized bed gasifier power plant. *Int. J. Hydrog. Energy* **2013**, *38*, 5857–5874. [CrossRef]
76. Perna, A.; Minutillo, G.; Jannelli, E.; Cigolotti, V.; Nam, S.W. Performance assessment of a hybrid SOFC/MGT cogeneration power plant fed by syngas from a biomass down-draft gasifier. *Appl. Energy* **2018**, *227*, 80–91. [CrossRef]
77. Isfahani, S.N.R.; Sedaghat, A. Hybrid micro gas turbine and solid state fuel cell power plant with hydrogen production and CO_2 capture. *Int. J. Hydrog. Energy* **2016**, *41*, 9490–9499. [CrossRef]
78. Zhen, Z.; Liu, T.; Liu, Q.; Lei, J.; Fang, J. A distributed energy system integrating SOFC-MGT with mid-and-low temperature solar thermochemical hydrogen fuel production. *Int. J. Hydrog. Energy* **2021**, *46*, 19846–19860. [CrossRef]
79. Karimi, M.H.; Chitgar, N.; Emadi, M.A.; Ahmadi, P.; Rosen, M.A. Performance assessment and optimization of a biomass-based solid oxide fuel cell and micro gas turbine system integrated with an organic Rankine cycle. *Int. J. Hydrog. Energy* **2020**, *45*, 6262–6277. [CrossRef]
80. Pantaleo, A.M.; Fordham, J.; Oyewunmi, J.F.; De Palma, P.; Markides, C.N. Integrating cogeneration and intermittent waste-heat recovery in food processing: Microturbines vs. ORC systems in the coffee roasting industry. *Appl. Energy* **2018**, *225*, 782–796. [CrossRef]
81. Kautz, M.; Hansen, U. The externally-fired gas-turbine (EFGT-Cycle) for decentralized use of biomass. *Appl. Energy* **2007**, *84*, 795–805. [CrossRef]
82. Aichmayer, L.; Spelling, J.; Laumert, B. Preliminary design and analysis of a novel solar receiver for a micro gas-turbine based solar dish system. *Sol. Energy* **2015**, *114*, 378–396. [CrossRef]
83. Yang, J.; Ziano, G.; Ghavami, M.; Al-Zaili, J.; Yang, T. Thermodynamic modelling and real-time control strategies of solar micro gas turbine system with thermochemical energy storage. *J. Clean. Prod.* **2021**, *304*, 127010. [CrossRef]
84. Cameretti, M.C.; Langella, G.; Sabino, S.; Tuccillo, R. Modeling of a hybrid solar micro gas-turbine power plant. *Energy Procedia* **2015**, *82*, 833–840. [CrossRef]
85. Wang, W.; Ragnolo, G.; Aichmayer, L.; Strand, T.; Laumert, B. Integrated design of a hybrid gas turbine-receiver unit for a solar dish system. *Energy Procedia* **2015**, *69*, 583–592. [CrossRef]
86. Giostri, A.; Binotti, M.; Sterpos, C.; Lozza, G. Small scale solar tower coupled with micro gas turbine. *Renew. Energy* **2020**, *147*, 570–583. [CrossRef]
87. Giostri, A.; Macchi, E. An advanced solution to boost sun-to-electricity efficiency of parabolic dish. *Sol. Energy* **2016**, *139*, 337–354. [CrossRef]
88. Aichmayer, L.; Spelling, J.; Laumert, B. Thermoeconomic analysis of a solar dish micro gas-turbine combined-cycle power plant. *Energy Procedia* **2015**, *69*, 1089–1099. [CrossRef]
89. Giovannelli, A.; Bashir, M.A. High-Temperature Cavity Receiver Integrated with a Short-Term Storage System for Solar MGTs: Heat Transfer Enhancement. *Energy Procedia* **2017**, *126*, 557–564. [CrossRef]
90. Xu, X.; Li, K.; Qi, F.; Jia, H.; Deng, J. Identification of microturbine model for long-term dynamic analysis of distribution networks. *Appl. Energy* **2017**, *192*, 305–314. [CrossRef]
91. Rist, J.F.; Dias, M.F.; Palman, M.; Zelazo, D.; Cukurel, B. Economic dispatch of a single micro-gas turbine under CHP operation. *Appl. Energy* **2017**, *200*, 1–18. [CrossRef]
92. Kanchev, H.; Lu, D.; Colas, F.; Lazarov, V.; Francois, B. Energy Management and Operational Planning of a Microgrid With a PV-Based Active Generator for Smart Grid Applications. *IEEE Trans. Ind. Electron.* **2011**, *58*, 4583–4592. [CrossRef]
93. Kanchev, H.; Colas, F.; Lazarov, V.; Francois, B. Emission Reduction and Economical Optimization of an Urban Microgrid Operation Including Dispatched PV-Based Active Generators. *IEEE Trans. Sustain. Energy* **2014**, *5*, 1397–1405. [CrossRef]
94. Rivarolo, M.; Greco, A.; Massardo, A.F. Thermo-economic optimization of the impact of renewable generators on poly-generation smart-grids including hot thermal storage. *Energy Convers. Manag.* **2013**, *65*, 75–83. [CrossRef]
95. Oh, S.J.; Yoo, C.H.; Chung, I.Y.; Won, D.J. Hardware-in-the-Loop Simulation of Distributed Intelligent Energy Management System for Microgrids. *Energies* **2013**, *6*, 3263–3283. [CrossRef]

96. Yan, X.; Abbes, D.; Francois, B. Development of a tool for urban microgrid optimal energy planning and management. *Simul. Model. Pract. Theory* **2018**, *89*, 64–81. [CrossRef]
97. Mohammadi, S.; Soleymani, S.; Mozafari, B. Scenario-based stochastic operation management of MicroGrid including Wind, Photovoltaic, Micro-Turbine, Fuel Cell and Energy Storage Devices. *Int. J. Electr. Power Energy Syst.* **2014**, *54*, 525–535. [CrossRef]
98. Cappelletti, A.; Martelli, F.; Bianchi, E.; Trifoni, E. Numerical Redesign of 100 kW MGT Combustor for 100% H2 fueling. *Energy Procedia* **2014**, *45*, 1412–1421. [CrossRef]
99. Devriese, C.; Penninx, G.; de Ruiter, G.; Bastiaans, R.; De Paepe, W. The CFD Design and Optimisation of a 100 kW Hydrogen Fuelled mGT. In Proceeding of the ASME Turbo Expo 2020, Online. 21–25 September 2020. Paper No. GT2020-14473.
100. Tuccillo, R.; Cameretti, M.C.; De Robbio, R.; Reale, F.; Chiariello, F. Methane-hydrogen blends in micro gas turbines: Comparison of different combustor concepts. In Proceeding of the ASME Turbo Expo 2019, Phoenix, AZ, USA, 17–21 June 2019. Paper No. GT2019-90229.
101. Reale, F.; Calabria, R.; Chiariello, F.; Pagliara, R.; Massoli, P. A micro gas turbine fuelled by methane hydrogen blends. *Appl. Mech. Mater.* **2012**, *232*, 792–796. [CrossRef]
102. Meziane, S.; Bentebbiche, A. Numerical study of blended fuel natural gas-hydrogen combustion in rich/quench/lean combustor of a micro gas turbine. *Int. J. Hydrog. Energy* **2019**, *44*, 15610–15621. [CrossRef]
103. Shih, H.Y.; Liu, C.R. A computational study on the combustion of hydrogen/methane blended fuels for a micro gas turbines. *Int. J. Hydrog. Energy* **2014**, *39*, 15013–15115. [CrossRef]
104. Calabria, R.; Chiariello, F.; Massoli, P.; Reale, F. Part Load Behavior of a Micro Gas Turbine Fed with Different Fuels. In Proceedings of the ASME Turbo Expo 2014, Düsseldorf, Germany, 16–20 June 2014. Paper No. GT2014-26631.
105. Bazooyar, B.; Darabkhani, H.G. The design strategy and testing of an efficient microgas turbine combustor for biogas fuel. *Fuel* **2021**, *294*, 120535. [CrossRef]
106. Ilbas, M.; Sahin, M. Effects of turbulator angle and hydrogen addition on a biogas turbulent diffusion flame. *Int. J. Hydrog. Energy* **2017**, *42*, 25735–25743. [CrossRef]
107. Cadorin, M.; Pinelli, M.; Vaccari, A.; Calabria, R.; Chiariello, F.; Massoli, P.; Bianchi, E. Analysis of a Micro Gas Turbine Fed by Natural Gas and Synthesis Gas: MGT Test Bench and Combustor CFD Analysis. *J. Eng. Gas Turbines Power* **2012**, *134*, 071401. [CrossRef]
108. Calabria, R.; Chiariello, F.; Massoli, P.; Reale, F. CFD Analysis of Turbec T100 Combustor at Part Load by Varying Fuels. In Proceedings of the ASME Turbo Expo 2015, Montreal, QC, Canada, 15–19 June 2015. Paper No. GT2015-43455.
109. Calabria, R.; Chiariello, F.; Massoli, P.; Reale, F. A Biogas Fuelled Micro Gas Turbine: Energetic performance and environmental impact analysis. In Proceedings of the ASME Turbo Expo 2019, Phoenix, AZ, USA, 17–21 June 2019. Paper No. GT2019-91483.

energies

Article

Numerical Modeling of Combustion and Detonation in Aqueous Foams

Alexey Kiverin * **and Ivan Yakovenko**

Joint Institute for High Temperatures of Russian Academy of Sciences, 125412 Moscow, Russia; yakovenko.ivan@bk.ru
* Correspondence: alexeykiverin@gmail.com

Abstract: Combustible aqueous foams and foamed emulsions represent prospective energy carriers. This paper is devoted to the overview of model assumptions required for numerical simulations of combustion and detonation processes in aqueous foams. The basic mathematical model is proposed and used for the analysis of the combustion development in the wet aqueous foam containing bubbles filled with reactive gas. The numerical results agree with the recent experimental data on combustion and detonation in aqueous foams containing premixed hydrogen–oxygen. The obtained results allowed for distinguishing the mechanisms of flame acceleration, transition to detonation, detonation propagation, and decay.

Keywords: combustion; flame acceleration; detonation; microfoam; numerical analysis

Citation: Kiverin, A.; Yakovenko, I. Numerical Modeling of Combustion and Detonation in Aqueous Foams. *Energies* **2021**, *14*, 6233. https://doi.org/10.3390/en14196233

Academic Editors: Michele Pinelli, Alessio Suman and Nicola Casari

Received: 27 August 2021
Accepted: 26 September 2021
Published: 30 September 2021

Publisher's Note: MDPI stays neutral with regard to jurisdictional claims in published maps and institutional affiliations.

1. Introduction

The development of new technologies for efficient and clean energy is one of today's topical challenges. Among the promising directions of development are: (1) the use of hydrogen, which represents one of the most promising renewable fuels due to its high energy density, low harmful emissions, and cost-efficiency [1,2]; (2) the improvement of fossil fuel combustion stability and efficiency via hydrogen addition [3]; and (3) fuel dilution with chemical inhibitors and inert heat carriers as a promising approach for reducing harmful emissions [4]. In the latter case, for example, water addition in the form of a spray [5] or within the emulsified fuel [6] defines the lower production of NO_x. In all the mentioned cases, one should deal with complex multiphase systems, and one needs precise instruments of experimental diagnostics and numerical analysis to obtain a clear understanding of the processes and methods of their control.

One of the most interesting combustible systems possessing all the above-mentioned features is aqueous foam, which can contain both liquid hydrocarbon fuels and/or gaseous fuels (such as hydrogen). Recently Kichatov et al. [7] introduced the foamed emulsion as a brand new type of complex fuel. Furthermore, Kichatov et al. considered the aqueous foam bubbled with a hydrogen–oxygen mixture [8] and the oil-in-water emulsion foamed with a hydrogen–oxygen mixture [9]. In the case of the hydrogen addition, the intense dynamic processes were registered and interpreted as detonation onset. Thus, in [8], the authors proposed an integral mechanism of flame acceleration based on the experimental data and traditional considerations about the positive feedback related to flame interaction with the flow induced by flame propagation through the foam. As a rule, such integral effects are determined by the local dynamics of the flame front. Thus, for example, in [10], the integral self-similar solution for the acceleration of the freely propagating flame is obtained based on the local dynamics of the flame subjected to the hydrodynamic instability. In a recent paper [11], the collective effect of water droplets suspended in a hydrogen–air mixture on the local flame dynamics and related flame intensification is studied. The local effects driving the evolution of the flame in the foamed emulsions and microfoams are still unresolved. Thus, it is of interest to investigate how the flame locally evolves and what physical mechanisms affect the flame dynamics in aqueous foams.

In the paper [8], the effect of energy focusing on the scales of a single bubble collapsing under compression was proposed as a mechanism driving the detonation propagation. These conclusions were also made based on an integral understanding of the process dynamics. However, the understanding of the local dynamics is of paramount importance here since the combustion arises and decays on the scales of individual gas bubbles. Generally speaking, similar features of the detonation evolution are observed in bubbly liquids and dry aqueous foams. In bubbly liquids [12], the bubbles are small, and the gas inside each bubble explodes as a whole, pushing the shock wave into the liquid. Subsequently, the shock wave compresses the neighboring bubbles causing explosions inside them, etc. In dry foams [13–15], the detonation wave decays when passing from one bubble to another due to the rupture of liquid films into fine water droplets inhibiting the reaction. As a result, the flame front enters the next bubble behind the shock wave, and the deflagration-to-detonation transition takes place on the scales of the single relatively large bubble, leading to the formation of a new detonation front. The local dynamics and particular mechanisms of flame acceleration, deflagration-to-detonation transition, and detonation propagation in wet foams are of interest and considered in this paper.

The paper is organized as follows. In Section 2, we propose a mathematical model describing the dynamic processes in combustible foams associated with flame propagation. In Section 3, the numerical results obtained with the use of the model formulated in Section 2 are discussed, and processes of flame acceleration and transition to detonation are interpreted on the basis of numerical analysis. We conclude in Section 4.

2. Mathematical Model and Problem Setup

The combustible wet foam considered recently in [8] represents a two-phase medium consisting of gas bubbles of $\sim$200 μm diameter separated by thin films of an aqueous solution of a surfactant or oil-in-water emulsion [9]. The hydrogen–oxygen mixture is used as a foaming gas.

Let us consider the following model for the dynamic processes in combustible wet foam. The foam represents two interacting continuums whose dynamics is governed by a two-velocity, two-temperature model written in a conventional form similar to that proposed in [16]. The equation of state for the gas phase is in the same form as for the mixture of ideal gases [17]. The liquid phase is treated as incompressible and interacting with the gas phase via momentum and heat exchange. Heating of the liquid phase by hot gas defines the evaporation from the interface between the liquid and the gas according to the heat balance, as in [18]. Momentum exchange between phases is modeled via an empiric coefficient $C^* \sim 1.0$–10.0 in the same way as it was proposed in [16].

In the process of combustion wave propagation, the liquid phase of the foam decays into the droplets via two basic mechanisms. The first one is due to the expansion of the gas inside the bubble either as a result of the foaming gas heating or liquid phase evaporation. The second mechanism is the dynamic action leading to the rupture of the liquid films. According to this, here, the following criteria for the foam decay are applied: (1) at $\rho_{gas} > \rho_{cr}$, (2) $\mathrm{We} > \mathrm{We}_{cr}$. Critical gas density ρ_{cr} corresponds to the gas volume fraction equal to 0.96, while the critical Weber number We_{cr} is chosen to be equal to 350, which is usually considered as a criterion for liquid droplet fragmentation (see, e.g., [19,20]).

After the foam decay, the two-phase medium is formed that represents an aerosol consisting of water droplets suspended in the hydrogen–oxygen gaseous mixture. The interaction between phases is now described via the same model but with different interaction source terms corresponding to the droplet–gas interaction. Herewith, the evaporation of droplets is modeled in the approximation of the thermal conductivity limit [21], while droplet fragmentation is described via the mechanism of fine droplets breaking away from the surface of the coarse droplets at We > 2700 [20]. According to [20], under such conditions, the diameter of fine droplets is estimated as 2 μm.

The hydrogen combustion kinetics is taken into account according to the detailed kinetic mechanism from [22]. The mechanism by Keromnes et al. is widely tested and provides very good performance in terms of reproduction of a wide range of experimental data on hydrogen–air combustion [23]. In [24], it was also shown that in a pressurized hydrogen-containing system, the considered mechanism from [22] provides results for the pressure range below 20 atm similar to those of another comprehensive detailed mechanism by Burke et al. [25].

2.1. Basic Two-Phase Model

The governing equations for gas phase in two-dimensions are written in the following way:

$$\frac{\partial \rho_g}{\partial t} + \frac{\partial \rho_g u_{g,i}}{\partial x_i} = G_w \rho_w$$

$$\frac{\partial \rho_g Y_k}{\partial t} + \frac{\partial \rho_g Y_k u_{g,i}}{\partial x_i} = \frac{\partial \rho_g Y_k V_{k,i}}{\partial x_i} + \dot{\omega}_k + \begin{cases} 0, & k \neq H_2O \\ G_w \rho_w, & k = H_2O \end{cases}$$

$$\frac{\partial \rho_g u_{g,i}}{\partial t} + u_{g,j} \frac{\partial \rho_g u_{g,i}}{\partial x_j} = -\frac{\partial p}{\partial x_i} + \frac{\partial \sigma_{ij}}{\partial x_j} + f_{b,i}$$

$$\frac{\partial \rho_g E_g}{\partial t} + \frac{\partial \rho_g E_g u_{g,i}}{\partial x_i} = \frac{\partial p u_{g,i}}{\partial x_i} - \frac{\partial}{\partial x_i}\left(\rho_g \sum_{k=1}^{N} h_{s,k} Y_k V_{k,i}\right) - \frac{\partial}{\partial x_i}\left(\kappa_g \frac{\partial T_g}{\partial x_i}\right) + \\ + \frac{\partial \sigma_{ij} u_{g,j}}{\partial x_i} - \sum_{k=1}^{N} \dot{\omega}_k \Delta h^0_{f,k} + \dot{q}$$

$$\sigma_{ij} = \mu_g \left(\frac{\partial u_{g,i}}{\partial x_j} + \frac{\partial u_{g,j}}{\partial x_i} - \frac{2}{3}\delta_{ij}\frac{\partial u_{g,l}}{\partial x_l} \right)$$

where ρ_g is the gas density, $u_{g,i}$ is the i-th component of the gas velocity, G_w is the evaporation rate of the water, Y_k is the mass fraction of the k-th specie, $V_{k,i}$ is the i-th diffusion velocity component of the k-th species, $\dot{\omega}_k$ is the source of the k-th component due to chemical reaction, E is the specific total energy of the gaseous mixture, T_g is the temperature of the gaseous mixture, p is the pressure of the gaseous mixture, κ_g is the thermal conductivity of the gaseous mixture, σ_{ij} is the viscous stress tensor of the gaseous mixture, μ_g is the viscosity of the gaseous mixture, and $\Delta h^0_{s,k}$ is the enthalpy of formation of the k-th species. $\dot{q}$ and $\vec{f}_b$ are energy and momentum sources related to the interphase interactions:

$$\dot{q} = \rho_d \left(\vec{u}_g \vec{F}_{St} + G_w \lambda_w - c_{p,w} \frac{(T_g - T_d)}{\tau_Q} \right)$$

$$\vec{f}_b = \vec{F}_{St} \rho_d + G_m \vec{u}_d \rho_d$$

where ρ_d is the mass density of the aqueous phase, T_d is the temperature of the aqueous phase, $\vec{F}_{St}$ is the drag force, λ_w is the specific latent heat of the vaporization of water, τ_Q is the characteristic time of heat exchange between liquid and gas phases, and $c_{p,w}$ is the isobaric specific heat capacity of water. For the droplets formed as a result of the foam decay, drag force is obtained from the conventional relation for spherical droplets:

$$\vec{F}_{St} = \frac{(\vec{u}_d - \vec{u}_g)}{\tau_{St}}$$

where τ_{St} is the characteristic time of momentum exchange between liquid and gas phases.

For the foam phase, drag force is given by the semi-empirical relation [16]

$$\vec{F}_{St} = \frac{C^*}{d_d}|\vec{u}_d - \vec{u}_g|(\vec{u}_d - \vec{u}_g)$$

with $C^* \sim 1.0$–10.0.

Characteristic time scales of the interphase momentum and heat exchange are

$$\tau_{St} = \frac{m_d}{3\pi d_d \mu_g}$$

$$\tau_Q = \frac{d_d^2 c_{p,w} \rho_d^0}{6\kappa_g \mathrm{Nu}}$$

where m_d is the mass of the water droplet, and $\mathrm{Nu} \approx 2$ is the Nusselt number.

For the liquid phase, governing equations are the following:

$$\frac{\partial \rho_d}{\partial t} + \frac{\partial \rho_d u_{d,i}}{\partial x_i} = -G_w \rho_d$$

$$\frac{\partial \vec{u}_d}{\partial t} + (\vec{u}_d \cdot \nabla)\vec{u}_d = -F_{St}$$

$$\frac{\partial T_d}{\partial t} + u_{d,i}\frac{\partial T_d}{\partial x_i} = \frac{T_g - T_d}{\tau_{St}}$$

The presented form of the hydrodynamic model is close to that used previously in [26]. It is assumed that the behavior of the liquid phase is determined by the compression of gas and its motion inside the bubbles or between the scattered droplets, while the liquid itself is incompressible. The foam phase and aerosol produced as a result of foam decay are assumed to differ in terms of different relations for the Stokes force (see relations for F_{St} above). Moreover, it is supposed that inner stresses within the liquid do not cause a significant effect on the flame propagation process. Here, it should be noted that such a model reproduces the dynamic processes in the foam quite well (see discussion in Section 2.2), in a way similar to the other models, which justifies the potential of its application. At the same time, such processes as the compaction of the liquid phase within the foam may play a significant role in the propagation of acoustic and shock waves in the foam [27]. Thus, one possible way to improve the model is to introduce compaction in the way it is implemented in, e.g., [28].

The main differences between the model presented here and one considered in [26] concern the models of heat transfer and momentum losses in the dispersed medium. In the gas-droplets region, the heating of the droplets is limited with the processes of phase transformation (evaporation). Here, the evaporation of the droplets is calculated in an approximation of evaporation limited by a heat transfer via the Spalding relation [21]:

$$G_m = \frac{2\pi d_d \kappa_g}{c_{V,w} m_d} \ln(1 + B_T)$$

$$B_T = \frac{c_{V,w}\Delta T}{\lambda_w}$$

where $c_{V,w}$ is the specific isochoric heat capacity of water.

The calculations are carried out using a conventional numerical technique called the "coarse particles method" and widely implemented for solving various problems in the field of gaseous and two-phase combustion (see [26], references within, and other papers by the authors). The numerical technique represents a finite-difference scheme with the second-order accuracy in space and first-order in time.

2.2. Dynamic Processes

It is of paramount importance to resolve the interactions between the gaseous and liquid phases of the foam to reproduce the dynamic processes in the aqueous foam accurately. According to the model formulation, there is a non-dimensional constant C^*, which determines the foam resistance and can be chosen to fit the experimental data. Here we used the data of the blast wave test carried out in [29] for the aqueous foam with the same expansion ratio ($f = 10$) and close bubbles size (50 μm). The results of the fitting routine are presented in Figure 1, where the dependencies of blast wave arrival time (a) and maximal pressure behind the shock (b) on the foam layer thickness are shown. One can see that the formulated model with $C^* = 10$ fits the experimental data with good agreement. Due to this, this value of C^* is used in further calculations.

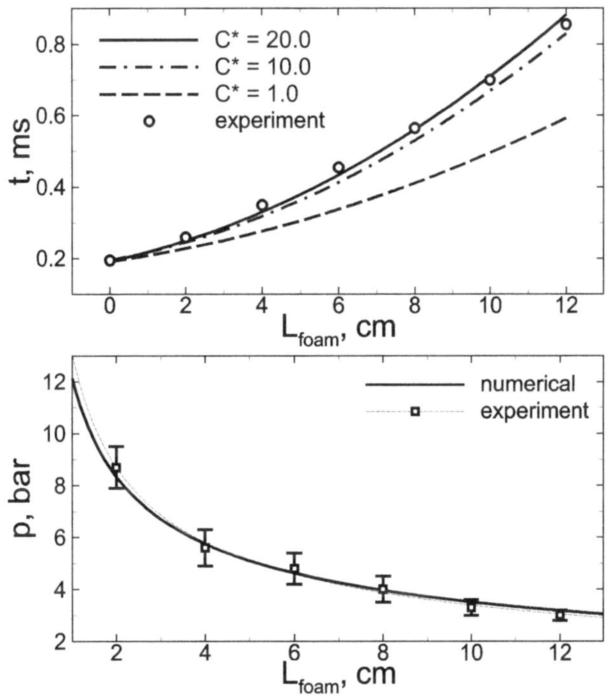

Figure 1. Comparison between the numerical and experimental data [29] on the blast wave attenuation by the foam layers of different thicknesses. Time of blast wave arrival (**top**) and pressure registered behind the wave (**bottom**) are plotted.

2.3. Peculiarities of the Problem Setup Related to the Foam Structure

As already mentioned in the Introduction, a crucial role in flame and detonation development belongs to the local effects. This is why the structure of the foam needs to be taken into account. Here, it is proposed to set a non-uniform initial distribution of the liquid fraction as shown in Figure 2a. In gray regions, representing the films between the bubbles and the plateau borders (regions of film intersections), the density of the liquid phase ($\rho_d = f(x,y)$) is maximal. In the white regions, the density of the liquid phase is minimal. It should be noted that here, the problem setup is two-dimensional, but there are no regions with zero liquid or gas fractions. This can be treated as a model representation of the two-dimensional projection of the real three-dimensional structure of the foam. The average value of ρ_d corresponds to the expansion ratio $f = 10$ and ~98.6 wt.% of water content in the foam.

The foam is ignited by a thin (1 mm) layer heated up to 3000 K at the top opened end of the channel where the outflow condition into air is imposed (Figure 2b). Solid wall conditions are imposed on the side walls and the bottom end of the channel. Walls are assumed to be isothermal ($T_{wall} = 300$ K) and non-slip. The initial temperature and pressure in the foam are equal to 300 K and 1 atm, correspondingly. After ignition, the flame is formed and propagates downward through the foam (Figure 2c). The characteristic spatial distribution ρ_d in the vicinity of the propagating flame is shown in Figure 2d.

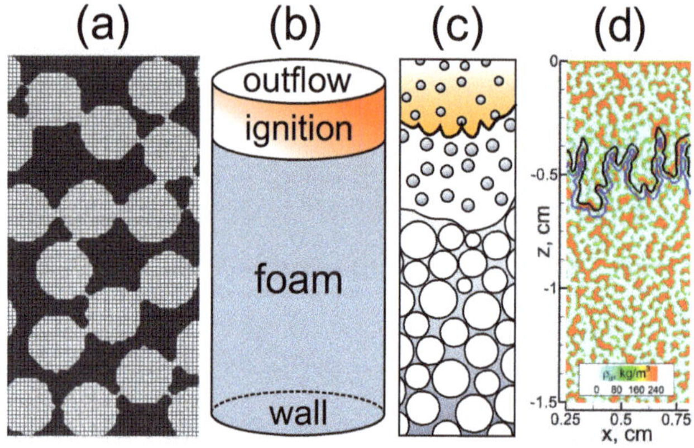

Figure 2. (**a**) The structure of the foam model: gray regions contain greater amount of liquid, white regions contain greater amount of gas, and grid corresponds to the numerical grid. (**b**) The geometry of computational domain: gray region—the foam at 300 K temperature, red region—the heated layer ($T = 3000$ K). The height of the channel is 14 cm, the width of the channel varies from 0.25 cm to 3.7 cm. (**c**) Schematic pattern of the downward flame propagation process in the foam (the same as was reported previously in [30]). (**d**) Calculated spatial distribution of liquid phase density ρ_d; black line—the position of the flame front at the considered time instant, blue line—foam surface; channel width is 1 cm.

Preliminary results of simulations for such a problem setup are presented in [30]. It is shown that the proposed model reproduces the experimentally obtained integral characteristics of the deflagration-to-detonation transition in stoichiometric hydrogen–oxygen microfoam such as pre-detonation length (L_{DDT}), maximal detonation speed achieved in the process of DDT, and quasi-steady value of the wave speed at the final stage of the process development. Here, let us provide the results of the convergence test for this solution. Table 1 illustrates the convergence of the solution for the pre-detonation length. As one can see, even rather coarse numerical grids reproduce the process with high accuracy. This is due to the features of the problem setup where the distribution $\rho_d = f(x, y)$ plays an important role and largely determines the development of the flame and detonation. The estimations of the exact solution and convergence order are calculated according to the standard routine proposed by Richardson [31]. The convergence rate is estimated as 1.87. Since the numerical scheme is of the first order, the solution is characterized by a super convergence that indicates a strong dependence of the solution on the features of the problem setup, including the non-uniform spatial distribution of ρ_d.

Table 1. Convergence test: numerical data on the cell size, L_{DDT}, and error. Asymptotic value of L_{DDT} and convergence order based on the estimation of L_{DDT} is also given in the bottom row of the table.

δx, μm	L_{DDT}, cm	Error, %
100	1.13	24
50	1.36	8
25	1.45	2
	Exact value estimate: $L_{DDT} = 1.48$ cm Convergence rate: 1.87	

In further numerical experiments, we used uniform meshes with the cell size $\delta x = 50$ μm, providing an 8% error in determination of pre-detonation distance. The time step is chosen dynamically to satisfy the Courant–Friedrichs–Lewy condition $\delta t < \frac{\delta x}{\max(|u_g|+c_g)}$, where c_g is the speed of sound in the gaseous phase. In particular, the criterion $\delta t = 0.5\frac{\delta x}{\max(|u_g|+c_g)}$ is used.

3. Analysis of Flame Dynamics

3.1. Flame Acceleration

Let us consider the flame evolution in the process of propagation through the foam. Figure 3 shows the time history of the flame speed defined as the velocity of the flame front leading point. It can be seen that the flame, once ignited, propagates with acceleration. Herewith, the acceleration proceeds via the exponential law (flame accelerates almost exponentially before transition to detonation at ~40–50 μs, which is shown with a green dotted line). This indicates that there is a mechanism with positive feedback responsible for the flame acceleration. It should also be noted that the numerical results agree well with the experiment that substantiates the proposed mathematical model of the flame propagation through the foam. A close but less accurate comparison was presented recently in [30]. Here the experimental points from [8] are reprocessed, and two different models are used to visualize the effect of droplet fragmentation.

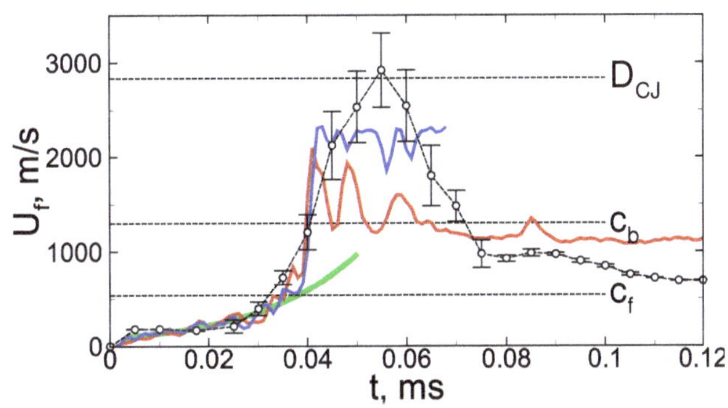

Figure 3. Time history of the flame speed in the microfoam bubbled with stoichiometric hydrogen–oxygen mixture. Signs connected with dashed lines—experimental data from [8], solid lines—numerical results with (red) and without (blue) account of droplet fragmentation. Green thick line shows the exponential approximation $U_f = 80 \times \exp(50 \times 10^{-3}t)$, illustrating the self-sustained flame acceleration within time interval from 3 μs to 40 μs before transition to detonation. Horizontal dashed lines show the characteristic velocity scales for hydrogen–oxygen: c_f—sonic speed in the fresh gas mixture, c_b—sonic speed in the combustion products, D_{CJ}—Chapman–Jouguet detonation speed.

The mechanism of flame acceleration is closely related to the flame–foam interaction. According to the proposed model and obtained numerical results, the flame–foam interaction proceeds via the following scenario. Under the thermal and dynamic action of the expanding combustion products on the foam, the foam decays into an aerosol consisting of droplets suspended in the hydrogen–oxygen mixture (Figure 1c). Figure 4 shows the multidimensional structure of the flame propagating into the foam. The red area represents the high-temperature combustion products, while the blue regions indicate the cold zone occupied by the foam and the suspension of gas and water droplets in the pre-flame area. Flame front structures in channels of different widths (3.7 cm, 1.0 cm, and 0.25 cm) are presented for the same time instant. Due to the initial non-uniform structure of the foam, there are zones in the pre-flame area with a larger and smaller volumetric content of droplets $\rho_d(x,y)$ (Figure 1d). This means that the flame propagates through an array of droplets scattered in space with an average distance between them equal to the initial bubble size (d_b). Herewith, the average diameter of droplets is equal to the size of the plateau borders, which are about $0.7d_b$.

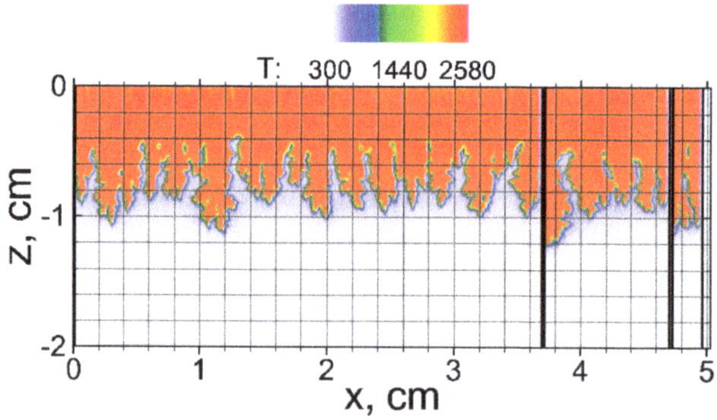

Figure 4. Flame structure in the stoichiometric hydrogen–oxygen microfoam in the tubes of different widths (from left to right, 3.7 cm, 1.0 cm, and 0.25 cm) at the same time instant.

Close conditions where the flame interacts with an array of equally spaced droplets were considered recently in [11]. In that paper, it was shown that the droplets affect the flame surface and cause excitation of the hydrodynamic instability development. As a result, the flame accelerates. Herewith, the most pronounced effect was observed in the case of rather coarse droplets (with diameter larger than 50 μm) scattered in space at the distance of the order of critical wavelength (λ_C) corresponding to the highest rate of hydrodynamic instability growth ($\gamma \propto 1/\lambda_C$) [32]. The same scenario is observed in the case of foam combustion. This exact mechanism plays the most important role in the process of flame acceleration in the hydrogen–oxygen microfoam. As Figure 4 clearly shows, the flame structure is almost independent of the channel width. Thus, it can be concluded that the momentum and heat losses to the walls play much less of a role in the flame acceleration compared with the effect of the flame–droplets interaction and instability development. Similar flame evolution patterns can be observed in obstructed spaces [33] or in porous media [34].

It is interesting to note that foam decay results in the formation of rather coarse droplets. Such droplets effectively excite perturbations on the flame surface due to high momentum losses, but the process of their evaporation is fairly slow, proceeds mainly in the combustion products area, and does not lead to flame quenching. In such a way, a natural separation between the combustion and evaporation zones is established. As a result, the foam with relatively high water content (~90 wt.% and greater) remains combustible. Moreover, the combustion in such a foam develops in the form of an accelerating flame.

With the account of the mechanism of flame propagation through the foam described above, the basic description of the flame acceleration mechanism is as follows. Foam decays under the action of the flame. Herewith, the intensity of foam decay increases with the increase in the flame speed and the flame surface. Thus, the faster and more developed flame causes more intense foam decay. On the other hand, foam decay defines additional flame acceleration via the mechanism described above. The flame perturbation by droplets and instability growth determines both flame acceleration and the increase in its surface. In such a way, a positive feedback loop is established, and the flame accelerates exponentially.

3.2. Detonation

The process of flame acceleration is always associated with the compression of the combustible medium. The flame propagating inside the channel acts like a piston on the unreacted medium ahead of the flame. When flame acceleration takes place in gases, peculiarities of flame dynamics on different stages of its acceleration define distinct modes of compression [35]. Thus, at the stage of exponential acceleration, the compression leads to shock wave generation at a certain distance ahead of the flame front. At the stage of subexponential acceleration (according to the power law with an exponent less than unity), the formation of the shock wave takes place in the immediate vicinity of the flame front [36]. Due to this, there is a coupling between the reaction front and compression zone, so the compression rate increases at this stage. In such a way, the accumulative effect results in a strong shock formation on the scales of the reaction zone, which finally causes the detonation onset. A detailed analysis of the flow patterns obtained in [36,37] shows that in the case of gas combustion in channels, this effect is related to the momentum losses on the channel side walls. Such an accumulative mechanism is realized at the later stages of flame acceleration as well when the flame achieves the maximal speed of CJ deflagration, which is of the order of sonic speed in the combustion products [38]. At this stage (of so-called chocked flame), the compression on the scales of the reaction zone is related to the peculiarities of the behavior of compression waves emitted from the reaction zone propagating with a transonic speed [35]. Depending on the rate of chemical kinetics and on the burning rate variation with compression, which is much more important, a further increase in the compression rate can take place. This finally leads to the detonation onset.

Let us now consider the peculiarities of the compression effect inherent to the process of the flame-accelerated propagation through the foam. The foam represents a two-phase medium in which the compression waves attenuate due to the momentum and heat losses. This feature of the foam, in particular, is widely used to suppress the dynamic loads on the solid constructions with foam barriers [29,39]. Due to this effect, the compression wave propagates through the foam with a lower speed that defines the same localization effect of the compression zone at a short distance ahead of the flame front (Figure 6), as discussed above. The coupling of the accelerating flame with a compression zone results in the intensification of the process, and the deflagration-to-detonation transition occurs via the mechanism proposed in [36]. Herewith, the foam decays in the pre-flame zone, and the detonation onset arises in the gas pockets between the droplets (Figure 5). Note that the temperature of the combustion product area behind the flame front is rather low. The evaporation of droplets located inside the combustion product area causes its cooling, but the temperature remains far beyond the temperature of water evaporation, so the expanding combustion products still can cause foam decay in the pre-flame zone. The boundary of the combustion products region propagates (solid black line in Figure 5) downwards together with the reaction front, which is scattered in space due to the heat and momentum losses in the process of detonation wave interaction with the foam.

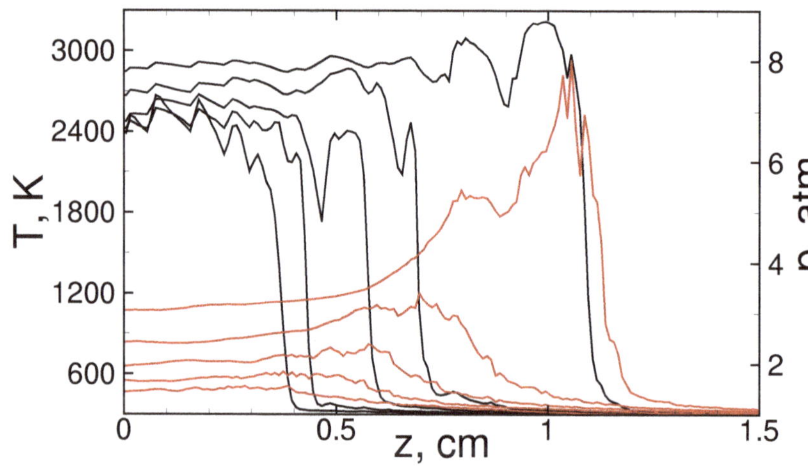

Figure 5. Temperature (black) and pressure (red) profiles in the fixed cross section at subsequent time instants (20, 25, 30, 35, and 40 μs).

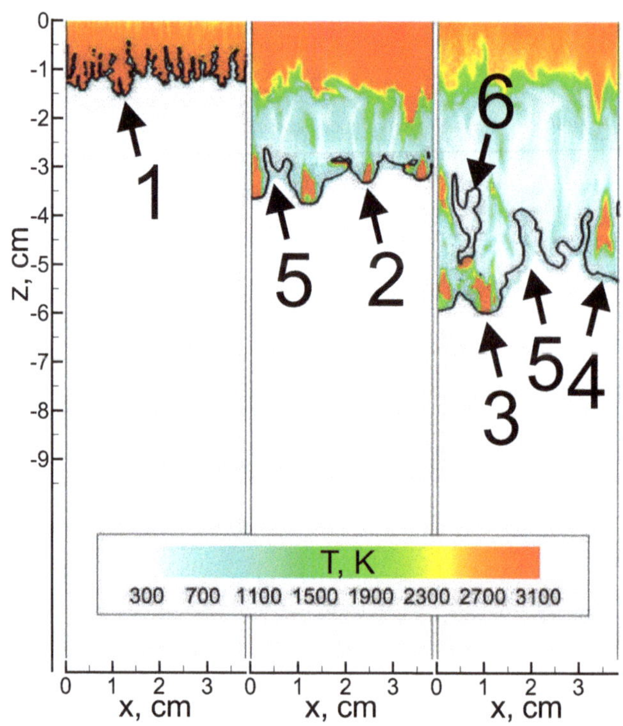

Figure 6. Numerically obtained flow pattern in the process of flame acceleration and detonation propagation in the foam bubbled with $2H_2 + O_2$ mixture in 37 mm channel. Time instants are: 40, 50, and 60 μs. 1—flame front, 2—local explosion, 3—decaying stage of the local explosion, 4—quenched kernel, 5—jet of combustion products, 6—unreacted pocket behind the leading front. Black line indicates the boundary of combustion products region. Calculation with account of droplet fragmentation.

The blue curve in Figure 3 shows the numerical solution with the formation of a self-sustained detonation wave. The self-sustained detonation was also observed experimentally. However, it was rather difficult to measure the exact value of the detonation speed since strong illumination did not allow for accurate discerning of the flow pattern (see Figure 4 in [8]). Here, the uncertainty in the determination of the detonation speed is shown via the error bars. The obtained detonation speed (Figure 3) is lower than the detonation speed in pure gas ($D_{CJ} \approx 2800$ m/s). Integrally, the shock wave driving the detonation loses its intensity when entering the foam, but locally, the process is more complex. The gas phase occurs as compressed even to a higher rate than in gaseous detonation. In the limit of the infinite thickness of the liquid films, the bubbles are collapsing, so extreme values of temperature and pressure can be achieved inside the bubble due to the energy focusing. In the case of the finite thickness of liquid films, they are also acting as soft shells on the gas inside, and their contraction also results in the gas compression limited by the time instant of the films' rupture. Nevertheless, the gas inside the bubbles (or regions between droplets after films rupture) occurs as compressed up to higher compression rates compared to the shock-compression of the pure gas. As a result, the lower intensity of the leading shock wave is sufficient for self-sustained detonation wave propagation through the foam. Due to this, the steady-state detonation speed in the wet foam occurs as lower than the detonation speed in gases.

As soon as the detonation wave is formed, it interacts with the foam. In particular, the detonation wave propagates from one bubble to the neighboring one. In the process of the detonation wave interaction with a liquid film between the bubbles, the film decays into an aerosol. Moreover, the intense dynamic loads cause the fragmentation of droplets into smaller ones. These small droplets take a significant part of the momentum and heat from the flow that leads to the deceleration of the leading shock wave. Thus, in Figure 5, one can observe the local decay of the detonation wave. Moreover, the detonation wave can decay as a whole. The solution shown by the red curve in Figure 3 was obtained with an account of droplet fragmentation, and contrary to the case without accounting for this factor (blue line), here, one can observe detonation deceleration. Finally, the detonation decays with a total quenching of the combustion. However, the combustion products continue to expand intensively, causing dynamic action on the foam in the bottom part of the channel below the area of detonation decay. The jet-like flow of the combustion products is formed, which destroys the remaining foam. Herewith, the formed droplets evaporate in the combustion products, and in the end, no foam remains in the channel.

It should be noted that there is good agreement in the dynamics of the process in the final stage observed experimentally and that obtained numerically (Figure 3). The available experimental diagnostics were limited, and the main difficulty is related to the fact that the foam is not an optically transparent medium. It impedes acquiring a clear pattern of the flow. Error bars on the experimental data in Figure 3 are based on the thickness of the light zone and indicate the estimation from below. The numerically obtained maximal speed is quite close to the experimental one. The largest visible divergence is about 300 m/s ($\sim$10% compared with the value of 3000 m/s). Given this, the representation of the process at this stage based on model assumptions and numerical results is quite plausible. The experimental diagnostics need to be improved to obtain a more thorough understanding of the process and, in particular, details of the detonation decay. In [8], only visible light was registered, while there are different possible reasons for illumination, including the luminosity of the collapsing bubbles.

4. Conclusions

In this paper, we propose a numerical model and present new results on flame acceleration and detonation onset, propagation, and decay in hydrogen–oxygen microfoam. It is demonstrated that the structure of the foam determines the features of deflagration and detonation waves propagating through the foam. In particular, it is shown that the leading role in flame propagation and acceleration belongs to the parameters of the aerosol formed as a result of foam decay. Rather coarse droplets of ∼140 μm diameter are scattered in space with an average distance of ∼200 μm, which corresponds to the regime of flame propagation under the permanent effect of excitation of the short-wavelength perturbations by droplets and intensification of the instability growth. The same phenomenon was recently described in detail in [11]. As a result, the flame acceleration proceeds in the same way, independent of the geometry of the reactor.

The detonation arises due to the compression wave localization directly ahead of the accelerated flame front. This localization takes place due to the high hydraulic resistance of the foam. As a result, the coupling of the reaction zone with the compression zone leads to an increased rate of flame acceleration as well as to a higher rate of compression on the scales of the reaction zone. Finally, this leads to the detonation wave formation. Herewith, it is initially formed in the aerosol and then propagates into the foam that leads to its deceleration. The detonation speed, as well as the magnitude of the leading shock wave, is lower in the foam compared with the pure gas. This can be explained by the structure of the reaction zone, representing the series of explosion kernels scattered in space behind the leading shock front. The positions of these kernels are associated with the bubbles in the foam, while the explosions are triggered by the shock wave focusing on the scales of each bubble due to the liquid shell collapse. Herewith, an important role belongs to the process of droplet fragmentation due to dynamic impact. Fine droplets formed in the zone of the detonation front cause both deceleration of the shock wave and quenching of the exothermic reaction. As a result, the detonation wave decays. The distinguished mechanisms of detonation onset, propagation, and decay widen the physical understanding of the detonation processes observed recently in [8].

The proposed model provides good qualitative results and allows the physical interpretation of the features of the flame-accelerated propagation and transition to detonation in hydrogen–oxygen microfoam. However, there is still room for improvement. The main strategy to enhance the performance of the elaborated model is to introduce an account of compaction phenomena in a way as it was implemented, e.g., by M.R. Baer and J.W. Nunziato in [28].

Author Contributions: Conceptualization, A.K.; methodology, A.K. and I.Y.; software, I.Y.; validation, A.K. and I.Y.; formal analysis, A.K. and I.Y.; investigation, A.K. and I.Y.; resources, A.K. and I.Y.; data curation, A.K. and I.Y.; writing—original draft preparation, A.K.; writing—review and editing, A.K. and I.Y.; visualization, A.K. and I.Y.; supervision, A.K.; project administration, A.K.; funding acquisition, A.K. Both authors have read and agreed to the published version of the manuscript.

Funding: This research was supported by The Ministry of Science and Higher Education of the Russian Federation (Agreement with Joint Institute for High Temperatures RAS No 075-15-2020-785 dated 23 September 2020).

Institutional Review Board Statement: Not applicable.

Informed Consent Statement: Not applicable.

Data Availability Statement: The data are available upon request from the corresponding authors.

Acknowledgments: We are grateful to Boris Kichatov and Alexey Korshunov for fruitful discussions. We acknowledge high-performance computing support from the Joint Supercomputer Center of the Russian Academy of Sciences and Supercomputing Center of Lomonosov Moscow State University.

Conflicts of Interest: The authors declare no conflict of interest.

References

1. Abe, J.; Popoola, A.; Ajenifuja, E.; Popoola, O. Hydrogen energy, economy and storage: Review and recommendation. *Int. J. Hydrog. Energy* **2019**, *44*, 15072–15086. [CrossRef]
2. Verhelst, S. Recent progress in the use of hydrogen as a fuel for internal combustion engines. *Int. J. Hydrog. Energy* **2014**, *39*, 1071–1085. [CrossRef]
3. Li, Y.; Bi, M.; Li, B.; Zhou, Y.; Gao, W. Effects of hydrogen and initial pressure on flame characteristics and explosion pressure of methane/hydrogen fuels. *Fuel* **2018**, *233*, 269–282. [CrossRef]

4. Taghavifar, H.; Anvari, S.; Parvishi, A. Benchmarking of water injection in a hydrogen-fueled diesel engine to reduce emissions. *Int. J. Hydrog. Energy* **2017**, *42*, 11962–11975. [CrossRef]
5. Anufriev, I.S.; Kopyev, E.P. Diesel fuel combustion by spraying in a superheated steam jet. *Fuel Process. Technol.* **2019**, *192*, 154–169. [CrossRef]
6. Huo, M.; Lin, S.; Liu, H.; Lee, C.f.F. Study on the spray and combustion characteristics of water–emulsified diesel. *Fuel* **2014**, *123*, 218–229. [CrossRef]
7. Kichatov, B.; Korshunov, A.; Kiverin, A.; Son, E. Foamed emulsion—Fuel on the base of water-saturated oils. *Fuel* **2017**, *203*, 261–268. [CrossRef]
8. Kichatov, B.; Korshunov, A.; Kiverin, A.; Yakovenko, I.; Gubernov, V.; Khomik, S.V.; Medvedev, S.P. Detonation in the hydrogen-oxygen microfoam on the aqueous base. *Int. J. Hydrog. Energy* **2019**, *44*, 31567–31578. [CrossRef]
9. Kichatov, B.; Korshunov, A.; Gubernov, V.; Kiverin, A.; Yakovenko, I. Combustion of heptane-in-water emulsion foamed with hydrogen-oxygen mixture. *Fuel Process. Technol.* **2020**, *198*, 106230. [CrossRef]
10. Gostintsev, Y.A.; Istratov, A.G.; Shulenin, Y.V. Self-similar propagation of a free turbulent flame in mixed gas mixtures. *Combust. Explos. Shock Waves* **1988**, *24*, 563–569. [CrossRef]
11. Yakovenko, I.; Kiverin, A. Intensification mechanisms of the lean hydrogen-air combustion via addition of suspended micro-droplets of water. *Int. J. Hydrog. Energy* **2021**, *46*, 1259–1272. [CrossRef]
12. Sychev, A.; Pinaev, A. Self-sustaining detonation in liquids with bubbles of explosive gas. *J. Appl. Mech. Tech. Phys.* **1986**, *27*, 119–123. [CrossRef]
13. Saint-Cloud, J.; Guerraud, C.; Moreau, M.; Manson, M. Experiences sur la propagation des detonations dans un milieu biphasique. *Acta Astronaut.* **1976**, *3*, 781–794. [CrossRef]
14. Segev, G.; Hasson, A.; Siman, M.; Burcat, A. Detonation waves through foam. In Proceedings of the 22nd Symposium (International) on Combustion, Seattle, Washington, USA, 14–19 August 1989; pp. 1751–1756.
15. Subbotin, V.; Usol'tsev, S. Study of the mechanism of the transfer of gaseous detonation through films of liquid. *Combust. Explos. Shock Waves* **1984**, *20*, 224–230. [CrossRef]
16. Faure, S.; Ghidaglia, J.M. Violent flows in aqueous foams I: Physical and numerical models. *Eur. J. Mech. B/Fluids* **2011**, *30*, 341–359. [CrossRef]
17. Kuo, K.K.; Acharya, R. *Fundamentals of Turbulent and Multiphase Combustion*, 1st ed.; John Wiley & Sons, Inc.: Hoboken, NJ, USA, 2012.
18. Shinjo, J.; Xia, J.; Ganippa, L.C.; Megaritis, A. Physics of puffing and microexplosion of emulsion fuel droplets. *Phys. Fluids* **2014**, *26*. [CrossRef]
19. Tonini, S.; Gavaises, M.; Theodorakakos, A. The role of droplet fragmentation in high-pressure evaporating diesel sprays. *Int. J. Therm. Sci.* **2009**, *48*, 554–572. [CrossRef]
20. Boiko, V.M.; Poplavski, S.V. Experimental study of two types of stripping breakup of the drop in the flow behind the shock wave. *Combust. Explos. Shock Waves* **2012**, *48*, 440–445. [CrossRef]
21. Spalding, D.B. *Combustion and Mass Transfer*, 1st ed.; Pergamon Press Inc.: Elmsford, New York, USA, 1979.
22. Kéromnès, A.; Metcalfe, W.K.; Heufer, K.A.; Donohoe, N.; Das, A.K.; Sung, C.J.; Herzler, J.; Naumann, C.; Griebel, P.; Mathieu, O.; et al. An experimental and detailed chemical kinetic modeling study of hydrogen and syngas mixture oxidation at elevated pressures. *Combust. Flame* **2013**, *160*, 995–1011. [CrossRef]
23. Olm, C.; Zsély, I.G.; Pálvölgyi, R.; Varga, T.; Nagy, T.; Curran, H.J.; Turányi, T. Comparison of the performance of several recent hydrogen combustion mechanisms. *Combust. Flame* **2014**, *161*, 2219–2234. [CrossRef]
24. Ahmed, S.F.; Santner, J.; Dryer, F.L.; Padak, B.; Farouk, T.I. Computational Study of NOx Formation at Conditions Relevant to Gas Turbine Operation, Part 2: NOx in High Hydrogen Content Fuel Combustion at Elevated Pressure. *Energy Fuels* **2016**, *30*, 7691–7703. [CrossRef]
25. Burke, M.P.; Chaos, M.; Ju, Y.; Dryer, F.L.; Klippenstein, S.J. Comprehensive H_2/O_2 kinetic model for high-pressure combustion. *Int. J. Chem. Kinet.* **2012**, *44*, 444–474. [CrossRef]
26. Ivanov, M.F.; Kiverin, A.D.; Liberman, M.A. Ignition of deflagration and detonation ahead of the flame due to radiative preheating of suspended micro particles. *Combust. Flame* **2015**, *162*, 3612–3621. [CrossRef]
27. Baer, M.R. A numerical study of shock wave reflections on low density foam. *Shock Waves* **1992**, *2*, 121–124. [CrossRef]
28. Baer, M.R.; Nunziato, J.W. A two-phase mixture theory for the deflagration-to-detonation transition (ddt) in reactive granular materials. *Int. J. Multiph. Flow* **1986**, *12*, 861–889. [CrossRef]
29. Sembian, S.; Liverts, M.; Apazidis, N. Attenuation of strong external blast by foam barriers. *Phys. Fluids* **2016**, *28*. [CrossRef]
30. Kiverin, A.D.; Yakovenko, I.S.; Kichatov, B.V.; Korshunov, A.M. Cumulative effect in foams and mechanism of detonation development. *J. Phys. Conf. Ser.* **2020**, *1686*, 012079. [CrossRef]
31. Roache, P.J. Perspective: A Method for Uniform Reporting of Grid Refinement Studies. *J. Fluids Eng.* **1994**, *116*, 405. [CrossRef]
32. Landau, L.D.; Lifshitz, E.M. *Fluid Mechanics*, 2nd ed.; Butterworth-Heinemann: Burlington, Massachusetts, USA, 1987; Volume 6.
33. Ogawa, T.; Gamezo, V.N.; Oran, E.S. Flame acceleration and transition to detonation in an array of square obstacles. *J. Loss Prev. Process Ind.* **2013**, *26*, 355–362. [CrossRef]
34. Babkin, V.S. Filtrational combustion of gases. Present state of affairs and prospects. *Pure Appl. Chem.* **1993**, *65*, 335–344. [CrossRef]
35. Kiverin, A.; Yakovenko, I.; Ivanov, M. On the structure and stability of supersonic hydrogen flames in channels. *Int. J. Hydrog. Energy* **2016**, *41*, 22465–22478. [CrossRef]

36. Ivanov, M.F.; Kiverin, A.D.; Liberman, M.A.; Fortov, V.E. The flame-acceleration mechanism and transition to detonation of a hydrogen-oxygen mixture in a channel. *Dokl. Phys.* **2010**, *55*, 480–484. [CrossRef]
37. Ivanov, M.; Kiverin, A.; Liberman, M.A. Flame acceleration and DDT of hydrogen–oxygen gaseous mixtures in channels with no-slip walls. *Int. J. Hydrog. Energy* **2011**, *36*, 7714–7727. [CrossRef]
38. Saif, M.; Wang, W.; Pekalski, A.; Levin, M.; Radulescu, M.I. Chapman–Jouguet deflagrations and their transition to detonation. *Proc. Combust. Inst.* **2016**, S1540748916303807. [CrossRef]
39. Britan, A.; Shapiro, H.; Liverts, M.; Ben-Dor, G. Macro-mechanical modeling of blast-wave mitigation in foams. Part III: Verification of the models. *Shock Waves* **2014**, *24*, 241–256. [CrossRef]

energies

Article

Estimation of Functional Form of Time-Dependent Heat Transfer Coefficient Using an Accurate and Robust Parameter Estimation Approach: An Inverse Analysis

Farzad Mohebbi [1,*] and Mathieu Sellier [2]

[1] Zienkiewicz Centre for Computational Engineering, Faculty of Science and Engineering, Swansea University, Swansea SA1 8EN, UK
[2] Department of Mechanical Engineering, University of Canterbury, Private Bag 4800, Christchurch 8140, New Zealand; mathieu.sellier@canterbury.ac.nz
* Correspondence: farzadmohebbi@yahoo.com

Abstract: This paper presents a numerical method to address function estimation problems in inverse heat transfer problems using parameter estimation approach without prior information on the functional form of the variable to be estimated. Using an inverse analysis, the functional form of a time-dependent heat transfer coefficient is estimated efficiently and accurately. The functional form of the heat transfer coefficient is assumed unknown and the inverse heat transfer problem should be treated using a function estimation approach by solving sensitivity and adjoint problems during the minimization process. Based on proposing a new sensitivity matrix, however, the functional form can be estimated in an accurate and very efficient manner using a parameter estimation approach without the need for solving the sensitivity and adjoint problems and imposing extra computational cost, mathematical complexity, and implementation efforts. In the proposed sensitivity analysis scheme, all sensitivity coefficients can be computed in only one direct problem solution at each iteration. In this inverse heat transfer problem, the body shape is irregular and meshed using a body-fitted grid generation method. The direct heat conduction problem is solved using the finite-difference method. The steepest-descent method is used as a minimization algorithm to minimize the defined objective function and the termination of the minimization process is carried out based on the discrepancy principle. A test case with three different functional forms and two different measurement errors is considered to show the accuracy and efficiency of the used inverse analysis.

Keywords: inverse heat transfer; steepest-descent method; sensitivity analysis; function estimation; parameter estimation; body-fitted grid generation; time-dependent heat transfer coefficient

Citation: Mohebbi, F.; Sellier, M. Estimation of Functional Form of Time-Dependent Heat Transfer Coefficient Using an Accurate and Robust Parameter Estimation Approach: An Inverse Analysis. *Energies* **2021**, *14*, 5073. https://doi.org/10.3390/en14165073

Academic Editor: Christopher Micallef

Received: 14 July 2021
Accepted: 12 August 2021
Published: 18 August 2021

Publisher's Note: MDPI stays neutral with regard to jurisdictional claims in published maps and institutional affiliations.

1. Introduction

In a heat transfer problem, the accuracy of thermophysical properties and boundary conditions is critical to obtain an accurate numerical simulation. As a boundary condition, convective heat transfer depends on different parameters such as time, surface geometry, and surface temperature, to name a few. The accurate determination of the convective heat transfer coefficient is a difficult task as convection is a very complicated phenomenon and expensive experiments with sophisticated instruments are required to appropriately unravel its dynamics [1].

The advent of high-speed and high-capacity computers and the development of different regularization methods over the past decades have played significant roles in successful applications of numerical inverse methods, as inexpensive alternatives to costly and time-consuming experiments with sophisticated instruments, to appropriately estimate unknown heat transfer quantities such as the heat transfer coefficient [2–10]. Inverse heat transfer problems are mathematically challenging problems because they are ill-posed and the accuracy of the estimation of an unknown quantity is very sensitive to measurement errors [11–13]. If the unknown quantity to be estimated (in this study, the heat transfer coefficient) can be expressed as a constant parameter [14,15] or represented by a few parameters [16], a parameter estimation approach may be used to estimate the parameters thereby estimating the unknown quantity. However, a function estimation approach should be used to estimate the unknown functional form of the unknown quantity when there is no information available on the functional form of the unknown quantity. In the function estimation approach, the *sensitivity* and *adjoint* problems are required to obtain the gradient of objective function with respect to unknown functional form which impose additional mathematical developments and computational costs on the inverse analysis. In this study, based on the numerical procedure employed in [17], a two-dimensional transient inverse heat conduction problem is considered. The thermophysical properties are assumed constant, the geometry of heat-conducting body is irregular, and the body is subject to Neumann and Robin boundary conditions at its boundary surface parts. Using the parameter estimation approach initially developed in [17] for the estimation of unknown functional form of a time-dependent heat flux (a boundary condition of second kind) imposed at a boundary surface, here the unknown functional form of a time-dependent heat transfer coefficient(a third-kind boundary condition) is estimated efficiently and accurately without involving the solution of the sensitivity and adjoint problems. Thus, the mathematical development effort and the computational cost are reduced significantly.

To do so, the heat-conducting body (the physical domain) is mapped onto a regular computational domain in order to take advantage of the ease of implementation of the finite-difference method to explicitly solve the transient heat conduction equation and the associated boundary conditions. As the body shape is irregular, a body-fitted (elliptic) grid generation method is used to mesh the irregular domain which makes the proposed method general and applicable to any irregular domain as long as it can be mapped onto a regular computational domain. Using the chain rule to relate the temperature at sensor place and the time-dependent heat transfer coefficient applied on the part of the body boundary, explicit expressions are derived to compute sensitivity coefficients during the solution of the transient heat conduction equation without the need for solving the sensitivity and adjoint equations. The steepest-descent method, as an iterative regularization method, with a stopping criterion specified by discrepancy principle is used to minimize the objective function and reach the solution accurately. A test case with three complicated functional forms of timewise variation of the heat transfer coefficient is presented to reveal the accuracy, efficiency, and robustness of the inverse analysis. Moreover, two different measurement errors are considered. It is shown that the inverse analysis is not strongly affected by the errors involved in the temperature measurements and the unknown functional forms of the timewise variation of the heat transfer coefficient can be recovered with excellent accuracy. As stated before, the objective of this study is to present a parameter estimation approach to estimate the unknown functional form of a time-dependent heat transfer coefficient efficiently and accurately.

2. Governing Equation

The body shown in Figure 1a is initially at the temperature T_0. At time $t > 0$, it is exposed to a time-dependent heat flux $\dot{q}(t)$ at boundary surface Γ_1 and convective heat transfer on boundary surfaces $\Gamma_i, i = 2, 3, 4$ with corresponding heat transfer coefficients $h_2(t)$, h_3, and h_4 and surrounding temperatures $T_{\infty_i}, i = 2, 3, 4$. The thermal conductivity, density, and specific heat of the body are k_T, ρ, and c, respectively.

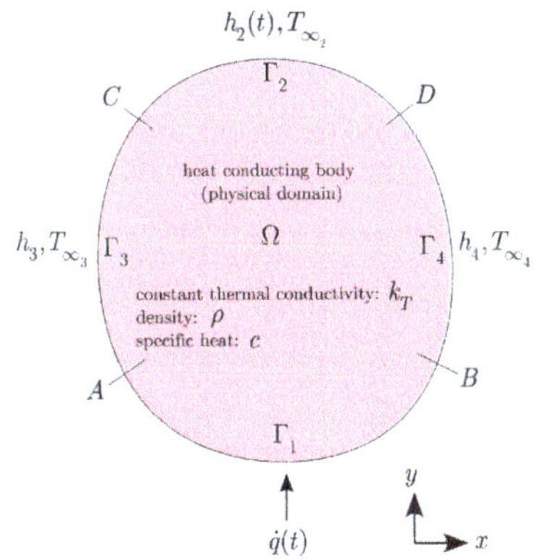

(a) Physical domain

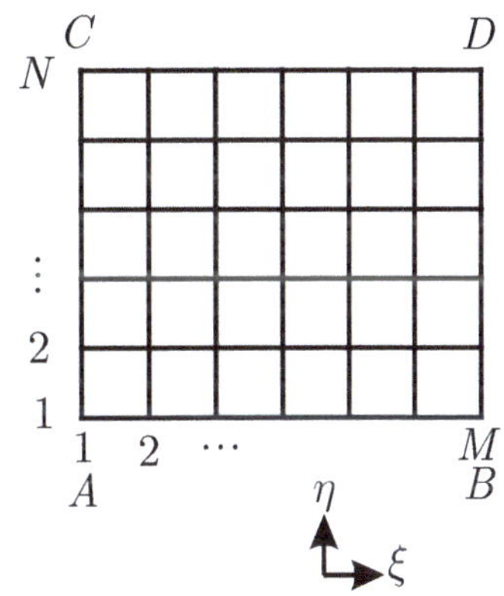

(b) Computational domain

Figure 1. Two-dimensional irregular (arbitrarily shaped) heat-conducting body (physical domain) subjected to a time-dependent heat flux $\dot{q}(t)$ on surface Γ_1 and convective heat transfer on surfaces $\Gamma_i, i = 2, 3, 4$ (**a**) and the corresponding computational domain (**b**).

The governing equation for a two-dimensional transient heat conduction problem with no heat generation can be expressed as [17,18]

$$k_T\left(\frac{\partial^2 T(x,y,t)}{\partial x^2} + \frac{\partial^2 T(x,y,t)}{\partial y^2}\right) = \rho c \frac{\partial T(x,y,t)}{\partial t} \text{ in physical domain } \Omega(x,y) \tag{1}$$

with the boundary and initial conditions

$$\frac{\partial T(x,y,t)}{\partial n_1} = \frac{\dot{q}(t)}{k_T} \quad \text{on boundary surface } \Gamma_1(x,y) \tag{2}$$

$$\frac{\partial T(x,y,t)}{\partial n_2} = -\frac{h_2(t)}{k_T}\left(T_{\Gamma_2}(x,y,t) - T_{\infty_2}\right) \quad \text{on boundary surface } \Gamma_2(x,y) \tag{3}$$

$$\frac{\partial T(x,y,t)}{\partial n_i} = -\frac{h_i}{k_T}\left(T_{\Gamma_i}(x,y,t) - T_{\infty_i}\right) \quad \text{on boundary surface } \Gamma_i(x,y),\ i = 3,4 \tag{4}$$

$$T(x,y,0) = T_0(x,y) \quad \text{in physical domain } \Omega(x,y) \tag{5}$$

where t is the time. Since the heat-conducting body is irregular, it (the x and y physical domain) can be mapped onto a regular one (the ξ and η computational domain). The elliptic grid generation method is employed to generate a grid over the physical domain. Then the heat conduction equation and its associated boundary and initial conditions can be transformed from the (x, y, t) to the (ξ, η, t) variables [12,18]. The transformation results in

$$\left(\frac{\alpha\frac{\partial^2 T(\xi,\eta,t)}{\partial \xi^2} - 2\beta\frac{\partial^2 T(\xi,\eta,t)}{\partial \xi \partial \eta} + \gamma\frac{\partial^2 T(\xi,\eta,t)}{\partial \eta^2}}{J^2} + (\nabla^2\xi)\frac{\partial T(\xi,\eta,t)}{\partial \xi} + (\nabla^2\eta)\frac{\partial T(\xi,\eta,t)}{\partial \eta}\right) = \frac{\rho c}{k_T}\frac{\partial T(\xi,\eta,t)}{\partial t} \tag{6}$$

where $\nabla^2\xi = P(\xi,\eta)$ and $\nabla^2\eta = Q(\xi,\eta)$ are grid control functions. If $P(\xi,\eta) = Q(\xi,\eta) = 0$, then a smooth grid over the physical domain is obtained. Therefore, Equation (6) becomes

$$\left(\frac{\alpha\frac{\partial^2 T(\xi,\eta,t)}{\partial \xi^2} - 2\beta\frac{\partial^2 T(\xi,\eta,t)}{\partial \xi \partial \eta} + \gamma\frac{\partial^2 T(\xi,\eta,t)}{\partial \eta^2}}{J^2}\right) = \frac{\rho c}{k_T}\frac{\partial T(\xi,\eta,t)}{\partial t} \quad \text{in } 1 < \xi < M,\ 1 < \eta < N, \quad \text{for } t > 0 \tag{7}$$

where

$$\alpha = x_\eta^2 + y_\eta^2$$
$$\beta = x_\xi x_\eta + y_\xi y_\eta$$
$$\gamma = x_\xi^2 + y_\xi^2$$
$$J = x_\xi y_\eta - x_\eta y_\xi \quad \text{(Jacobian of transformation)} \tag{8}$$

The transformed boundary and initial conditions can be expressed as

$$\left(\frac{-1}{J\sqrt{\gamma}}\left(\gamma\frac{\partial T(\xi,\eta,t)}{\partial \eta} - \beta\frac{\partial T(\xi,\eta,t)}{\partial \xi}\right)\right)_{\Gamma_1} = \frac{\dot{q}(t)}{k_T} \quad \text{at } 1<\xi<M,\ \eta=1, \quad \text{for } t>0 \tag{9}$$

$$\left(\frac{1}{J\sqrt{\gamma}}\left(\gamma\frac{\partial T(\xi,\eta,t)}{\partial \eta} - \beta\frac{\partial T(\xi,\eta,t)}{\partial \xi}\right)\right)_{\Gamma_2} = -\frac{h_2(t)}{k_T}(T(\xi,\eta,t) - T_{\infty_2}) \quad \text{at } 1<\xi<M,\ \eta=N, \quad \text{for } t>0 \tag{10}$$

$$\left(\frac{-1}{J\sqrt{\alpha}}\left(\alpha\frac{\partial T(\xi,\eta,t)}{\partial \xi} - \beta\frac{\partial T(\xi,\eta,t)}{\partial \eta}\right)\right)_{\Gamma_3} = -\frac{h_3}{k_T}(T(\xi,\eta,t) - T_{\infty_3}) \quad \text{at } 1<\eta<N,\ \xi=1, \quad \text{for } t>0 \tag{11}$$

$$\left(\frac{1}{J\sqrt{\alpha}}\left(\alpha\frac{\partial T(\xi,\eta,t)}{\partial \xi} - \beta\frac{\partial T(\xi,\eta,t)}{\partial \eta}\right)\right)_{\Gamma_4} = -\frac{h_4}{k_T}(T(\xi,\eta,t) - T_{\infty_4}) \quad \text{at } 1<\eta<N,\ \xi=M, \quad \text{for } t>0 \tag{12}$$

$$T(\xi,\eta,0) = T_0^*(\xi,\eta) \quad \text{in } 1<\xi<M,\ 1<\eta<N, \quad \text{for } t=0 \tag{13}$$

where the initial condition $T_0(x,y)$ is rewritten as $T_0^*(\xi,\eta)$ in terms of the variables ξ and η. Now the finite-difference method can be employed to discretize the derivatives present in the above equations in the regular computational domain, as follows (assuming $\Delta\xi = \Delta\eta = 1$)

$$f_\xi = \frac{1}{2}(f_{i+1,j} - f_{i-1,j})$$
$$f_\eta = \frac{1}{2}(f_{i,j+1} - f_{i,j-1})$$
$$f_{\xi\xi} = f_{i+1,j} - 2f_{i,j} + f_{i-1,j}$$
$$f_{\eta\eta} = f_{i,j+1} - 2f_{i,j} + f_{i,j-1}$$
$$f_{\xi\eta} = \frac{1}{4}(f_{i+1,j+1} - f_{i-1,j+1} - f_{i+1,j-1} + f_{i-1,j-1}) \tag{14}$$

where $f \equiv x, y, T$. One-sided forward and one-sided backward relations are used to discretize the boundary condition equations. The explicit method can be used to solve the resulting transient heat conduction equation, Equation (7). Using forward-time-central-space (FTCS) discretization and the relations in Equation (14), we get

$$\begin{aligned}\frac{1}{J^2}\Big(\alpha(T^n_{i+1,j} - 2T^n_{i,j} + T^n_{i-1,j}) - 2\beta\frac{1}{4}(T^n_{i+1,j+1} - T^n_{i-1,j+1} - T^n_{i+1,j-1} + T^n_{i-1,j-1}) + \gamma(T^n_{i,j+1} - 2T^n_{i,j} + T^n_{i,j-1})\Big) \\ = \frac{\rho c}{k_T}\frac{T^{n+1}_{i,j} - T^n_{i,j}}{\Delta t}, \quad i = 2,\ldots,M-1,\ j = 2,\ldots,N-1 \text{ for } t>0\end{aligned} \tag{15}$$

where Δt is the time step. Taking into account the stability criterion, the time-marching procedure can be used to solve Equation (15) and obtain $T^{n+1}_{i,j}$. That is, the nodal temperatures at the time level $n+1$, $T^{n+1}_{i,j}$, can be determined from the knowledge of nodal temperatures at the previous time level n, $T^n_{i,j}$, as follows

$$\begin{aligned}T^{n+1}_{i,j} = T^n_{i,j} + \\ \frac{k_T\Delta t}{\rho c J^2}\Big(\alpha(T^n_{i+1,j} - 2T^n_{i,j} + T^n_{i-1,j}) - 2\beta\frac{1}{4}(T^n_{i+1,j+1} - T^n_{i-1,j+1} - T^n_{i+1,j-1} + T^n_{i-1,j-1}) + \gamma(T^n_{i,j+1} - 2T^n_{i,j} + T^n_{i,j-1})\Big)\end{aligned} \tag{16}$$

3. The Inverse Analysis

3.1. Objective Function

The inverse heat transfer problem of interest deals with the estimation of the time-dependent heat transfer coefficient $h_2(t)$ applied at the time t_i and at the surface Γ_2, $h_2(t_i), i = 1, \ldots, r$ (r is the number of time steps) using the transient readings of a single sensor S placed at the point (Si, Sj) inside the heat conducting body. In inverse analysis, the aim is to minimize the mismatch between the estimated temperatures at the sensor place, $T_e(Si, Sj, t_i)$, computed from the solution of the direct transient heat conduction problem using the estimated heat transfer coefficient $h_2(t_i)$ and the measured temperatures $T_m(Si, Sj, t_i)$ over the time domain $0 < t < t_r$. This can be mathematically expressed as a least-squares minimization as follows

$$\min \left\{ \underset{\mathbf{h}_2 \text{ at } \Gamma_2}{\mathrm{J}} := \|\mathbf{T}_e(Si, Sj, t) - \mathbf{T}_m(Si, Sj, t)\|^2 : \text{Equation (1) in } \Omega, \text{ BCs and IC in Equations (2)–(5)} \right\} \tag{17}$$

where $\mathbf{h}_2 = [h_2(t_1), h_2(t_2), h_2(t_3), \ldots, h_2(t_r)]^{\mathrm{T}}$. Therefore, the objective function can be expressed as

$$\mathrm{J} = \sum_{i=1}^{r} [T_e(Si, Sj, t_i) - T_m(Si, Sj, t_i)]^2 \tag{18}$$

3.2. Sensitivity Analysis

The calculation of the gradient of the objective function J defined by Equation (18) with respect to $h_2(t_i)$, $i = 1, \ldots, r$ is required in gradient-based minimization methods. Therefore, it can be written

$$\frac{\partial \mathrm{J}}{\partial h_2(t_{i'})} = 2\sum_{i=1}^{r} [T_e(Si, Sj, t_i) - T_m(Si, Sj, t_i)] \frac{\partial T_e(Si, Sj, t_i)}{\partial h_2(t_{i'})} \tag{19}$$

The sensitivity coefficients $\frac{\partial T_e(Si,Sj,t_i)}{\partial h_2(t_{i'})}$ $(i = 1, \ldots, r,\ i' = 1, \ldots, r)$ can be explicitly expressed using the chain rule (using the constant thermal conductivity k_T) as follows

$$\frac{\partial T_e(Si, Sj, t_i)}{\partial h_2(t_{i'})} = \frac{\dfrac{\partial T_e(Si, Sj, t_i)}{\partial k_T}}{\dfrac{\partial h_2(t_{i'})}{\partial k_T}} \tag{20}$$

The expression in the numerator of Equation (20), $\frac{\partial T_e(Si,Sj,t_i)}{\partial k_T}$, can be obtained by taking derivative of $T_{i,j}^{n+1}$ in Equation (16) with respect to k_T, as follows

$$\begin{aligned} \frac{\partial T_e^{n+1}(Si, Sj, t_i)}{\partial k_T} &= \frac{\Delta t}{\rho c J^2} \Big(\alpha (T_{Si+1,Sj}^n - 2T_{Si,Sj}^n + T_{Si-1,Sj}^n) \\ &-2\beta \frac{1}{4} (T_{Si+1,Sj+1}^n - T_{Si-1,Sj+1}^n - T_{Si+1,Sj-1}^n + T_{Si-1,Sj-1}^n) + \gamma (T_{Si,Sj+1}^n - 2T_{Si,Sj}^n + T_{Si,Sj-1}^n) \Big) \end{aligned} \tag{21}$$

The expression in the denominator of Equation (20), $\frac{\partial T_e(Si,Sj,t_i)}{\partial k_T}$, can be obtained from the boundary condition involving the heat transfer coefficient $h_2(t)$, Equation (9), as follows

$$\left(\frac{1}{J\sqrt{\gamma}} \left(\gamma \frac{\partial T(\xi, \eta, t)}{\partial \eta} - \beta \frac{\partial T(\xi, \eta, t)}{\partial \xi} \right) \right)_{\Gamma_2} = -\frac{h_2(t)}{k_T} (T(\xi, \eta, t) - T_{\infty 2}) \tag{22}$$

therefore, we get

$$\frac{\partial h_2(t_{i'})}{\partial k_T} = \frac{-1}{T(\xi, \eta, t) - T_{\infty 2}} \left(\frac{1}{J\sqrt{\gamma}} \left(\gamma \frac{\partial T(\xi, \eta, t)}{\partial \eta} - \beta \frac{\partial T(\xi, \eta, t)}{\partial \xi} \right) \right)_{\Gamma_2} \tag{23}$$

where the terms T_ξ, T_η, J, γ and β are computed using the finite-difference expressions associated with the surface Γ_2. That is,

$$f_\xi = \frac{1}{2}(f_{i+1,N} - f_{i-1,N})$$

$$f_\eta = \frac{1}{2}(3f_{i,N} - 4f_{i,N-1} + f_{i,N-2})$$

where $f \equiv x, y, T$. Therefore, the sensitivity coefficients in Equation (20) can be calculated by dividing the term in Equation (21) by the one in Equation (23).

$$\frac{\partial T_e(Si,Sj,t_i)}{\partial h_2(t_{i'})} = \frac{1}{\left(\frac{-1}{T(\xi,\eta,t)-T_{\infty_2}}\left(\frac{1}{J\sqrt{\gamma}}\left(\gamma\frac{\partial T(\xi,\eta,t)}{\partial \eta} - \beta\frac{\partial T(\xi,\eta,t)}{\partial \xi}\right)\right)_{\Gamma_2}\right)_{t_{i'}}} \frac{\Delta t}{\rho c J^2}\Big(\alpha(T^{t_i}_{Si+1,Sj} - 2T^{t_i}_{Si,Sj} + T^{t_i}_{Si-1,Sj}) \\ -2\beta\frac{1}{4}(T^{t_i}_{Si+1,Sj+1} - T^{t_i}_{Si-1,Sj+1} - T^{t_i}_{Si+1,Sj-1} + T^{t_i}_{Si-1,Sj-1}) + \gamma(T^{t_i}_{Si,Sj+1} - 2T^{t_i}_{Si,Sj} + T^{t_i}_{Si,Sj-1})\Big) \quad (24)$$

It can be seen that all sensitivity coefficients $\frac{\partial T_e(Si,Sj,t_i)}{\partial h_2(t_{i'})}$ can be computed during the transient solution of the direct heat conduction equation. The sensitivity matrix **Ja** can be explicitly written as

$$\mathbf{Ja}_{h_2(t)} = \begin{bmatrix} \frac{\partial T_e(Si,Sj,t_1)}{\partial h_2(t_1)} & \frac{\partial T_e(Si,Sj,t_1)}{\partial h_2(t_2)} & \frac{\partial T_e(Si,Sj,t_1)}{\partial h_2(t_3)} & \cdots & \frac{\partial T_e(Si,Sj,t_1)}{\partial h_2(t_r)} \\ \frac{\partial T_e(Si,Sj,t_2)}{\partial h_2(t_1)} & \frac{\partial T_e(Si,Sj,t_2)}{\partial h_2(t_2)} & \frac{\partial T_e(Si,Sj,t_2)}{\partial h_2(t_3)} & \cdots & \frac{\partial T_e(Si,Sj,t_2)}{\partial h_2(t_r)} \\ \frac{\partial T_e(Si,Sj,t_3)}{\partial h_2(t_1)} & \frac{\partial T_e(Si,Sj,t_3)}{\partial h_2(t_2)} & \frac{\partial T_e(Si,Sj,t_3)}{\partial h_2(t_3)} & \cdots & \frac{\partial T_e(Si,Sj,t_3)}{\partial h_2(t_r)} \\ \vdots & \vdots & \vdots & \ddots & \vdots \\ \frac{\partial T_e(Si,Sj,t_r)}{\partial h_2(t_1)} & \frac{\partial T_e(Si,Sj,t_r)}{\partial h_2(t_2)} & \frac{\partial T_e(Si,Sj,t_r)}{\partial h_2(t_3)} & \cdots & \frac{\partial T_e(Si,Sj,t_r)}{\partial h_2(t_r)} \end{bmatrix}_{r\times r} \quad (25)$$

We know that the temperature estimated at any time is independent of a yet-to-occur future heat transfer coefficient component [11,19] which gives rise to a lower-triangular sensitivity matrix. That is, for $i' > i$ (the terms above the main diagonal of the sensitivity matrix), $\frac{\partial T_e(Si,Sj,t_i)}{\partial h_2(t_{i'})} = 0$. Thus, we get

$$\mathbf{Ja}_{h_2(t)} = \begin{bmatrix} \frac{\partial T_e(Si,Sj,t_1)}{\partial h_2(t_1)} & 0 & 0 & \cdots & 0 \\ \frac{\partial T_e(Si,Sj,t_2)}{\partial h_2(t_1)} & \frac{\partial T_e(Si,Sj,t_2)}{\partial h_2(t_2)} & 0 & \cdots & 0 \\ \frac{\partial T_e(Si,Sj,t_3)}{\partial h_2(t_1)} & \frac{\partial T_e(Si,Sj,t_3)}{\partial h_2(t_2)} & \frac{\partial T_e(Si,Sj,t_3)}{\partial h_2(t_3)} & \cdots & 0 \\ \vdots & \vdots & \vdots & \ddots & \vdots \\ \frac{\partial T_e(Si,Sj,t_r)}{\partial h_2(t_1)} & \frac{\partial T_e(Si,Sj,t_r)}{\partial h_2(t_2)} & \frac{\partial T_e(Si,Sj,t_r)}{\partial h_2(t_3)} & \cdots & \frac{\partial T_e(Si,Sj,t_r)}{\partial h_2(t_r)} \end{bmatrix}_{r\times r} \quad (26)$$

3.3. The Steepest-Descent Method

The steepest-descent optimization method is used in this study to minimize the objective function given by Equation (18) by searching along the direction of steepest descent $\mathbf{d}^{(k)}$ using a search step length $\beta^{(k)}$.

$$\mathbf{h}_2{}^{(k+1)} = \mathbf{h}_2{}^{(k)} + \beta^{(k)}\mathbf{d}^{(k)} \quad (27)$$

The negative of the gradient direction $\nabla J^{(k)}$ denotes the direction of steepest descent $\mathbf{d}^{(k)}$. Thus,

$$\mathbf{d}^{(k)} = -\nabla J^{(k)} \quad (28)$$

In the expression $\mathbf{h}_2{}^{(k+1)} = \mathbf{h}_2{}^{(k)} + \beta^{(k)}\mathbf{d}^{(k)} = \mathbf{h}_2{}^{(k)} - \beta^{(k)}\nabla J^{(k)}$, the search step-length $\beta^{(k)} > 0$ is given as follows [12]

$$\beta^{(k)} = -\frac{\left[\mathbf{Ja}^{(k)}\mathbf{d}^{(k)}\right]^{T}\left[\mathbf{T}_c - \mathbf{T}_m\right]}{\left[\mathbf{Ja}^{(k)}\mathbf{d}^{(k)}\right]^{T}\left[\mathbf{Ja}^{(k)}\mathbf{d}^{(k)}\right]} \tag{29}$$

Optimization Algorithm

The proposed numerical procedure to estimate the time-dependent heat transfer coefficient at the boundary surface Γ_2, $h_2(t)$, can be summarized as follows:

1. Measure the temperatures at the sensor place $S_{Si,Sj}$ and the time t_i $(i = 1, \ldots, r)$, $T_m(Si, Sj, t_i)$.
2. Solve the direct problem to obtain the temperature values at the sensor place and the time t_i $(i = 1, \ldots, r)$, $T_e(Si, Sj, t_i)$, through solving Equations (7)–(13).
3. Compute the objective function value ($J^{(k)}$) using Equation (18).
4. If value of the objective function obtained in step 3 is less than the specified stopping criterion, the optimization is finished. Otherwise, go to step 5.
5. Compute the sensitivity matrix $\mathbf{Ja}_{h_2(t)}$ from Equation (26), the gradient direction $\nabla J^{(k)}$ from Equation (19), the direction of descent $\mathbf{d}^{(k)}$ from Equation (28), and the search step-length $\beta^{(k)}$ from Equation (29).
6. Update $\mathbf{h}_2$ from Equation (27) and set the next iteration ($k = k + 1$) and return to the step 2.

3.4. Stopping Criterion

The minimization process can be terminated if

$$J^{(k)} < \varepsilon \tag{30}$$

where ε is chosen based on obtaining stable and appropriate results. In this study, when there are no measurement errors, $\varepsilon = 10^{-6}$. However, if the temperature measurements contain errors, then the *discrepancy principle* is used to stop the iterative procedure and obtain stable results. In this case,

$$\varepsilon = r\sigma^2 \tag{31}$$

where σ is the standard deviation of the measurement errors [17] and is assumed constant in this study ($\sigma = 0.001$ and $\sigma = 0.005$). Here, the measured temperatures containing random errors, $T_{\text{meas}}(Si, Sj, t_i)$, $(i = 1, \ldots, r)$, are generated by adding an error term $\omega\sigma$ to the exact temperatures $T_{\text{exact}}(Si, Sj, t_i)$ to give

$$T_{\text{meas}}(Si, Sj, t_i) = T_{\text{exact}}(Si, Sj, t_i) + \omega\sigma \tag{32}$$

where ω is a random variable with normal (Gaussian) distribution, zero mean, and unitary standard deviation. Assuming 99% confidence for the measured temperature, ω lies in the range $-2.576 \leq \omega \leq 2.576$ and it is randomly generated by using MATLAB.

4. Results

A test case with three different complicated functional forms of variation of the heat transfer coefficient with time is presented to investigate the accuracy, efficiency, and robustness of the proposed sensitivity analysis method to estimate the time-dependent heat transfer coefficient on part of the boundary of a heat conducting body. Initially the heat transfer coefficient is assumed to be known, the transient heat conduction problem is then solved to calculate the temperature at the sensor place at times t_i $(i = 1, \ldots, r)$. Then, the calculated temperatures are used as *simulated measured* ones to recover the initially used heat transfer coefficient. The three different forms of timewise variation of the heat transfer coefficient considered are as follows (Figures 2–4)

$$h_2^{(1)}(t) = \begin{cases} 12, & 0 \le t < 100\,\text{s} \\ 8, & 100 \le t < 200\,\text{s} \\ \dfrac{16-8}{300-200}(t-200)+8, & 200 \le t \le 300\,\text{s} \\ \dfrac{10-16}{400-300}(t-300)+16, & 300 < t \le 400\,\text{s} \\ 10, & 400 < t \le 500\,\text{s} \\ 11, & 500 < t \le 600\,\text{s} \end{cases}$$

$$h_2^{(2)}(t) = 10 + \log(t^{0.5})\sin(\frac{t}{6}) + t^{0.45}$$

and $h_2^{(3)}(t)$ which is an arbitrary waveform generated in MATLAB used here to model the variation of the heat transfer coefficient with time.

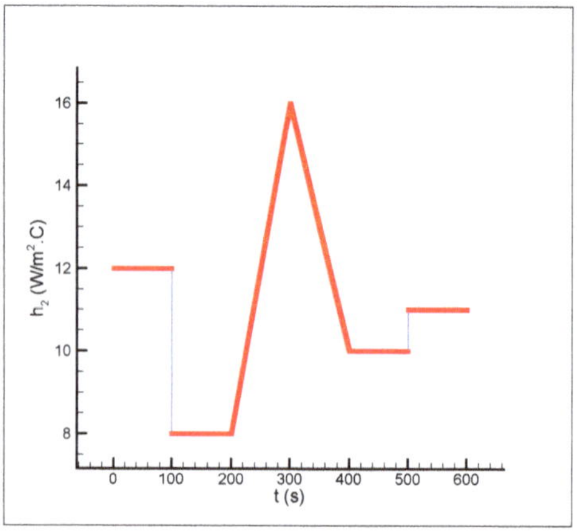

Figure 2. Timewise variation of the heat transfer coefficient $h_2^{(1)}$.

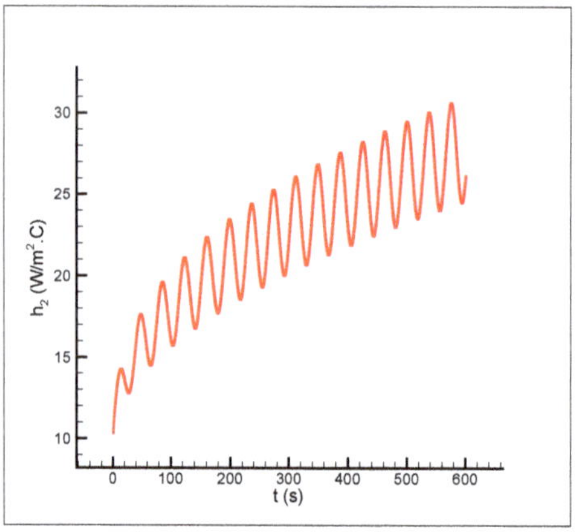

Figure 3. Timewise variation of the heat transfer coefficient $h_2^{(2)}$.

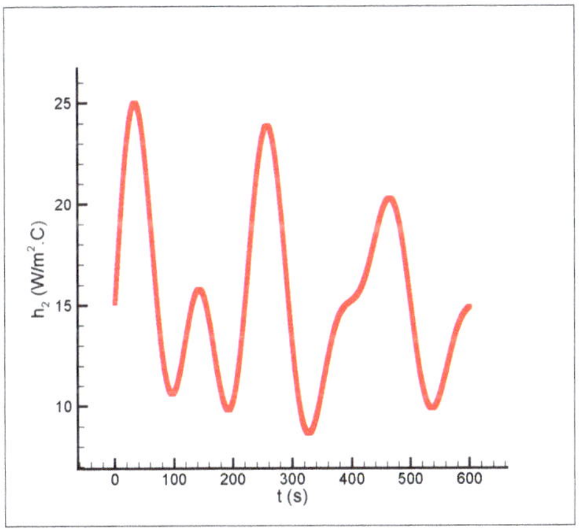

Figure 4. Timewise variation of the heat transfer coefficient $h_2^{(3)}$.

Assuming the heat conducting body is made of stainless steel (type 304), the numerical values of the coefficients involved in the test case are listed in Table 1.

Table 1. Data used for the heat conduction problem (the body is made of stainless steel (type 304) [20]).

$\dot{q}\,(\frac{W}{m^2})$	$k_T\,(\frac{W}{m\cdot°C})$	$\rho\,(\frac{kg}{m^3})$	$c\,(\frac{J}{kg\cdot°C})$	$h_2\,(\frac{W}{m^2.°C})$	$h_i\,(\frac{W}{m^2.°C})$, $i = 3,4$	T_{∞_i} (°C), $i = 2,3,4$
$2000 + 1000\sin(\frac{\pi t}{180})$	14.9	7900	477	$h_2^{(1)}(t), h_2^{(2)}(t), h_2^{(3)}(t)$	5	30

In all simulations in this study, the heat conducting body is meshed using a grid size of $M \times N = 40 \times 40$, the temperature measurement sensor is placed at node $(Si, Sj) = (\frac{M}{2}, N-3) = (20, 37)$ (close to the boundary subject to convective heat transfer with the convective heat transfer coefficient h_2 to obtain sensible sensitivity coefficients) (Figure 5), the initial temperature is $T(x, y, 0) = T_0(x, y) = 20$ °C, the final time is $t_r = 600$ s, and the time step is $\Delta t = 0.1$ s. Thus, the number of transient readings of the single sensor S is $r = \frac{t_r}{\Delta t} = \frac{600\,\text{s}}{0.1\,\text{s}} = 6000$. This means that that the number of unknown parameters is 6000. Thus, the estimation of such a large number of unknown parameters using the parameter estimation approach commonly used in the literature is not feasible. However, using the proposed sensitivity analysis, one can handle the estimation of the large number of unknown parameters accurately and efficiently. In this study, two different measurement errors of $\sigma = 0.001$ and $\sigma = 0.005$ are considered. The stopping criteria for the test case with the following measurement errors are

$$\sigma = 0.001 \Rightarrow \varepsilon = \sigma^2 r = 0.001^2(6000) = 0.006$$

$$\sigma = 0.005 \Rightarrow \varepsilon = \sigma^2 r = 0.005^2(6000) = 0.15$$

As the size of Jacobian matrix is $r \times r$, we will deal in this study with a Jacobian matrix of size 6000×6000. Once the temperature at the sensor place is obtained at the time t_i, the elements of the Jacobian matrix can be calculated during the transient solution using the obtained expression for the sensitivity coefficients; that is, during the solution of the direct problem, the terms $T_{k_T}(i,1) = \frac{\partial T_e(Si, Sj, t_i)}{\partial k_T}$, $i = 1, \ldots, r$ and $h_{2_{k_T}}(j,1) = \frac{\partial h_2(t_j)}{\partial k_T}$,

$j = 1, \ldots, r$ are computed from Equations (21) and (23), respectively, and then the sensitivity coefficients can be obtained using the following pseudo-code

```
do i = 1, r
   do j = 1, r
     if(i.LT.j)then
     Ja(i,j) = 0.0
     else
     Ja(i,j) = T_kT(i,1) / h2_kT(j,1)
     endif
   enddo
enddo
```

Initially, the implementation of the direct problem solver is validated with the results obtained from the commercial finite element software COMSOL. To do so, using the data given in Table 1, $h_2^{(2)}$, and the body shown in Figure 5, the temperature distribution in the body is calculated by the two methods (our finite-difference explicit code, Equation (16), using two different time steps of 0.1 and 0.001 s and the finite element software COMSOL) which is shown in Figure 6. Moreover, the temperature history of the place of the sensor, $S(\frac{M}{2}, N-3)$, obtained by both methods is shown in Figure 7. The comparison between the results reveals a very good agreement hereby verifying the correct implementation of the explicit solver.

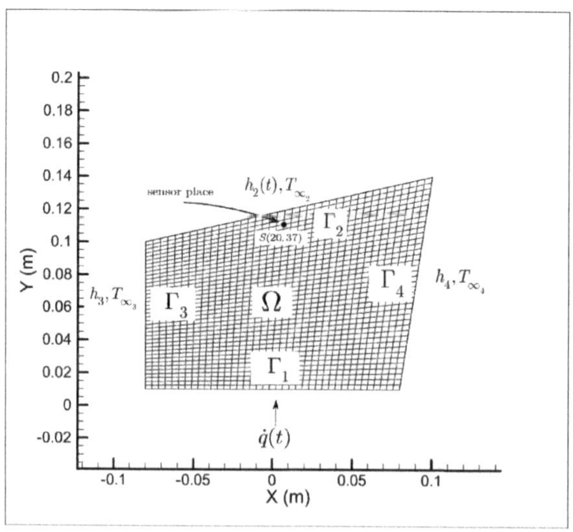

Figure 5. The location of the sensor S, $S(\frac{M}{2}, N-3) = S(20,37)$, used to measure the temperature at time t_i.

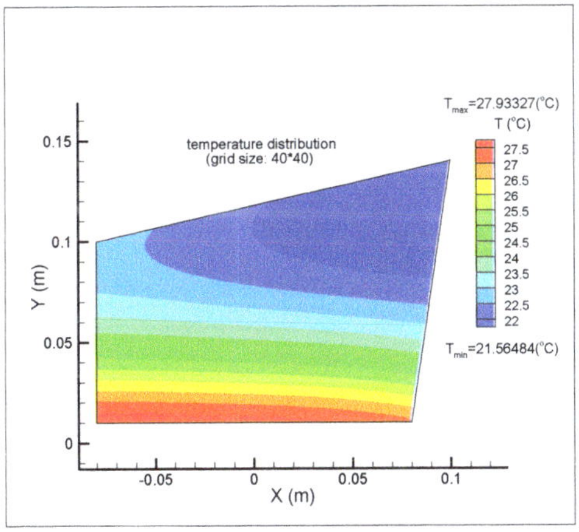

(**a**) time step = 0.1 s

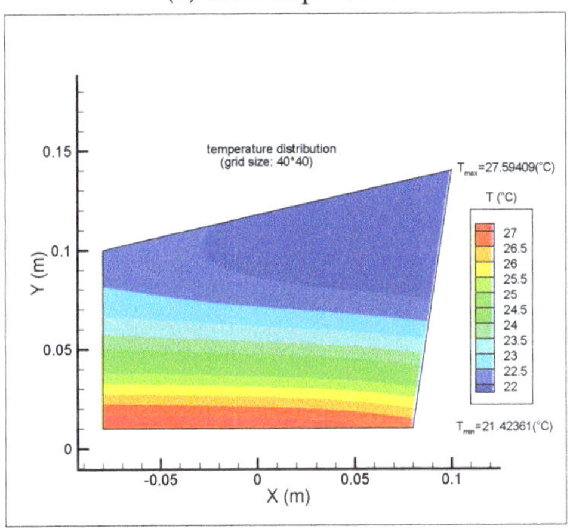

(**b**) time step = 0.001 s

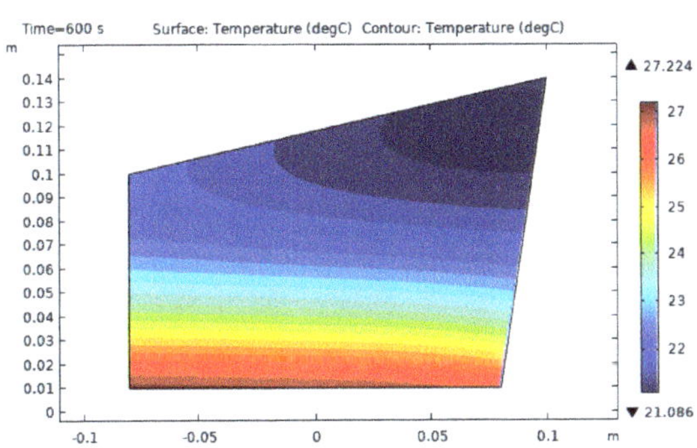

(**c**)

Figure 6. Validation of the direct problem solver using the finite element software COMSOL. The result obtained by using our explicit code (**a**) using the time step of 0.1 s and (**b**) using the time step of 0.001 s, and the result obtained by using COMSOL (**c**).

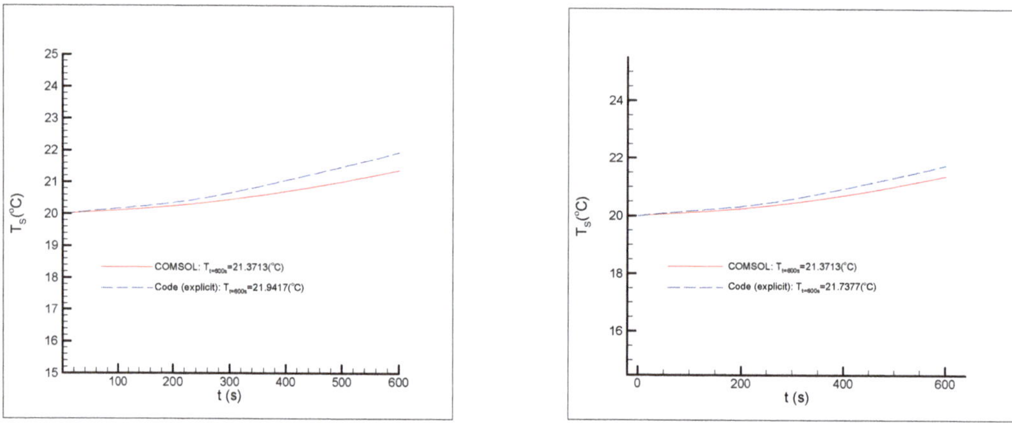

(**a**) time step = 0.1 s (**b**) time step = 0.001 s

Figure 7. Comparison of temperature history at the sensor location, $S(\frac{M}{2}, N-3)$, obtained from the explicit solver using two different time steps and the finite-element software COMSOL. The temperature history using the time step of 0.1 s is used as simulated measured temperatures in inverse analysis.

In this inverse heat conduction problem, three different and complicated functional forms of timewise variations (including the variations which are difficult to be recovered by an inverse analysis such as discontinuities and sharp corners [12]) for the heat transfer coefficient are chosen to examine the accuracy, efficiency, and robustness of the inverse analysis presented in this study. A comparison of the initial (guessed), final, and desired heat transfer coefficients is shown in Figures 8a, 9a and 10a (for the case of no measurement error, $\sigma = 0$), Figures 8c, 9c and 10c (for the measurement error of $\sigma = 0.001$), and Figures 8e, 9e and 10e (for the measurement error of $\sigma = 0.005$). By comparing the desired and final functional forms shown in the above figures, it can be seen that the desired functional forms are recovered accurately which implies that the inverse analysis is not strongly affected by the errors involved in the temperature measurements due to the accuracy of the proposed sensitivity analysis scheme. When measurement errors exist, some oscillatory behaviors are observed around the exact values due to the ill-posed nature of the inverse heat transfer problem. The convergence histories of the objective function for the three functional forms of interest are shown in Figures 8b, 9b and 10b (for the case of no measurement error, $\sigma = 0$), Figures 8d, 9d and 10d (for the measurement error of $\sigma = 0.001$), and Figures 8f, 9f and 10f (for the measurement error of $\sigma = 0.005$). The details of the results, including the initial and desired values for the unknown time-dependent heat transfer coefficient, the initial and final values of the objective function, and the number of iterations required to reach the solutions are given in Table 2. The computation time for each iteration (the direct and inverse solutions) is about 4 s. In spite of large unknown variables (6000 in the test case) and large final time, $t_f = 600$ s, this short computation time confirms that the employed inverse analysis based on the proposed sensitivity analysis is very efficient. The results are obtained by a FORTRAN compiler and computations are run on a PC with Intel Core i5 and 6G RAM.

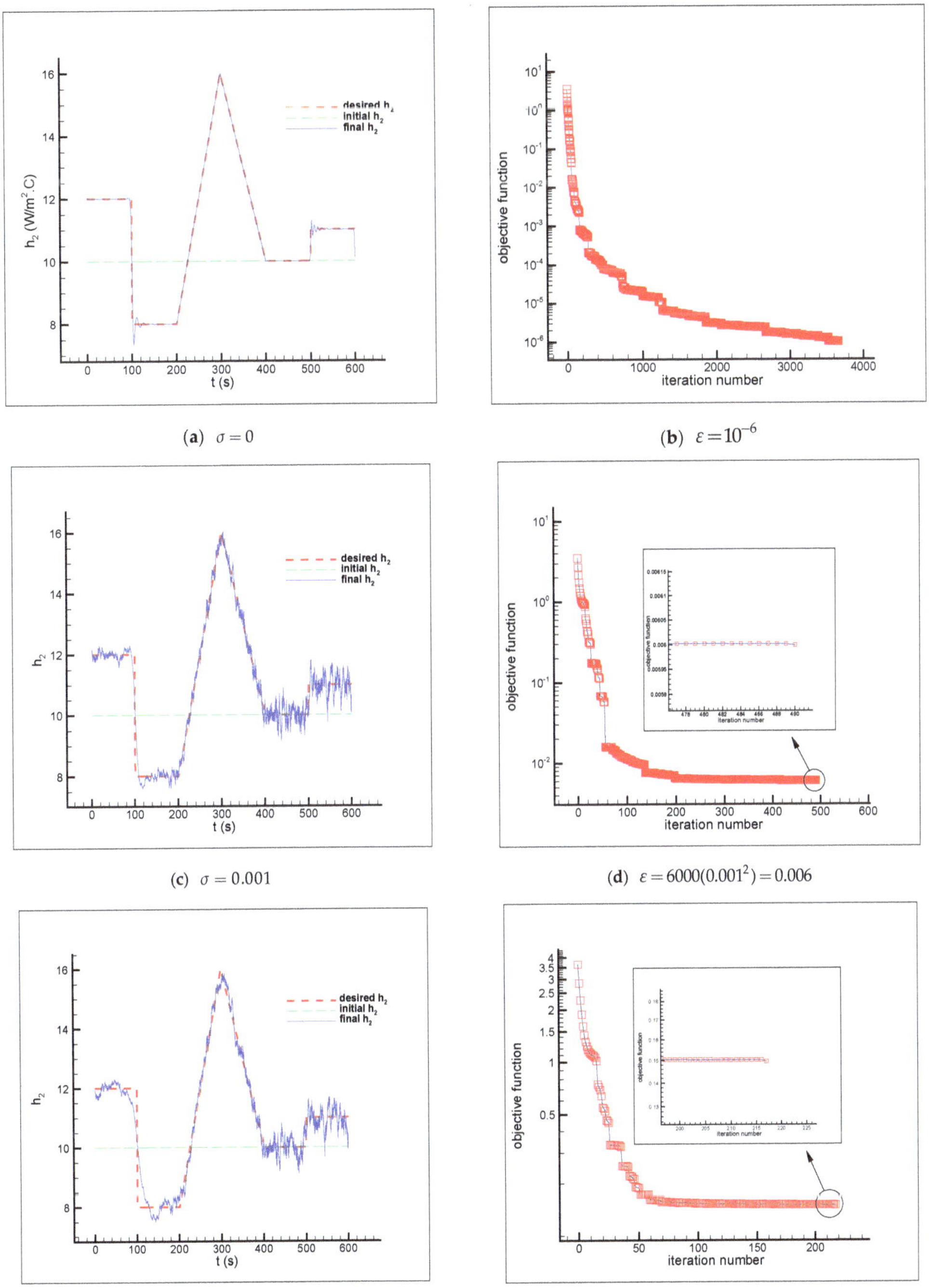

(**a**) $\sigma = 0$ (**b**) $\varepsilon = 10^{-6}$

(**c**) $\sigma = 0.001$ (**d**) $\varepsilon = 6000(0.001^2) = 0.006$

(**e**) $\sigma = 0.005$ (**f**) $\varepsilon = 6000(0.005^2) = 0.15$

Figure 8. Estimation of the time-dependent heat transfer coefficient $h_2^{(1)}$ using an initial guess $h_{2_{\text{initial}}}(t) = 10.0\,(\frac{\text{W}}{\text{m}^2\cdot{}^\circ\text{C}})$ and objective function versus iteration number for cases of no measurement error (**a**,**b**) and measurement error of $\sigma = 0.001$ (**c**,**d**), and $\sigma = 0.005$ (**e**,**f**).

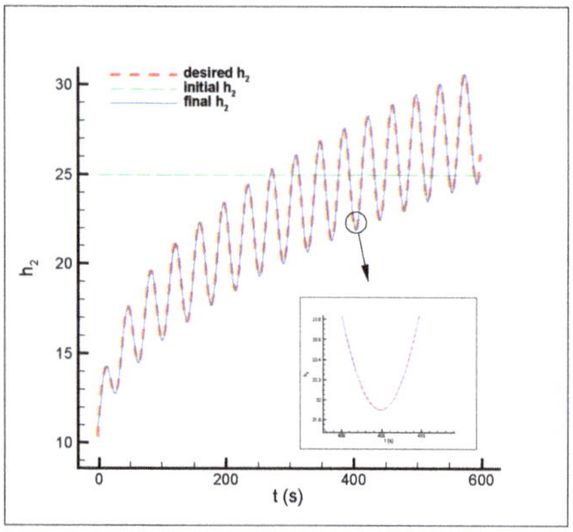

(a) $\sigma = 0.0$

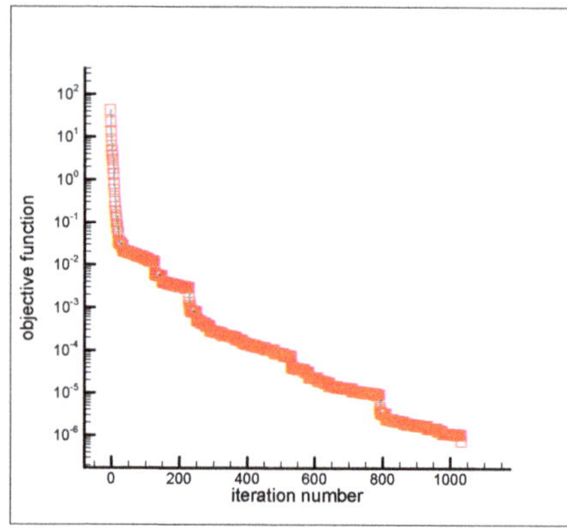

(b) $\varepsilon = 10^{-6}$

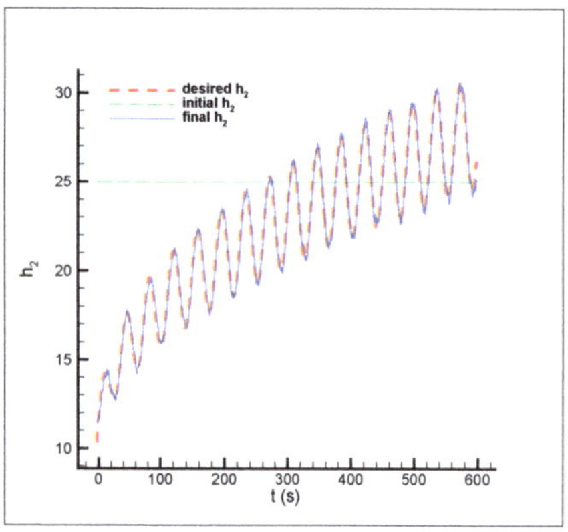

(c) $\sigma = 0.001$

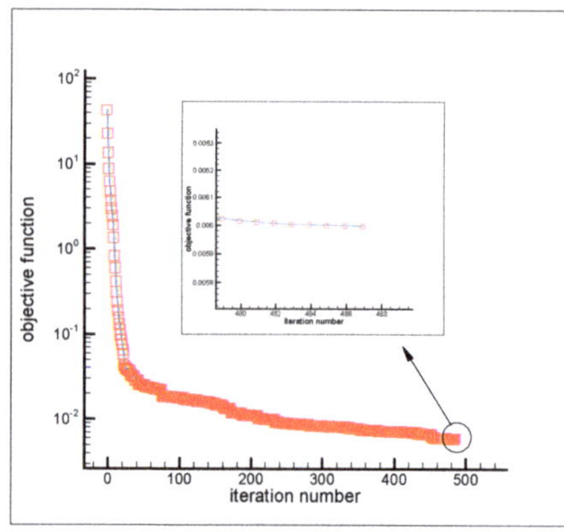

(d) $\varepsilon = 6000(0.001^2) = 0.006$

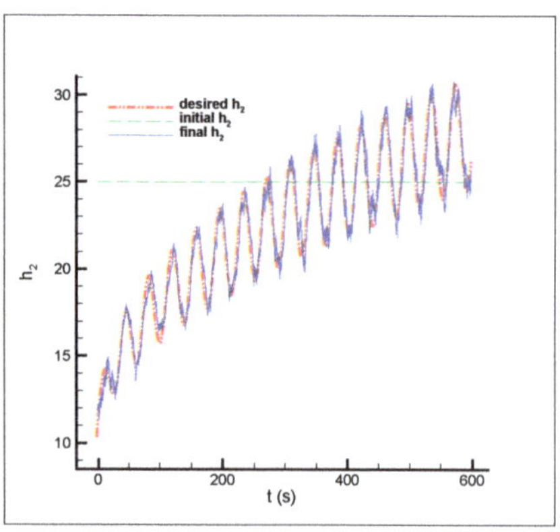

(e) $\sigma = 0.005$

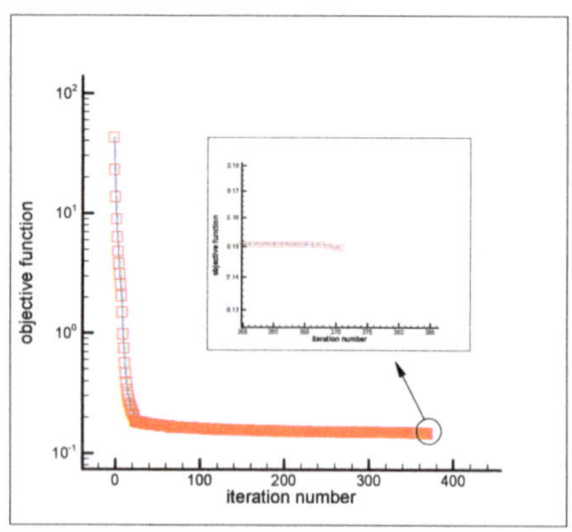

(f) $\varepsilon = 6000(0.005^2) = 0.15$

Figure 9. Estimation of the time-dependent heat transfer coefficient $h_2^{(2)}$ using an initial guess $h2_{\text{initial}}(t) = 25.0\,(\frac{\text{W}}{\text{m}^2 \cdot {}^\circ\text{C}})$ and objective function versus iteration number for cases of no measurement error (**a**,**b**) and measurement error of $\sigma = 0.001$ (**c**,**d**), and $\sigma = 0.005$ (**e**,**f**).

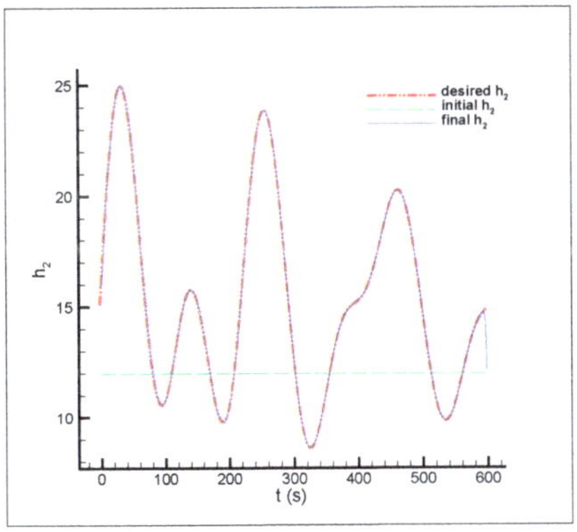

(a) $\sigma = 0.0$

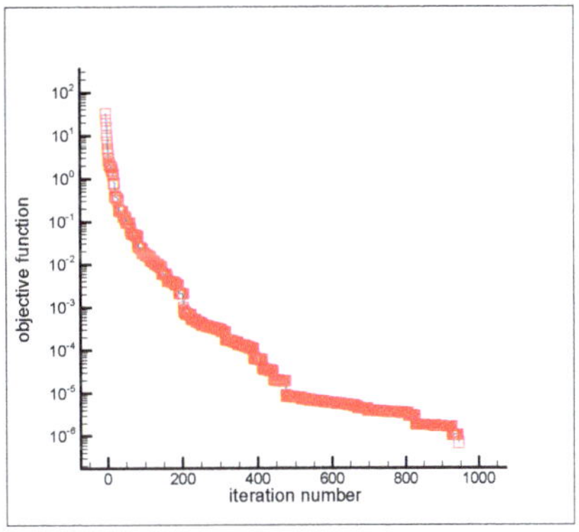

(b) $\varepsilon = 10^{-6}$

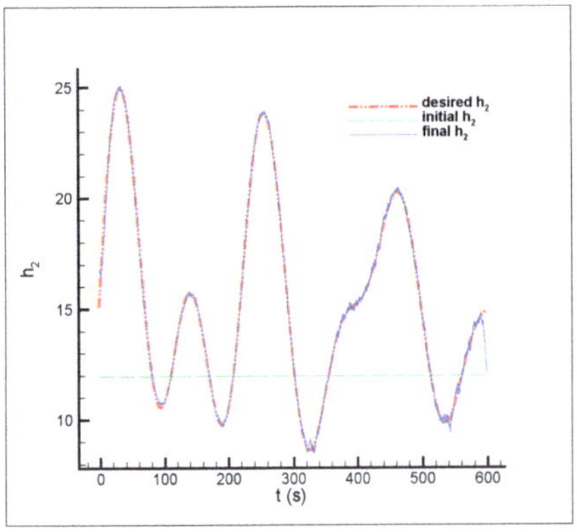

(c) $\sigma = 0.001$

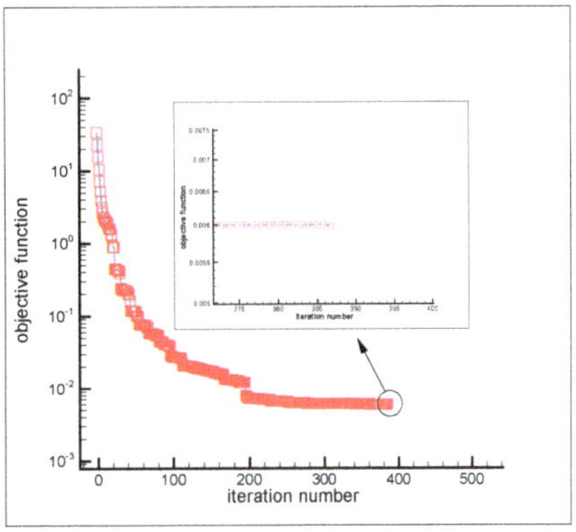

(d) $\varepsilon = 6000(0.001^2) = 0.006$

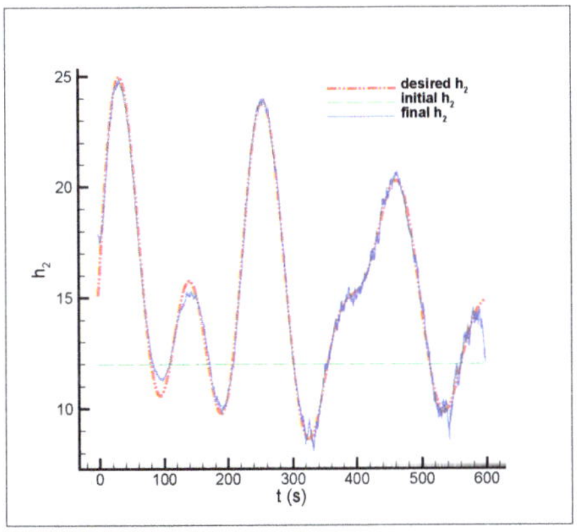

(e) $\sigma = 0.005$

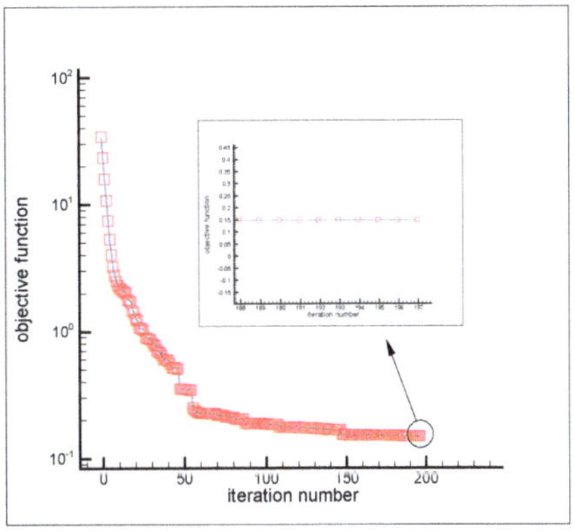

(f) $\varepsilon = 6000(0.005^2) = 0.15$

Figure 10. Estimation of the time-dependent heat transfer coefficient $h_2^{(3)}$ using an initial guess $h_{2_{\text{initial}}}(t) = 12.0\ (\frac{\text{W}}{\text{m}^2 \cdot {}^\circ\text{C}})$ and objective function versus iteration number for cases of no measurement error (**a**,**b**) and measurement error of $\sigma = 0.001$ (**c**,**d**), and $\sigma = 0.005$ (**e**,**f**).

Table 2. A summary of results for the estimation of the time-dependent heat transfer coefficient.

Functional Form	Initial Guess, $h_{2_{\text{initial}}}$	Temperature Measurement Error, σ	Initial Value of Objective Function	Final Value of Objective Function	Number of Iterations
$h_2^{(1)}$	10	$\sigma = 0.0$	3.50	9.99×10^{-7}	3654 (~244 min)
$h_2^{(1)}$	10	$\sigma = 0.001$	3.51	0.006	490 (~33 min)
$h_2^{(1)}$	10	$\sigma = 0.005$	3.65	0.15	217 (~14 min)
$h_2^{(2)}$	25	$\sigma = 0.0$	43.14	7.90×10^{-7}	1037 (~69 min)
$h_2^{(2)}$	25	$\sigma = 0.001$	43.14	0.006	487 (~32 min)
$h_2^{(2)}$	25	$\sigma = 0.005$	43.26	0.15	371 (~25 min)
$h_2^{(3)}$	12	$\sigma = 0.0$	34.28	6.95×10^{-7}	948 (~63 min)
$h_2^{(3)}$	12	$\sigma = 0.001$	34.28	0.006	387 (~26 min)
$h_2^{(3)}$	12	$\sigma = 0.005$	34.42	0.15	197 (~13 min)

From the above figures, we can also see that the estimated heat transfer coefficient deviates from the exact one in a neighborhood of t_r and approaches the initially guessed heat transfer coefficient. The mathematical reason is that by approaching the final time t_r, the number of the zero elements in the column vectors of the sensitivity matrix **Ja** also increases so that there exists only one nonzero element in the last column vector because the sensitivity matrix is a lower-triangular matrix (see Equation (26)). Thus, the last column vector can be written as

$$\begin{bmatrix} 0 \\ 0 \\ 0 \\ \vdots \\ 0 \\ \dfrac{\partial T_e(Si,Sj,t_r)}{\partial h_2(t_r)} \end{bmatrix} \tag{33}$$

From Equation (19), ($\dfrac{\partial J}{\partial \mathbf{h_2}} = 2\mathbf{Ja}^{\mathrm{T}}[\mathbf{T}_e - \mathbf{T}_m]$), we can write the gradient of the objective function J with respect to $\mathbf{h}_2$ at the final time t_r as

$$\frac{\partial J}{\partial h_2(t_r)} = 2\begin{bmatrix} 0 \\ 0 \\ 0 \\ \vdots \\ 0 \\ \dfrac{\partial T_e(Si,Sj,t_r)}{\partial h_2(t_r)} \end{bmatrix}^{\mathrm{T}} \begin{bmatrix} T_e(Si,Sj,t_1) - T_m(Si,Sj,t_1) \\ T_e(Si,Sj,t_2) - T_m(Si,Sj,t_2) \\ T_e(Si,Sj,t_3) - T_m(Si,Sj,t_3) \\ \vdots \\ T_e(Si,Sj,t_{r-1}) - T_m(Si,Sj,t_{r-1}) \\ T_e(Si,Sj,t_r) - T_m(Si,Sj,t_r) \end{bmatrix} = 2\frac{\partial T_e(Si,Sj,t_r)}{\partial h_2(t_r)}[T_e(Si,Sj,t_r) - T_m(Si,Sj,t_r)] \tag{34}$$

which is a very small number. Substituting a very small value for $\nabla J^{(k)}_{h_2(t_r)}$ into Equation (28), $d^{(k)}_{h_2(t_r)} = -\nabla J^{(k)}_{h_2(t_r)}$, results in a very small value for $d^{(k)}_{h_2(t_r)}$. Likewise, substituting a very small number for $d^{(k)}_{h_2(t_r)}$ into Equation (27), $h_2(t_r)^{(k+1)} = h_2(t_r)^{(k)} + \beta^{(k)} d^{(k)}_{h_2(t_r)}$, results in $h_2(t_r)^{(k+1)} \approx h_2(t_r)^{(k)}$, as observed. In other words, by approaching the final time t_r, there is no significant modification in the value of h_2 during the minimization process and the heat transfer coefficient retains its initially guessed value until the end of the minimization process.

5. Conclusions

Based on explicit expressions, an accurate and efficient sensitivity analysis scheme was proposed to estimate the unknown functional form of a time-dependent heat transfer coefficient applied on part of the boundary of an irregular heat conducting body subjected to specified initial and boundary conditions through transient readings of a single sensor placed inside the irregular body. Since there was no prior information available on the functional form of the variable to be estimated, the commonly used method to address this inverse heat conduction problem was based on the function estimation approach. However, here a parameter estimation approach was proposed to estimate the unknown functional form accurately and efficiently. As the body geometry was general (irregular), the physical domain was mapped onto a regular computational domain in order to take advantage of the ease of implementation of the finite-difference method to explicitly solve the transient heat conduction equation and associated boundary conditions. The chain rule was used to relate the sensor temperature and the time-dependent heat transfer coefficient applied on the part of the body boundary, the two ingredients of the sensitivity coefficients. Formulating this way, all sensitivity coefficients could be computed during the transient solution of the direct heat conduction problem without the need for solving the sensitivity and adjoint equations. The steepest-descent method with a stopping criterion specified by the discrepancy principle was used to minimize the objective function and reach the solution accurately. The accuracy, efficiency, and robustness of the inverse analysis were presented through considering three different complicated functional forms. As a future study, more challenging problems of heat conduction in materials with temperature-dependent or space-dependent thermal conductivity (in functionally graded materials, for instance) may be considered. In this study, the transient heat transfer equation (direct solution) was solved by the explicit method; the feasibility of derivation of the sensitivity coefficients using an implicit method needs to be investigated.

Author Contributions: Conceptualization, F.M. and M.S.; data curation, F.M.; formal analysis, F.M.; funding acquisition, F.M.; investigation, F.M. and M.S.; methodology, F.M.; software, F.M.; validation, F.M.; writing—original draft, F.M.; writing—review and editing, F.M. and M.S. Both authors have read and agreed to the published version of the manuscript.

Funding: This research was supported by funding from the European Union's Horizon 2020 research and innovation program under the Marie Skłodowska-Curie grant agreement No 663830.

Institutional Review Board Statement: Not applicable.

Informed Consent Statement: Not applicable.

Data Availability Statement: The data presented in this study are available on request from the corresponding author.

Conflicts of Interest: The authors declare no conflict of interest.

Nomenclature

c	specific heat ($\frac{\text{J}}{\text{kg}\cdot{}^\circ\text{C}}$)
$\mathbf{d}^{(\text{k})}$	direction of steepest descent
h	heat transfer coefficient ($\frac{\text{W}}{\text{m}^2\cdot{}^\circ\text{C}}$)
k_T	thermal conductivity of the solid body ($\frac{\text{W}}{\text{m}\cdot{}^\circ\text{C}}$)
J	objective function
J	Jacobian of transformation
Ja	Jacobian (sensitivity) matrix

$\dot{q}$	heat flux ($\frac{W}{m^2}$)
S	sensor
T	temperature (°C)
T_e	estimated temperature (°C)
T_m	measured temperature (°C)
T_0	initial temperature (°C)
T_∞	ambient temperature (°C)
t	time (s)
x, y	Cartesian coordinates in the physical domain
α, β, γ	metric coefficients in two-dimensional elliptic grid generation
$\beta^{(k)}$	search step-length at iteration k
ξ, η	Cartesian coordinates in the computational domain
Γ	boundary
Ω	domain (heat-conducting body)
ρ	density ($\frac{kg}{m^3}$)
σ	standard deviation of the measurement error
ε	stopping criterion
ω	random variable
r	number of time steps
k	iteration number

References

1. Kreith, F.; Manglik, R.M.; Bohn, M.S. *Principles of Heat Transfer*, 7th ed.; Cengage Learning: Stamford, CT, USA, 2010.
2. Louahlia-Gualous, H.; Panday, P.; Artioukhine, E. Inverse determination of the local heat transfer coefficients for nucleate boiling on a horizontal cylinder. *J. Heat Transf.* **2003**, *125*, 1087–1095. [CrossRef]
3. Beck, J.; Osman, A. Nonlinear inverse problem for the estimation of time-and-space-dependentheat-transfer coefficients. *J. Thermophys. Heat Transf.* **1989**, *3*, 146–152. [CrossRef]
4. Taler, J. Determination of local heat transfer coefficient from the solution of the inverse heat conduction problem. *Forsch. Ing.* **2007**, *71*, 69–78. [CrossRef]
5. Martin, T.; Dulikravich, G. Inverse determination of steady heat convection coefficient distributions. *Trans. Am. Soc. Mech. Eng. J. Heat Transf.* **1998**, *120*, 328–334. [CrossRef]
6. Rainieri, S.; Pagliarini, G. Data filtering applied to infrared thermographic measurements intended for the estimation of local heat transfer coefficient. *Exp. Therm. Fluid Sci.* **2002**, *26*, 109–114. [CrossRef]
7. Maillet, D.; Degiovanni, A.; Pasquetti, R. Inverse heat conduction applied to the measurement of heat transfer coefficient on a cylinder: Comparison between ananalytical and a boundary element technique. *J. Heat Transf.* **1991**, *113*, 549–557. [CrossRef]
8. Su, J.; Hewitt, G.F. Inverse heat conduction problem of estimating time-varying heat transfer coefficient. *Numer. Heat Transf. Part A Appl.* **2004**, *45*, 777–789. [CrossRef]
9. Hào, D.N.; Thanh, P.X.; Lesnic, D. Determination of the heat transfer coefficients in transient heat conduction. *Inverse Probl.* **2013**, *29*, 095020. [CrossRef]
10. Skubisz, P.; Adrian, H. Estimation of heat transfer coefficient of forced-aircooling and its experimental validation in controlled processing of forgings. *Numer. Heat Transf. Part A Appl.* **2018**, *73*, 163–176. [CrossRef]
11. Beck, J.V.; Blackwell, B.; Clair, C.R.S. *Inverse Heat Conduction: Ill-Posed Problems*; Wiley: Hoboken, NJ, USA, 1985.
12. Özisik, M.; Orlande, H. *Inverse Heat Transfer: Fundamentals and Applications*; Taylor & Francis: Abingdon, UK, 2000.
13. Alifanov, O.M. *Inverse Heat Transfer Problems*; Springer: Berlin/Heidelberg, Germany, 1994.
14. Mohebbi, F.; Sellier, M. Parameter estimation in heat conduction using a two-dimensional inverse analysis. *Int. J. Comput. Methods Eng. Sci. Mech.* **2016**, *17*, 274–287. [CrossRef]
15. Mohebbi, F.; Sellier, M. Estimation of thermal conductivity, heat transfer coefficient, and heat flux using a three dimensional inverse analysis. *Int. J. Therm. Sci.* **2016**, *99*, 258–270. [CrossRef]
16. Mohebbi, F.; Sellier, M. Identification of space-and temperature-dependent heat transfer coefficient. *Int. J. Therm. Sci.* **2018**, *128*, 28–37. [CrossRef]
17. Mohebbi, F. Function Estimation in Inverse Heat Transfer Problems Based on Parameter Estimation Approach. *Energies* **2020**, *13*, 4410. [CrossRef]
18. Özişik, M.N.; Orlande, H.R.B.; Colaço, M.J.; Cotta, R.M. *Finite Difference Methods in Heat Transfer*; CRC Press: Boca Raton, FL, USA, 2017.
19. Özisik, M. *Heat Conduction*; John Wiley & Sons: Hoboken, NJ, USA, 1993.
20. Pioro, I.L.; Mahdi, M.; Popov, R. Heat Transfer Media and Their Properties. In *Handbook of Thermal Science and Engineering*; Kulacki, F.A., Ed.; Springer International Publishing: Cham, Switzerland, 2017; pp. 1–100.

energies

MDPI

Article

Numerical Investigation of the Performance of a Submersible Pump: Prediction of Recirculation, Vortex Formation, and Swirl Resulting from Off-Design Operating Conditions

Virgel M. Arocena, Binoe E. Abuan, Joseph Gerard T. Reyes, Paul L. Rodgers and Louis Angelo M. Danao *

Department of Mechanical Engineering, University of the Philippines Diliman, Quezon City 1101, Philippines; vmarocena@up.edu.ph (V.M.A.); beabuan@up.edu.ph (B.E.A.); jtreyes2@up.edu.ph (J.G.T.R.); paul112464@yahoo.com (P.L.R.)

* Correspondence: louisdanao@up.edu.ph

Abstract: Like any other turbomachinery, it is essential that the hydraulic behavior and performance of mixed-flow pumps are evaluated way in advance prior to manufacturing. Pump performance relies heavily on the proper design of the intake structure. Intake structures should be accurately designed in order to minimize and avoid unnecessary swirl and vortex formations. Ensuring the optimum performance condition as well as predicting how a particular intake structure affects the efficiency of the pump often requires either physical model studies or theoretical evaluations. Unfortunately, physical models are costly, time-consuming, and site-specific. Conversely, design and performance predictions using a theoretical approach merely gives performance values or parameters, which are usually unable to determine the root cause of poor pump performance. This study evaluates the viability of using Computational Fluid Dynamics (CFD) as an alternative tool for pump designers and engineers in evaluating pump performance. A procedure for conducting CFD simulations to verify pump characteristics such as head, efficiency, and flow as an aid for preliminary pump design is presented. Afterwards, a multiphase simulation using the VOF approach is applied to compare the fluid dynamics between four different pump intake structures. A full-sized CFD model of the pump sump complete with the pump's active components was used for the intake structure analysis in order to avoid scaling issues encountered during the reduced-scale physical model test. The results provided a clear illustration of the hydraulic phenomena and characteristic curves of the pump. A performance drop in terms of reduction in TDH was predicted across the various intake structure designs. The CFD simulation of intake structure provided a clear insight on the varying degree of swirl, flow circulation, and effect on pump efficiency between all four cases.

Keywords: mixed-flow pumps; CFD; performance; *Q-H*; intake structure; sumps

Citation: Arocena, V.M.; Abuan, B.E.; Reyes, J.G.T.; Rodgers, P.L.; Danao, L.A.M. Numerical Investigation of the Performance of a Submersible Pump: Prediction of Recirculation, Vortex Formation, and Swirl Resulting from Off-Design Operating Conditions. *Energies* **2021**, *14*, 5082. https://doi.org/10.3390/en14165082

Academic Editors: Michele Pinelli, Alessio Suman and Nicola Casari

Received: 16 July 2021
Accepted: 16 August 2021
Published: 18 August 2021

1. Introduction

A few of the specific hydraulic conditions that greatly affect the performance of any pump are swirl, vortices, and circulation. These are just some of the undesirable hydraulic phenomena that often lead to performance degradation, vibration, noise, structural damage, and even catastrophic failure. These phenomena are also highly influenced by flow conditions in the intake structure. Poorly designed intake structures are those that fail to control the possible harmful formation of swirls, free-surface vortices, and submerged vortices. These swirls and vortices can induce cavitation growth if not fully suppressed. Despite this fact, the proper design of these intake structures is amongst the most overlooked aspects when designing a pumping station.

Theoretical means of evaluating a pump's performance is a highly intensive process. Aside from the numerous geometric parameters involved, the hydraulic conditions present in any pump are complex, and the physics of which are not fully understood. It is for this reason that most pump engineers often rely on designs based on rule-of-thumb or standards established from previously proven designs. This practice, along with some financial constraints, often limits design iteration and restricts the development of new products.

Most major hydraulic projects are model tested to optimize the design. For intake structure design, reduced-scale physical model tests had remained a primary mandatory requirement in keeping with existing codes and standards. ANSI/HI 9.8 [1] states that model tests shall be conducted when the sump geometry deviates from standardized designs, when approach flow is non-uniform or non-symmetric, or when the discharge rate is high (greater than 2.5 m^3/s per pump or 6 m^3/s per station). These tests provide design engineers with a visual image of what the flow will be during actual running conditions. They represent reality at a certain scale. Among the advantages in performing a physical model tests are that most parameters and operating boundaries involved are controllable, and the measurements, in general, are easy to perform. Unfortunately, aside from being expensive, site-specific, and time-consuming, results from physical model tests are highly influenced by scaling issues, incomplete modeling, and laboratory effects.

On the other hand, the implementation of Computational Fluid Dynamics (CFD) in fluid flow applications had grown in parallel with the advances in computer technology. CFD coupled with stress analysis had been actively used in the design of various pump components such as shafts, seals, impellers, diffusers, and casings, among others. For numerical studies, both the accuracy and robustness of the results vary heavily based on various turbulence parameters and modeling methodologies. Particularly for pump intake structure design, prediction of both free-surface and subsurface vortices using CFD had been a continuous focus. Constantinescu and Patel [2] were among the first to simulate the three-dimensional flow field on a simple water-intake bay. Their work solves the Reynolds-averaged Navier-Stokes equation with a two-layer k-ϵ turbulence model. The numerical solution was able to predict the expected vortex formation. However, it was emphasized that the resulting symmetric vortices caused by idealized flow conditions in the numerical solution are unlikely to exist in reality. These results were later verified by Rajendran and Patel [3] by constructing a 0.003 m^3/s rectangular pump sump model. Using particle-image velocimetry (PIV) to measure velocity fields, they were able to confirm that the previous CFD results for the position, number, and overall structure for both the free-surface and subsurface vortices were in good agreement with the physical model.

While there are numerous numerical and experimental studies concerning vortex prediction and flow phenomena in pump intake structures [4–6], most do not consider any impeller-induced flow. This is because most physical model experiments also do not include the impeller geometry in the scale model. Instead, a vertical pipe is used to represent the pump structure. Most studies only focus mainly on the flow condition upstream of the pump. One slight exception is the study made by Nagahara et al. [7]. In their study, they included the shape profile of the pump impeller and guidevane. They aimed to identify the relationship between the subsurface vortex strength and the size of the vapor cavity in the blades. The impeller remained stationary during the test, and no impeller-induced flow was considered. Instead, only the interaction between the vortex strength and the radial and axial forces on the blades were evaluated.

In factory test facilities, physical model studies, and even on existing numerical simulations, pump and intake structures are often evaluated separately. Constructing a test facility to verify the performance of the pump and sump prototype is not ideal for even the largest factory. Similarly, no specific code and test procedure are established in conducting a reduced-scale hydraulic model test that incorporates the pump geometry to the model. In this regard, this study presents a methodology to evaluate both pump and intake structure performance using CFD. To establish this, pump performance characteristics in terms of head, flow, and efficiency are first simulated. Then, flow dynamics, vortex prediction, and pump performance are conducted using a full-scale numerical model of the entire pump and intake structure. To analyze the effect of pump intake geometry, several intake structure designs are presented in order to compare the performance of each type. The focus is on a method that would result in reduced computational efforts in order to merit its use as an engineering design tool.

2. Pump Performance Prediction

This section focuses on the CFD simulation of a mixed flow pump predicting various performance characteristics such as head, capacity, efficiency, and flow dynamics. The analysis aims to define through numerical simulation the performance curves of the mixed-flow pump, whose specifications are outlined in Table 1. The KSP-CV pump model is among the long-range of high-quality pumps manufactured by Tsuchiura Works, Machinery System Division of Hitachi, Ltd. These pumps are designed for various applications such as steelworks, industrial and chemical plants, irrigation, waterworks and sewage systems, power stations, and nuclear power plants. The KSP-CV pump used in this study is a single-stage vertical-shaft mixed flow submersible pump. These particular pump models have a typical head range between 3 m to 90 m and are designed for large-capacity operations. The active components for this pump are the rotating impeller and the stationary guidevane. The impeller has 5 blades and is rotating at N = 600 RPM, while the stationary guidevane has 12 vanes. The pump is driven by a 2500 kW motor, which proves that it is a high-energy pump. To protect proprietary information, only the basic data about the pump (i.e., Expected Performance Curves [8], Ratings) will be presented in this paper. The curves derived from the simulation for this pump will serve as the basis for the full-scale model analysis that will be presented in the succeeding sections of this study. In order to not exceed the available computational resources, disk friction losses, mechanical losses, leakage losses, and tip clearance effects are not included in this calculation. In addition, the walls were assumed to be smooth, and any disturbances due to surface roughness were neglected.

Table 1. Pump Specifications.

Specification	Value
Type Form	KSP-CV
Specific Speed [a]	83
Suction Bore (m)	1.257
Discharge Bore (m)	1.25
Stages	single
Capacity (m^3/h)	16,900
Total Dynamic Head (m)	39
NPSHa (m)	13
Speed (rpm)	600
Driver Type	motor-driven
Driver Rating (kW)	2500
Design Efficiency (%)	85
Service	CWP

[a] Dimensions are not used. Values shown are based on rate of flow at optimum efficiency in m^3/s, total head in m, and rotative speed in rpm.

2.1. Computational Domain

The computational domain includes the inlet or suction pipe, impeller, guidevane, column pipe, and 90° discharge elbow. The hexahedral meshing scheme is used for the whole domain. The 3D shape of the impeller was generated using ANSYS CFX-BladeGen, while the O-type grid was generated using ANSYS CFX-TurboGrid. The domain is separated into 4 regions in order to achieve a reasonable number of grid elements. The impeller and guidevane regions were modelled with a high-density mesh since these areas were expected to have a high solution gradient. The inlet pipe, column pipe, and discharge elbow regions, on the other hand, have lower density mesh count. The inlet pipe for this particular pump model incorporates four (4) 90° cross plates regulating pre-whirl and impeller backflow under partial flow conditions. This feature, together with the inlet extension, ensures that the flow is fully developed as it enters the domain. Similarly, the column pipe length is designed with enough length to prevent flow from recirculating back to the guidevane. The front and rear cavities of the impeller were partially included in the control volume but with zero leakage flow rates. The 3D model and the mesh model of the mixed flow impeller and guidevane are shown in Figure 1. Similarly, a schematic of the whole computational domain, as well as the mesh for the whole assembly, is shown in Figure 2.

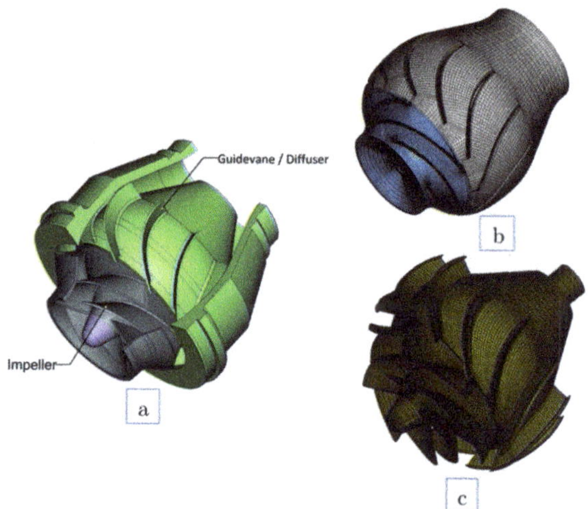

Figure 1. (**a**) The active components for the KSP-CV: 5-blade rotating impeller and 12-blade stationary guidevane/diffuser. (**b**) The generated hexahedral mesh used for the numerical analysis. (**c**) Impeller and guidevane blade walls.

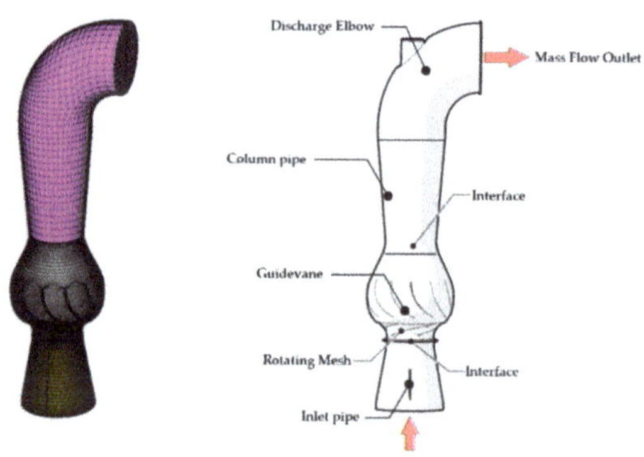

Figure 2. A schematic of the whole computational domain (**right**) and the mesh for the whole assembly (**left**).

2.2. Boundary Conditions and Solution Schemes

To solve the rotor-stator interaction problem in the domain, the multiple reference frame (MRF) approach is applied. The MRF approach is one of several models available in ANSYS Fluent in dealing with flows in systems containing translating or rotating components. For mixed flow pumps, all parts connected to the rotating shaft are moving with a certain angular velocity, while the other parts, such as the casing, guidevane, suction bell, etc., remain stationary. In general, the flows in such turbomachinery are unsteady in nature. This unsteadiness becomes even more apparent when viewed by a stationary observer. On the contrary, if viewed by an observer standing on the rotor, the flow would appear to be in a steady state. Using this perspective, the MRF model treats an unsteady problem (with respect to the absolute reference frame) as a steady-state problem with respect to the moving reference frame. When using the MRF model, additional source terms such as centripetal and Coriolis acceleration are added to the equation of motion to account for the transformation from the stationary to the moving reference frame. Note that, during the analysis, the mesh is fixed or frozen on a specified position, and the instantaneous flow field is observed with the rotor in that position. It is because of this analogy that the MRF is often referred to as the "frozen rotor approach". Although this approach is clearly an approximation of steady-state conditions, the MRF model simplifies the problem and consequently requires less time and computational effort. For preliminary design and development, this makes the MRF approach an attractive model for industrial applications such as stirred tanks, wind turbines, fans, and other turbomachinery. For this study, the impeller region is assigned to a moving reference frame, while the rest of the region is assigned to a stationary frame. Calculations were carried out under single-phase, steady-state conditions with water at 25° C as the working medium. Flow is incompressible and isothermal with constant fluid properties. Turbulence was modeled using the Realizeable k-ε since it shows better agreement with pump test data as compared to the other two k-ε models (i.e., RNG and standard) [9]. A pressure-based coupled solver was applied. The second-order discretization scheme was used for pressure, momentum, density, and turbulence equations. Static pressure is defined at the inlet, while a mass flow rate was prescribed at the outlet. A 5% turbulent intensity was specified for the inlet, and the hydraulic diameter for both inlet and outlet was set equivalent to the respective pipe diameters. The flow was specified as subsonic, and all walls were treated as no-slip.

2.3. Grid Independence Study

To optimize the computational time and to study the effects of grid resolutions, a preliminary grid independence study at 90% rated flow was conducted to compare the results for three CFD models. The grid independence study ensures that the results are due to the boundary conditions and physics used and not the mesh resolution. Three hexahedral mapped meshed models were generated, with element counts of 0.9, 3.6, and 5.2 million cells to represent models with coarse, medium, and fine mesh, respectively. For the model used in this paper, the mesh refinement did not follow the usual half/double element size. This is because refining the mesh by a factor of 2 will result in an 8-fold increase in problem size and is unacceptable for engineering design purposes. Since the analysis is concerned with flow dynamics in the impeller and guidevane, the mesh variation for all three CFD models is focused on these two regions. There is no significant need to refine the mesh for the inlet pipe and discharge elbow because the velocity gradients in these areas are expected to be low. Likewise, knowledge of the flow patterns in these areas is not key in the analysis. Non-overlapping mesh interfaces were adapted in ANSYS Fluent to combine the four regions. The independence between areas with different mesh densities was improved by refining 2 to 3 mesh layers near the interface. A focused view of the coarse, medium, and fine meshed models are shown in Figure 3, while a tabulated data of the mesh count and the respective solution time using an 8-core 4.6 GHz Intel Core i7 9700 K desktop with 32 Gb of memory is shown in Table 2.

Table 2. The cell count for each region and simulation time.

	Coarse Mesh	Medium Mesh	Fine Mesh
Inlet Pipe	208,880	208,880	208,880
Impeller	374,300	2,994,400	2,994,400
Guidevane	317,160	317,160	2,537,280
Discharge Elbow	37,904	37,904	47,088
Total Cells	938,244	3,558,344	5,787,648
Solution Time	15 min	47 min	7 h. 8 min.

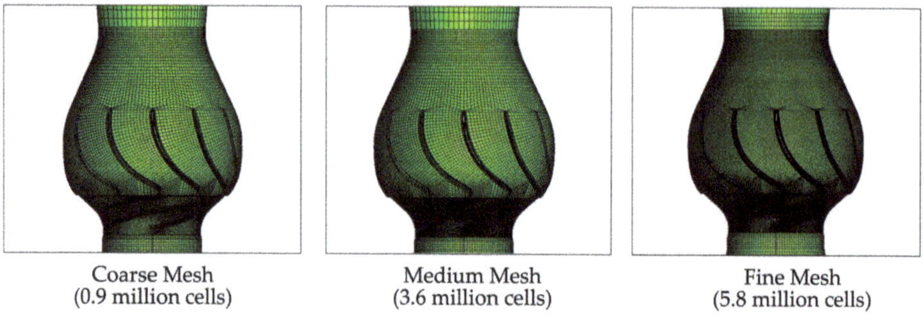

Figure 3. The mesh used for the grid independence study.

For all three numerical models, a steady-state simulation for the problem was conducted, where the mean velocity and the amplitude of the fluctuating field do not vary for more than 1% for each iteration. The target residual errors are kept below 10^{-4}. The pump head, as monitored throughout the entire calculation period, can be seen in Figure 4. For the three mesh densities, it can be hypothesized that the solution is very stable and that the calculated pump heads are nearly identical.

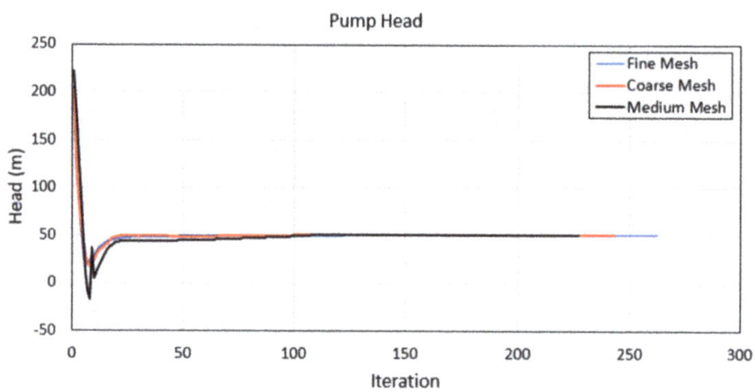

Figure 4. The pump head at 90% rated capacity as monitored during numerical simulation.

To qualitatively evaluate the grid independence, the axial velocity (z) at several points along a line just below the impeller eye and perpendicular to the flow direction is plotted, as shown in Figure 5. Additionally, in Figure 6, the total pressure contour on the center of the pump assembly is plotted for all three numerical models. Based on these figures, the simulation results show, in general, a close agreement between the three results. Specifically, for the velocity plots, a maximum variation of 1.4% was observed, particularly between the results for the fine mesh and the coarse mesh models. Additionally, the evaluation shows that even though the cell count for the medium mesh model is about 25% less than that of the fine mesh model and with a 90% reduction in calculation time, the difference in the results between these two models is almost negligible at 0.01%. The pressure distribution within the impeller and the guidevane regions on all three models are identical. The resemblance between the results from all three models shows that these simulations can be considered relatively grid-independent based on the presented mesh densities. Generally, for such cases, the coarse mesh is the best alternative for succeeding simulations in terms of minimizing computational time. However, in this paper, considering how close the results are between the fine mesh and the medium size mesh, and to more fully resolve the other phenomena of interest (e.g., vortex formation and air entrainment), the medium-sized mesh model was used.

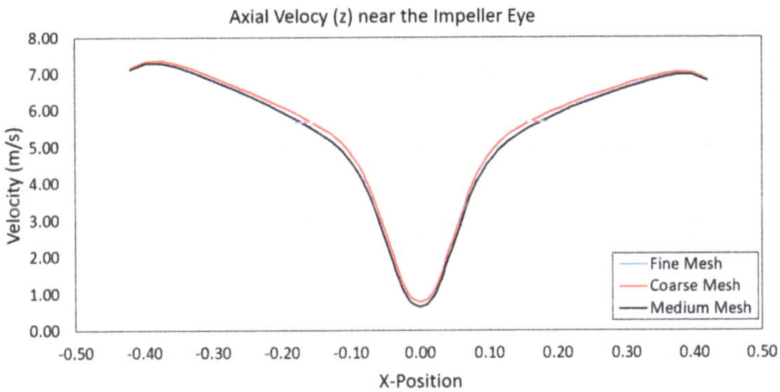

Figure 5. The axial velocity distribution near the impeller eye at 90% rated capacity.

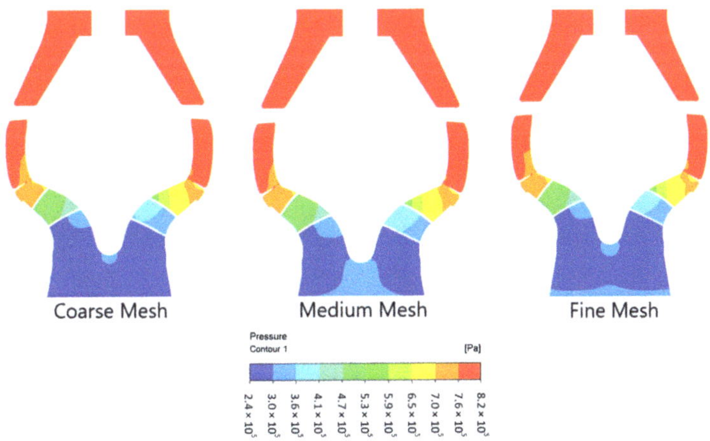

Figure 6. The pressure gradient across the pump assembly at 90% rated capacity.

3. Results and Discussions

Performance Curve

For the performance prediction, numerical simulations were performed from 10% to 150% of the nominal rated capacity with 10% increments, while shut-off (zero flow) may be interpolated. For all simulations, the rotor speed is kept constant at 600 RPM. The pump head, shaft power, and efficiency were calculated as follows:

$$\text{Head}: \; H = \frac{P_{out}}{\rho g} - \frac{P_{in}}{\rho g} + \Delta z, \tag{1}$$

$$\text{Shaft Power}: \; P = T \times \omega, \tag{2}$$

$$\text{Efficiency}: \; \eta = \frac{\rho g Q H}{P}, \tag{3}$$

wherein P_{in}, P_{out} are the area weighted average of the total pressure taken at a cross-section at the pipe inlet and at the end of the pipe column (just before entering the discharge elbow), respectively. Q is the fluid flow rate, g the acceleration due to gravity, and ρ is the fluid density. The tabulated results for all analysis points are shown in Table 3. Figure 7 shows the graph of the numerical results for the head, efficiency, and power as a function of the flow rate superimposed on the expected pump performance curve as provided by the manufacturer [8]. It can be seen from this figure that all three parameters follow a consistent trend as compared with the manufacturer's data. This further shows that for this particular pump, doing the grid independence study at only one flow condition, for this case at 90% rated flow, is enough to consider the flow relatively grid-independent.

Table 3. Tabulated performance results for mixed-flow pump at 600 RPM.

Flow Rate (m^3/h)	Shaft Power (kW)	Pump Head (m)	Efficiency (%)
1519	2288	78	14
3038	2260	75	27
4556	2118	70	41
6075	2007	65	54
7594	1963	62	66
9113	2080	62	74
10,632	2164	60	80
12,151	2230	57	85
13,669	2285	54	88
15,188	2328	50	89
16,707	2318	45	89
18,226	2262	39	87
19,745	2163	33	82
21,264	2017	25	71
22,782	1824	16	54

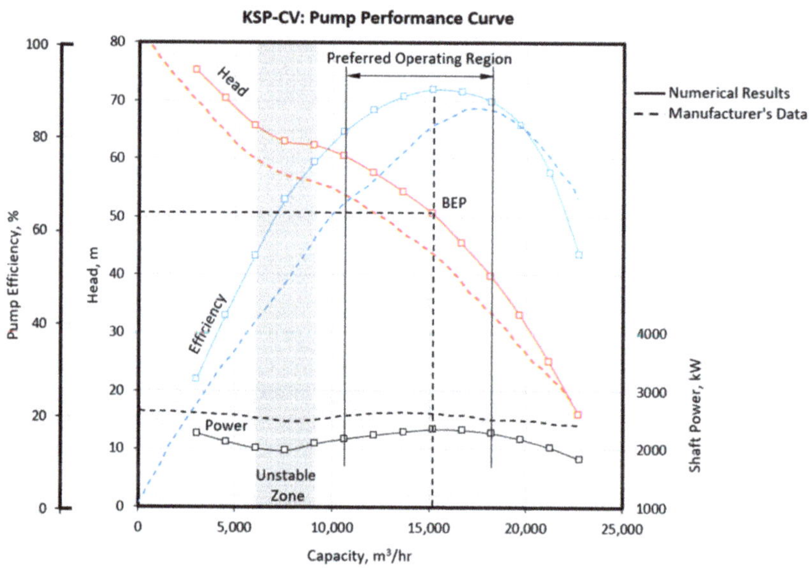

Figure 7. The pump performance curve based on numerical results as compared with the expected performance curve acquired from the manufacturer [8].

Based on the relationship between the head, flow rate, power, and pump efficiency (Figure 7), it can be seen that the power requirement is much flatter and does not vary too much along the pump's operating range. This type of characteristic curve is typical of the mixed-flow pump and is, in turn, different from those of radial-type pumps, where a larger flow requires higher power. This is because of the very large area of flow against the comparatively small impeller diameter in axial and mixed flow pumps. Hydraulics performance, however, differs appreciably, and the head capacity (*Q-H*) curve tends to be steep, with the point of maximum efficiency shifted towards maximum capacity. This characteristic can be verified in the chart where the best efficiency point (BEP) is located at high flow rates as expected.

Figure 8 shows the flow pattern and pressure distribution when operating the pump at Q_{BEP}. It can be seen that at Q_{BEP}, flow is perfectly aligned with the blade profile. However, it is evident that the leading edge of the blade acts as an obstruction for incoming flow, creating an area having a low static pressure on both sides of the blades. This shows that even at the best efficiency point where the flow direction is perfectly aligned with the blade, cavitation can still take place once there is a significant drop in the inlet pressure. Specifically, for the pump in this study, it can be assumed based on the same figure that because of the lower pressure, cavitation will most likely start to form on the suction side of the blade.

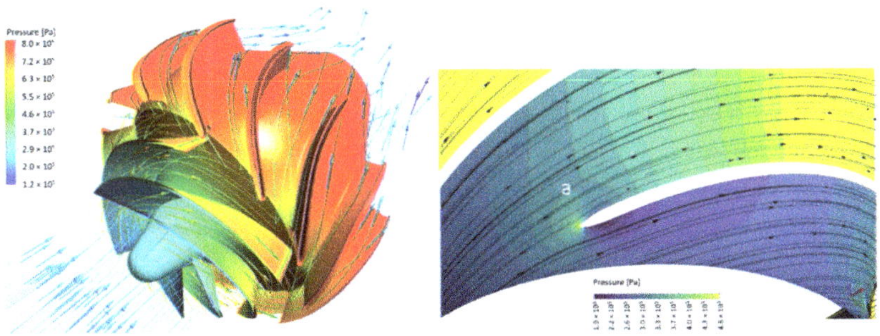

Figure 8. Velocity streamlines and pressure distribution at $Q_{BEP.}$ Even when flow direction is perfectly aligned with the blade, the blockage effect (a) still exists on the blade tip causing a decrease in pressure on both sides of the impeller blade.

All rotodynamic vertical pumps have limitations on the minimum flow at which they should be operated continuously or for an extended period of time. For specialized high energy pumps, flow limitations may be set as high as 70% of Q_{BEP}, while for low energy pumps, the allowable flow may be set as low as 20% of Q_{BEP} [10]. These flow limitations vary across various manufacturers and depend heavily on the type of application and impeller design. The next two figures show the flow pattern inside the pump when it is operated within the unstable region.

In Figure 9, it can be seen that at the rate of flow equivalent to 50% of Q_{BEP}, the fluid tends to return from the guidevane back towards the impeller. This phenomenon is called discharge recirculation, which primarily occurs during part-load or at low capacities. Recirculation often results in strong noise, vibration, and rotor stall. Further reduction in flow beyond the unstable region (30% of Q_{BEP}) will cause recirculation to shift towards the suction side of the impeller, forming a vortex region between the impeller blades, as shown in Figure 10. This low-pressure area can lead to cavitation accompanied by intense pressure pulsations which ultimately damages the impeller because of the undue stress to the rotor components. Additionally, Alpan and Peng [11] reported that fluid recirculation in the suction nozzle (suction recirculation) is the main source of power loss at low flow rates.

Generally, the effect of recirculation is still questionable and sometimes trivial at best. In some applications, recirculation is a constructive phenomenon and may even be advantageous to the design (e.g., regenerative pumps). In these cases, a desirable effect of recirculation is to have the same (recirculating) volume of fluid to be acted upon by the impeller multiple times, thereby increasing fluid energy. Therefore, it is easy to conclude that recirculation can contribute towards higher heads compared to conditions without recirculation. However, there are also some instances where recirculation can lead to an unfavorable dip in the *Q-H* curve (Figure 7). At least for the mixed flow pumps covered in this study, recirculation is an unfavorable phenomenon, and the additional head caused by recirculation is meaningless since mixed-flow pumps cannot operate in these regions. Operating at such low flows will cause these types of pumps to experience full cavitation.

Based on the results of the CFD analysis, for this pump, it can be concluded that continuous operation below 60% of Q_{BEP} should be avoided.

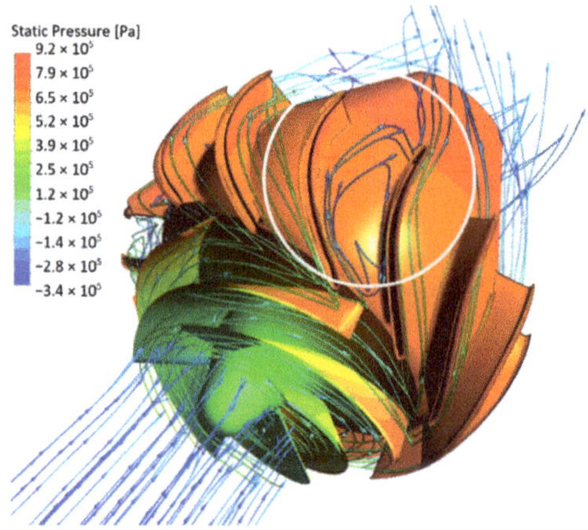

Figure 9. Discharge recirculation at a flow equivalent to 50% of QBEP. The fluid tends to recirculate from the guidevane back to the impeller. This circulatory flow can cause large forces on the impeller shrouds resulting in unbalanced axial forces and high thrust.

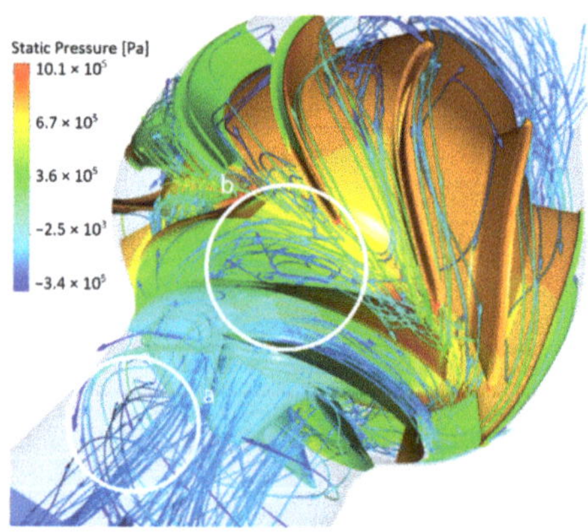

Figure 10. Suction recirculation at 30% of QBEP. The pump operating at this reduced flow causes the flow to separate from the blade and circulate back upstream, and in this case, even back to the suction pipe. Undesirable phenomena such as eddies and pre-rotation are observed to form within the inlet pipe (a) as well as strong, high-velocity vortices between the impeller blades (b).

The MRF mesh calculations led to some overprediction of the head as compared to the expected pump performance data. The overprediction increases with decreasing flow rate. This is primarily due to the leakage flow in the pump, which was not accounted for in the present calculations since the gap between the impeller and the labyrinth seal is considered non-existent. Applying the necessary correction for leakage losses would certainly bring the calculated results close to the manufacturer's data.

If the overprediction is to be disregarded or considered acceptable, these results would mean that the published rated capacity of 16,900 m^3/h and TDH of 39 m, as previously shown in Table 1, is 10% above the BEP.

4. Full-Size Pump Sump Numerical Analysis

Numerous design suggestions and guidelines regarding reliable optimal intake structures can be found in various pump standards, such as those published by the Hydraulic Institute [1] and the British Hydromechanics Research Association. Ideally, pump sumps should be simple and easy to construct. Although it is not always possible to build an intake structure that conforms to these guidelines. In certain cases, space and cost limitations are a primary concern. Conversely, in others, such as during site rehabilitation, replacing an aging pump with a new one poses a daunting challenge, especially because hydraulic conditions on the existing sump are unknown.

4.1. Sump Geometry Variations

Figure 11a represents the geometry of the rectangular sump, which served as the baseline for the analysis (Case 1). All dimensions are according to recommendations set by the Hydraulic Institute [1]. No splitters, fillet, curtain wall, or any form of anti-vortex device was installed in order to obtain a clear baseline result. ANSI/HI 9.8 [1] outlines the recommended dimensions for a rectangular sump, but oftentimes, due to site-specific flow conditions, additional modifications are necessary. This is to prevent any vortex formation that could affect the performance of the pump. In such cases, various pump publications outline design recommendations that aim to aid engineers in selecting the proper design. Some designers prefer to further modify this design by providing the sump with two semi-circular walls. With the notion that this would eliminate the vortices that appear in the corners of the sump. In practice, the advantage of this modification is trivial. In fact, some installations have experienced vortices along the center of the semicircles. Nearer the suction pipe as compared to sumps with corner vortices. For this study, a variation of this design is presented in Figure 11b (Case 2), wherein a floor cone and splitter are added in order to try to alleviate the side effect mentioned above.

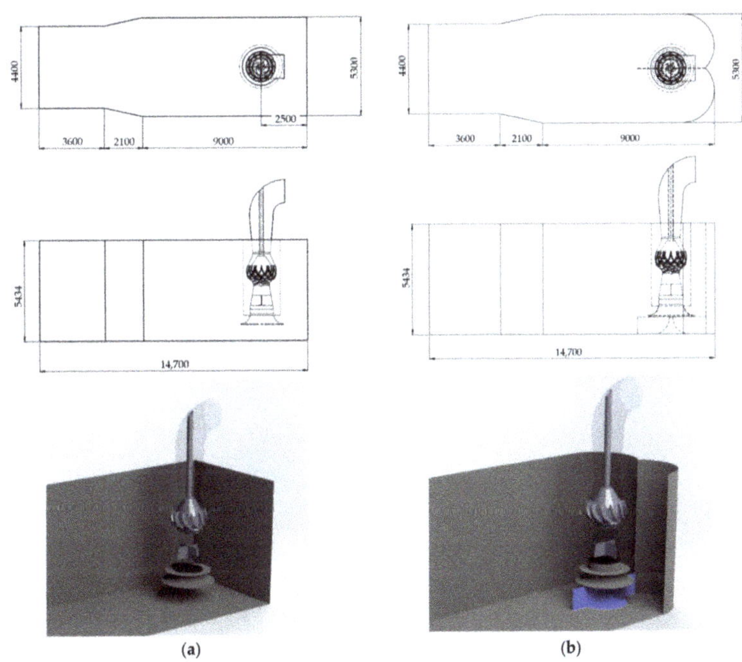

Figure 11. (**a**) A rectangular pump sump model with mixed-flow pump and guidevane (Case 1); (**b**) a pump sump utilizing semicircular arches as a backwall (Case 2). In addition, a floor cone and a splitter are installed to prevent floor sub-surface vortices.

Another popular variation on the recommended sump design is one proposed by Ingersoll-Rand [12]. Their proposed design uses tetrahedral floor splitters instead of the conventional fillet. A slight variation on the design, as shown in Figure 12a (Case 3), was used as the third model for this study. For this model, the central floor splitter was replaced with a floor cone and plate combination in order to increase the area underneath the bellmouth. The aim is to reduce the head loss observed in the unmodified sump (Case 1). Lastly, Figure 12b shows a formed suction intake (FSI) structure used for stations with adverse inflow conditions. It is said that these designs are relatively insensitive to the direction of approach and skewed velocity distribution at the suction bell mouth. Such configurations are usually used where it is difficult to design standard inlet structures due to space limitations. An advantage that an FSI has is that it can normalize any erratic or swirling flow by re-directing the flow vertically into the pump impeller. The increase in the fluid's velocity as it enters the FSI reduces the sump's susceptibility for sub-surface vortex formation. The FSI model presented in this study (Case 4) is a variation of the "TYPE 10" FSI design developed from the results of the research conducted by the US Army Engineer Waterways Experiment Station (WES). Specifications for FSI types 1 and 10 can be seen in USACE engineering and design manual EM 1110-2-3105 [13], referencing ETL 1110-2-327.

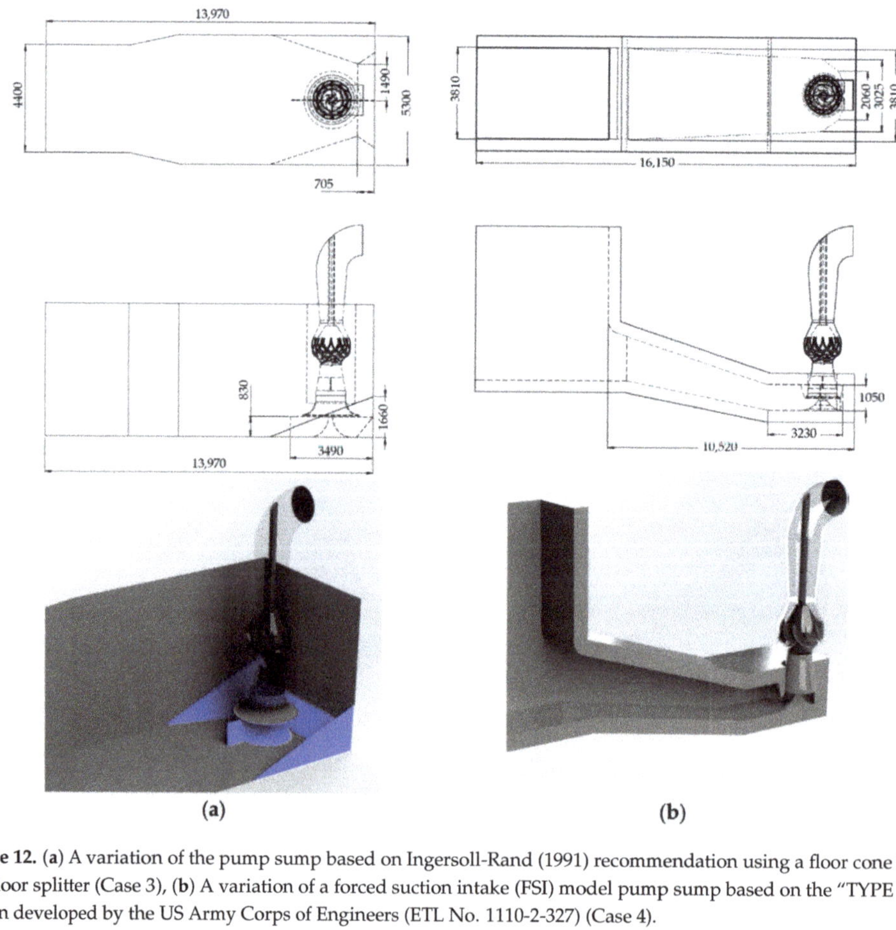

Figure 12. (**a**) A variation of the pump sump based on Ingersoll-Rand (1991) recommendation using a floor cone instead of a floor splitter (Case 3), (**b**) A variation of a forced suction intake (FSI) model pump sump based on the "TYPE 10" FSI design developed by the US Army Corps of Engineers (ETL No. 1110-2-327) (Case 4).

4.2. Computational Domain and Boundary Conditions

For all the cases, the geometry and mesh were created using an in-house pre-processing software CADAS, and the numerical simulations were run using ANSYS Fluent. Similar approaches and methodologies with those discussed in Arocena et al. [14] are used throughout this section. Grid size and element lengths for Cases 1, 2, and 3 were derived from the grid independence study conducted on Case 1 of the same publication. The pump sump is modelled using a high-resolution hexahedral mapped mesh near the pump bellmouth (pump bay) and a coarser mesh upstream of the pump (forebay). The mesh of the pump used in the analysis in the previous sections is then added to complete the model. Figure 13 shows an overview of the mesh scheme for all four cases, with each hydraulic zone represented by a different color. All boundary conditions were calculated based on the rated capacity of 16,700 m^3/h, which is about 110% of Q_{BEP} and with a low water level (LWL) of 5.484 m. A velocity flow inlet with negative velocity magnitude (outflow) was prescribed as outlet boundary condition at the end of the discharge pipe, while the rectangular section serving as the entry point for the sump was prescribed as a pressure inlet boundary condition. The boundary condition for the air surface 2 m over the water surface was also specified as pressure outlet boundary with zero backflow volume fraction. Indicating that only air can pass through this boundary.

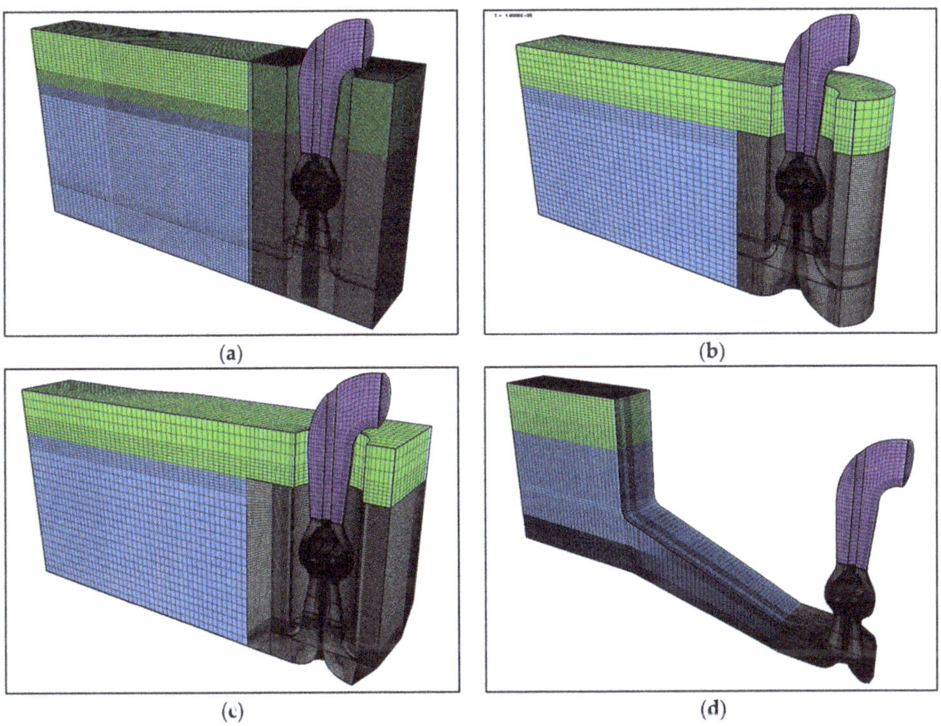

Figure 13. A cross-sectional view of meshed geometries for the 4 sump models: (a) Case 1, (b) Case 2, (c) Case 3, and (d) Case 4. Each model was separated into regions where high-density mesh count is created for regions with high velocity gradients.

At this point, it is important to highlight that the forebay for Case 1 represents the full-scale prototype for the reduced-scaled sump used in the physical model test presented in [14]. As an overview, it should be noted that for open channel flow, gravity and inertial forces play a more dominant role than viscous or turbulent shear forces. Keeping the

Froude number in both the model and the prototype constant is a good approximation of dynamic similitude [15]. As such, the 1:10 undistorted scale selected for the physical model test in [14] is based on a constant Froude number ($Fr = 0.38$) as computed across the suction bell for both the model and the prototype. Additionally, it was verified that the Reynolds number ($Re = 1.44 \times 10^5$) across the 260 mm diameter suction bell of the model is way above the minimum criteria of 6×10^4 as set by ANSI/HI [1]. This ensures that any scale effects will be minimized and that the flow in the model will be as turbulent as that of the prototype.

From the results of the physical model test, it can be expected that the selected forebay length for the numerical model is sufficient in providing a stable velocity gradient from the inlet of the domain. For pump intake structures, this stable approach flow is important in order to prevent any unnecessary turbulence, which could influence any free surface or subsurface vortex formation downstream of the forebay.

To solve the rotor-stator interaction problem in the domain, the multiple reference frame (MRF) scheme is applied to the domain. The impeller region is assigned to a moving reference frame, while the rest of the region is assigned to a stationary frame. Calculations were carried out under multiphase VOF transient conditions with water at 25 °C as the secondary phase and air as the primary phase. Turbulence was modelled using the k-ω shear stress transport (SST). SST k-ω had been found to be suitable for numerical modelling of free-surface vortices [16,17]. This turbulence model exhibits better performance in predicting flows at walls and adverse pressure gradients as compared to other eddy-viscosity models [18]. The transient formulation is second-order implicit. The converged solution from a steady-state simulation was used for the initial conditions.

5. Results and Discussions

Isosurface plots and surface streamline plots for the first three cases (Figure 14a–c) show no free-surface vortex formation stronger than type 2 (ARL's classification [19]) near the pump compartment. Cases 1 and 2 show type 1 (surface swirl) to type 2 (surface dimple) on the free surface of the water. As discussed earlier, the semicircular arcs added to the backwall magnified the surface swirl and directed the swirl nearer the pump column. If the semicircular arcs were constructed from two corner fillets instead of the current design, these surface swirls would have developed into full type 3 or type 4 dye core vortices. Instead, the sharp intersection of the two semicircular arcs served as a vertical backwall splitter that prevented any surface swirl from organizing and developing to a much stronger surface vortex.

For Case 1, strong type 2 (dye core) vortices attached to the sidewall and the sump floor were observed as the fluid began to separate from the wall and flow towards the bellmouth (Figure 14a). In contrast, for Case 2 and Case 3, there were no flow separation and the fluid remain attached to the fillets and splitters. This indicates that aside from using the recommended sump dimensions as published in various pump standards, it is still more advisable to check and minimize any sharp corners where eddies or dead zones would occur. The floor cone was effective in eliminating the floor vortex, but care should be given to not restrict the area under the bell too much that it would result in the pump experiencing additional head loss.

Although formed structure intakes (Case 4) are generally less sensitive to adverse inlet flow conditions, poorly designed FSI structures may still suffer from surface vortices in the forebay and subsurface vortices near the FSI channel. Particularly for Case 4, swirl and recirculation in the forebay area can be observed from the streamlines in Figure 14d. Large surface swirl/depression can be observed on the free surface area of the forebay. These are weak vortices whose cores do not reach the pump. Such phenomena can be avoided by the installation of curtain walls upstream. For FSI, however, such vortices, if left unchecked, can cause catastrophic damage to the structure. Noise, cavitation, and structural damage are among the most common results of such phenomena for these types of structures. Additionally, air-entrained pockets were observed, causing dispersed air bubbles to constantly reach the intake bellmouth. This can be attributed to the same swirling action on the surface of the forebay.

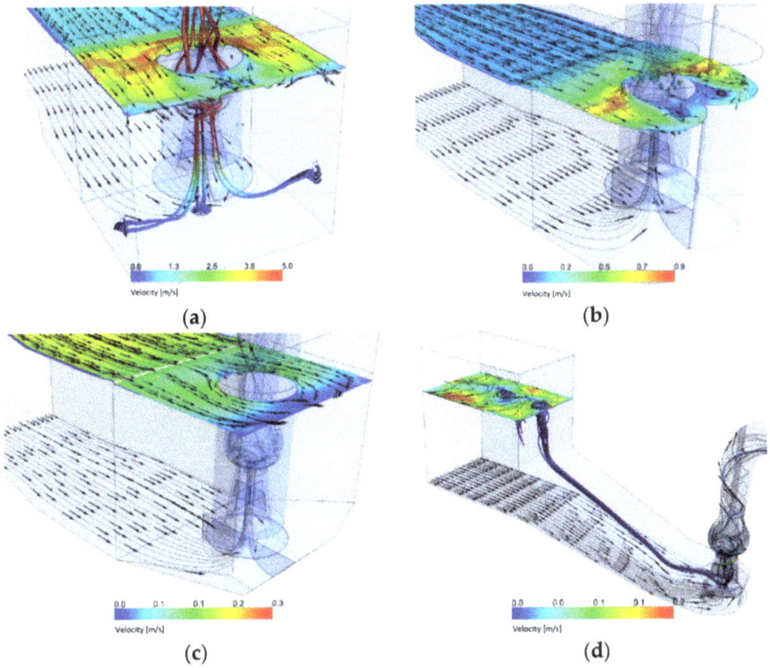

Figure 14. Plot of volume fraction isosurface with velocity contour overlay, approach flow streamlines, and vortex streamline: (a) case 1, (b) case 2, (c) case 3, and (d) case 4.

The CFD results (Figure 15) showed that for Cases 1 to 3, the swirl angles at the impeller eye are well within the 5° acceptance criteria for the short-term 30-s maximum set by ANSI/HI [1]. For these two cases, the average swirl angle was around 3.5°, while Case 2 showed a slightly higher average swirl of 3.8°. This negligible difference may be attributed to the slight turbulence present at the backwall for Case 2 caused by the semicircular arcs. Case 4 showed a very high swirl angle with an average of 10.1, which exceeds the maximum allowable value. This is in part due to the high fluid rotation as water enters the bell. The protrusions made by the floor cone (Figure 12b) into the bellmouth caused the fluid to rotate excessively as it enters the pump. This phenomenon caused additional issues such as uneven velocity distribution (Figure 16d), which usually results in unbalanced loading at the impeller shaft. Conversely, Figure 16a–c shows uniform velocity distribution for the first three cases.

Figure 17 shows point velocities as measured from the intake bellmouth throat. Cases 1–3 all show acceptable conditions based on ANSI/HI [1] criteria. The improvement in the velocity variation can be clearly seen across the three cases, with Case 3 having the most stable plot with a standard deviation of 0.1. On the other hand, Case 4 showed a large variation in point velocity again, proving that the flow is highly turbulent as the fluid enters the bellmouth.

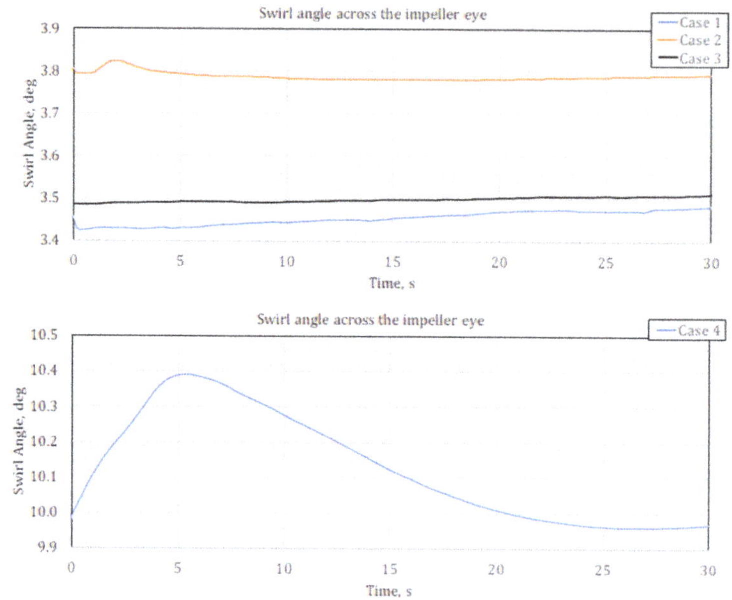

Figure 15. Swirl angle across the impeller.

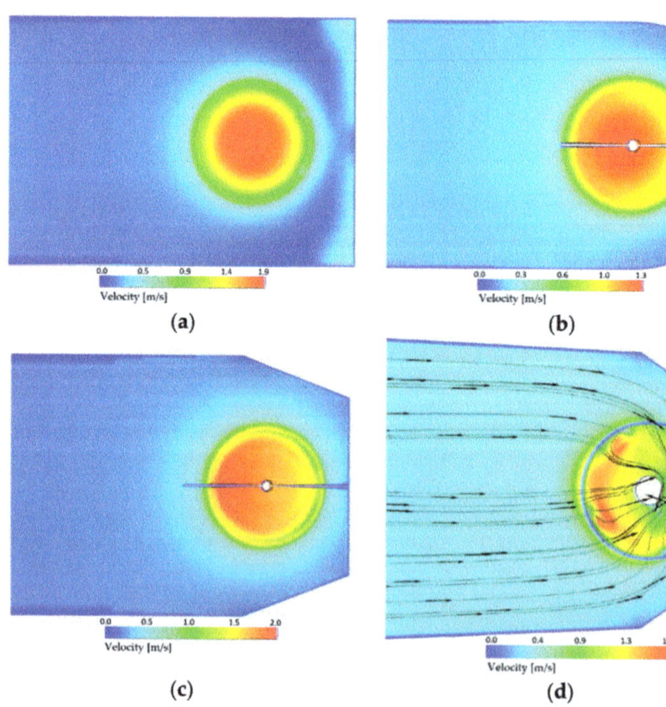

Figure 16. Velocity distribution at the suction bellmouth for all test cases: (**a**) case 1, (**b**) case 2, (**c**) case 3, (**d**) case 4. Notice the uneven velocity distribution for case 4.

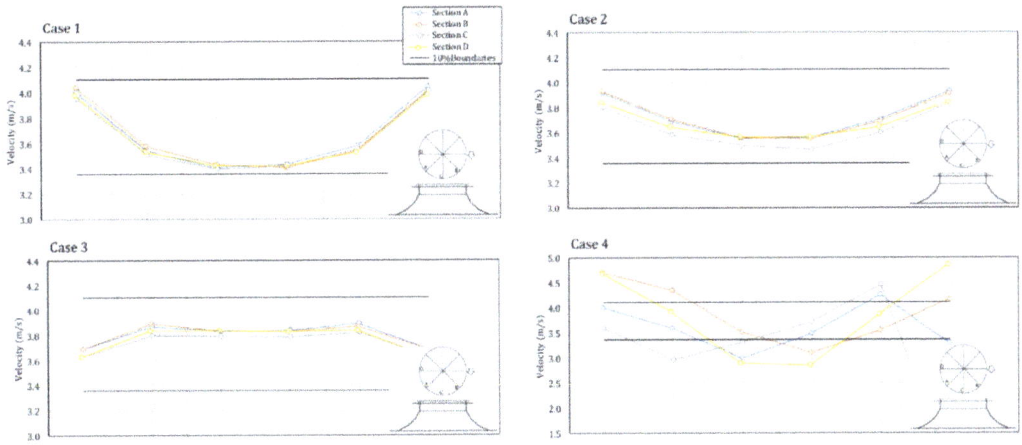

Figure 17. Point velocities at bellmouth throat.

An evaluation of the pump's performance for all the three cases, as shown in Table 4, further verifies that Case 3 presents the best hydraulic condition providing about 1% efficiency gain from the baseline geometry. Additionally, the table shows that although Case 2 was able to prevent the formation of subsurface vortices, the pump's efficiency suffered by 1% from baseline. This is due to the condition that swirl is higher in Case 2 as compared to the other two cases. This high swirl, as mentioned previously, may be attributed to the circulation at the back of the pump.

Table 4. Pump performance as installed in different sump geometries.

Case No.	Shaft Power (kW)	Pump Head (m)	Efficiency (%)
1	2165	37.4	78.6
2	2219	37.7	77.5
3	2196	38.2	79.3
4	2164	64.2	-

For the FSI model, CFD results showed a very high head. This is caused by the recirculation created by the excessive swirl at the intake of the pump, as shown in Figure 18. This increase in the head is primarily caused by having the same (recirculating) volume of fluid being acted upon by the impeller multiple times, thereby increasing fluid energy. This is desirable for regenerative pumps but not in mixed-flow pumps since, for Case 4, the pump would suffer severe cavitation.

Figure 18. Suction and discharge recirculation caused by excessive swirl at the pump inlet bellmouth.

6. Conclusions

This paper presented numerical solutions to investigate the hydraulic performance of mixed flow submersible pumps under varying pump intake geometry and conditions. The focus of the paper is to develop a suitable method for setting up a numerical model and simulation procedure that could be utilized as an engineering tool during product development. To achieve this, the analysis should provide a balance between numerical accuracy and the efficient use of computational resources. In this study, a numerical model of the pump was used to establish the pump performance characteristics. Head, flow, and efficiency were compared to data provided by the pump manufacturer. Afterwards, CFD simulations of a full-scale pump and intake structure were conducted to predict the formation of free-surface and submerged vortices during operation. Several cases were presented in order to compare the effect of various geometries and vortex splitters on pump performance. The models were created in full-scale so as to avoid the scaling effects normally encountered during reduced-scale physical model tests. The results showed suitable accuracy in predicting the flow dynamics, intake structure performance, and pump characteristic curves.

From this study, it can be implied that significant room for improvements in terms of increasing accuracy of the results can easily be achieved through the inclusion of disk friction losses, mechanical losses, leakage losses, and the tip clearance effect, which were previously neglected in this study. The grid independence study made in this paper as conducted at only one flow condition, for this case at 90% of Q_{BEP}, is proven to be sufficient since the CFD results for the pump's characteristic curve were able to closely match the supplied performance curve. Conversely, if calculations show significant deviation from expected data, then it is strongly advised that the user conduct additional grid independence studies, particularly at flows where unsteadiness and circulation are apparent.

Additionally, note that although the MRF approach has found use in various industrial applications, this approach, being the simplest, is a steady-state approximation in which individual cell zones move at different rotational and/or translational speeds [20]. For purposes of product development and preliminary design, the MRF approach can provide reasonable insight into the flow. For turbomachinery applications with strong rotor-stator interaction or where large transient effects are expected, a more robust solution such as the sliding mesh model should be considered.

On the other hand, it was observed that CFD could provide results within a shorter period of time with a lower financial impact. Operating parameters, such as suction pressure, rotational speed, and fluid density, can easily be modified to verify pump performance under such loads. Additionally, through CFD, the ease of conducting geometric and parametric revisions, such as the number of blades, impeller attack angle, and impeller diameter on the numerical model, can provide pump designers with enough insight on the effect of these parameters, aiding in developing more efficient and reliable pumps. Various factors can still be developed to obtain more accurate results, such as consideration of cavitation growth in unsteady flow to analyze noise, vibration, and flow instabilities. However, for the purposes of this study, CFD may prove to be a viable option in developing optimum pumps, thereby reducing the need for extensive physical model experiments.

Author Contributions: Conceptualization, V.M.A.; methodology, V.M.A. and L.A.M.D.; software, V.M.A., B.E.A. and L.A.M.D.; validation, V.M.A. and L.A.M.D.; formal analysis, V.M.A. and L.A.M.D.; investigation, V.M.A.; resources, V.M.A.; data curation, B.E.A., P.L.R. and J.G.T.R.; writing—original draft preparation, V.M.A. and L.A.M.D.; writing—review and editing, B.E.A., P.L.R. and J.G.T.R.; visualization, V.M.A. and L.A.M.D.; supervision, B.E.A., P.L.R., J.G.T.R. and L.A.M.D.; funding acquisition, L.A.M.D. All authors have read and agreed to the published version of the manuscript.

Funding: This research was funded by the Department of Science and Technology (DOST) through the Engineering Research and Development for Technology (ERDT) Program—Local Graduate Scholarships. The APC was funded by DOST-ERDT Faculty Research Dissemination Grant.

Acknowledgments: The authors would like to thank Hitachi Plant Technologies, Ltd., Philippine Branch Office for granting permission to use the pump physical model test data and pump performance data presented as well as the meshing software used in this paper.

Conflicts of Interest: The authors declare no conflict of interest.

References

1. American National Standard. *Pump Intake Design*; ANSI/HI 9.8-1998; Hydraulic Institute: Parsippany, NJ, USA, 2012.
2. Constantinescu, G.; Patel, V. Numerical Model for Simulation of Pump-Intake Flow and Vortices. *J. Hydraul. Eng.* **1998**, *124*, 123–134. [CrossRef]
3. Rajendran, V.P.; Constantinescu, G.S.; Patel, V.C. Experimental Validation of Numerical Model of Flow in Pump-Intake Bays. *J. Hydraul. Eng.* **1999**, *125*, 1119–1125. [CrossRef]
4. Okamura, T.; Kamemoto, K.; Matsui, J. CFD Prediction and model Experiment on Suction Vortices in pump sump. In Proceedings of the 9th Asian Conference on Fluid Machinery, Jeju, Korea, 16–19 October 2007; pp. 1–10.
5. Rajendran, V.P.; Patel, V.C. *Patel, Characterization of Vortices in Model Pump-Bay Using Particle Image Velocimetry*; IIHR Technical Report No. 396; University of Iowa: Iowa City, IA, USA, 1998.
6. Lai, Y.G.; Weber, L.J.; Patel, V.C. A non-hydrostatic three-dimensional numerical model for hydraulic flow simulation-Part II: Validation and application. *J. Hydraul. Eng.* **2003**, *129*, 206–214. [CrossRef]
7. Nagaha, T.; Sato, T.; Kawabata, S.; Okamura, T. Effect of submerged vortex cavitation in pump suction intakes on mixed flow pump impeller. *Turbomach. Soc. Jpn.* **2002**, *30*, 70–75. (In Japanese)
8. Hitachi Plant Technologies. *PBO, Internal Technical Report CWP No. MR-110519-15-114-001*; 2012; Unpublished.
9. Li, X.-H.; Zhang, S.-J.; Zhu, B.-L.; Hu, Q.-B. The study of the k-ε turbulence model for numerical simulation of centrifugal pump. In Proceedings of the 7th International Conference on Computer-Aided Industrial Design and Conceptual Design, Hangzhou, China, 17–19 November 2006; pp. 1–5.
10. American National Standard. *Rotodynamic Vertical Pumps of Radial, Mixed, and Axial Flow Types*; ANSI/HI 2.3-2013; Hydraulic Institute: Parsippany, NJ, USA, 2013.
11. Alpan, K.; Peng, W.W. Suction reverse flow in an axial-flow pump. *J. Fluids Eng. ASME Trans.* **1991**, *113*, 90–97. [CrossRef]
12. Ingersoll-Rand Company. *Test Standards for Modelling Inlets to Pumps*; Ingersoll-Rand Company: Phillipsburg, NJ, USA, 1991.
13. U.S. Army Corps of Engineers. *Mechanical and Electrical Design of Pumping Stations EM 1110-2-3105*; U.S. Army Corps of Engineers: Washington, DC, USA, 1999.
14. Arocena, V.M.; Abuan, B.E.; Reyes, J.G.T.; Rodgers, P.L.; Danao, L.A.M. Reduction of Entrained Vortices in Submersible Pump Suction Lines Using Numerical Simulations. *Energies* **2020**, *13*, 6136. [CrossRef]
15. Hecker, G.E. Model-prototype comparison of free surface vortices. *J. Hydraul. Div.* **1981**, *107*, 1243–1259. [CrossRef]
16. Qian, Z.; Wu, P.; Guo, Z.; Huai, W.-X. Numerical simulation of air entrainment and suppression in pump. *Sci. China Technol. Sci.* **2016**, *59*, 1847–1855. [CrossRef]
17. Ahn, S.-H.; Xiao, Y.; Wang, Z.; Zhou, X.; Luo, Y. Numerical prediction on the effect of free surface vortex on intake flow characteristics for tidal power stations. *Renew. Energy* **2017**, *101*, 617–628. [CrossRef]
18. Menter, F. Review of the shear-stress transport turbulence model experience from an industrial perspective. *Int. J. Comput. Fluid Dyn.* **2009**, *23*, 305–316. [CrossRef]
19. Knauss, J. Swirling flow problems at intakes. In *IAHR Hydraulic Structures Design Manual*; A.A. Balkema: Rotterdam, The Netherlands, 1987; Volume 1.
20. ANSYS, Inc. *ANSYS Fluent Theory Guide*; ANSYS, Inc.: Canonsburg, PA, USA, 2018.

energies

Article

Prediction of Abrasive and Impact Wear Due to Multi-Shaped Particles in a Centrifugal Pump via CFD-DEM Coupling Method

Cheng Tang [1], You-Chao Yang [2], Peng-Zhan Liu [1] and Youn-Jea Kim [3,*]

1 Graduate School of Mechanical Engineering, Sungkyunkwan University, Suwon 16419, Korea; tangcheng@skku.edu (C.T.); pengzhan@skku.edu (P.-Z.L.)
2 Chongqing Pump Industry Co., Ltd., Chongqing 400033, China; lzlgdxyyc@163.com
3 School of Mechanical Engineering, Sungkyunkwan University, Suwon 16419, Korea
* Correspondence: yjkim@skku.edu

Abstract: Since solid particles suspended in the fluid can cause wear in centrifugal pumps, intensive attention has been focused on the numerical prediction for the wear of flow parts in centrifugal pumps. However, most numerical studies have focused on only one wear model and a sphere particle model. The impact of particle shape on the wear of flow parts in centrifugal pumps is under-studied, particularly considering abrasive and impact wear simultaneously. In this work, the Computational Fluid Dynamics (CFD)-Discrete Element Method (DEM) coupling method with an abrasive and impact wear prediction model was adopted to study the wear characteristics of a centrifugal pump. Moreover, four regular polyhedron particles and a sphere particle with the same equivalent diameter but different sphericity were mainly analyzed. The results demonstrate that more particles move closer to the blade pressure side in the impeller passage, and particles tend to cluster in specific areas within the volute as sphericity increases. The volute suffers the principal wear erosion no matter what the shapes of particles and wear model are. Both the impact and abrasive wear within the impeller occur primarily on the blade leading edge. The pump's overall impact wear rate decreases first and then increases with particle sphericity rising, while the pump's overall abrasive wear rate grows steadily.

Keywords: centrifugal pump; CFD-DEM coupling method; abrasive and impact wear; particle shape

Citation: Tang, C.; Yang, Y.-C.; Liu, P.-Z.; Kim, Y.-J. Prediction of Abrasive and Impact Wear Due to Multi-Shaped Particles in a Centrifugal Pump via CFD-DEM Coupling Method. *Energies* **2021**, *14*, 2391. https://doi.org/10.3390/en14092391

Academic Editors: Michele Pinelli and Elia Distaso

Received: 8 March 2021
Accepted: 21 April 2021
Published: 23 April 2021

1. Introduction

The centrifugal pump has been widely used as essential and effective equipment in the hydraulic transportation of solids. In the pumping process, the solid particles carried by the fluid stream would cause considerable erosion due to abrasive and impact wear on the flow parts' surface [1], which determines the operational life and the reliability of slurry pumps. Theoretically, abrasive wear occurs when particles slide against equipment surfaces, while impact wear occurs when particles' impact equipment surfaces. For many years, the results of the experiment and numerical simulation have been carried out to predict and reveal the erosion due to wear in centrifugal pumps. The wear distribution and rates were assessed in a centrifugal pump impeller through a wear test rig by Tao et al. [2]. They concluded that the areas with severe wear in impeller passages are located at the blade leading edge, hub surface, and pressure side. Ahmad et al. [3] developed a computer program with an impact wear model to study the wear of a centrifugal pump impeller and validate the results by the paint wear experiment. They pointed out that the maximum wear happened on the blade leading edge and hub. Lei et al. [4] numerically analyzed the wear on blade surfaces by a Eulerian-Lagrangian method with an impact wear model. Results showed that the impact wear of the blade pressure side is more severe than that of the suction side. Noon et al. [5] utilized a similar numerical method to compare the numerical wear with experimental data in a centrifugal pump and demonstrated that the volute would be the most affected regions of impact wear damage. In addition, the effects of particle size, slurry concentration, and particle velocity on the wear rate were investigated. Unlike the numerical methods above, Huang et al. [6] applied a Computational Fluid Dynamics (CFD)-Discrete Element Method (DEM) coupling method with an abrasive wear model to study the wear in a centrifugal pump quantitatively. Results claimed that the number of particle collisions with flow parts mainly proportionally followed the wear loss on each part. Generally, these numerical methods treated the particles as a sphere and solely adopted one wear model to obtain the wear. However, the particle shapes have an essential influence on the wear characteristics [7]. Moreover, no previous study has investigated the abrasive and impact wear simultaneously in the centrifugal pump, which cannot provide a reliable prediction for the wear law. Therefore, this paper aims to compare the abrasive and impact wear characteristics of flow parts in a centrifugal pump and explore the relationship between wear and particle shape.

As an alternative, the DEM is a competent numerical approach to analyze particle systems. Moreover, Tsuji et al. [8] and Kafui et al. [9] developed a CFD-DEM coupling method in order to enhance the numerical precision in calculating multiphase flow. More importantly, with increasing attention on particle shape relevance in collision behavior and flow interactions [10–12], the multi-shaped particle modeling has been developed into various forms in the DEM. On the other hand, the abrasive and impact wear, which are two broad categories of wear erosion due to particle interactions, can be numerically achievable in the DEM. Hence, the results of abrasive and impact wear can be obtained simultaneously and analyzed separately for comparison.

According to these above, a comparative analysis of abrasive and impact wear on flow parts of a centrifugal pump was examined for the first time via the CFD-DEM coupling method, which takes full account of the particle shapes. A DEM code analyzed the particle distribution and motion combined with two wear erosion models based on a commercial platform STAR-CCM+. Furthermore, four regular polyhedral particles (tetrahedron, hexahedron, octahedron, and dodecahedron) and one sphere particle were mainly modeled to investigate the impact of particle shapes on the wear rate. These results would provide new insights into the wear research in the centrifugal pump, which may benefit optimization design for a centrifugal pump.

2. Methodology

2.1. Geometric Model and Mesh

In this paper, a centrifugal pump geometric model involving an inlet pipe, impeller, and volute was built for calculating the wear, as shown in Figure 1a. For validating the numerical method, the computational domain holds the same geometry as the original pump provided by one of the pump companies in China. The design flow rate of the pump is 68.4 m^3/h with a head of 8.3 m, under a 1450 rpm rotation speed. The diameters of the pump inlet, impeller, and pump outlet are 115 mm, 190 mm, and 100 mm, respectively. The polyhedron meshes of the computational domain and five prism layers at the wall boundaries (see Figure 1b) were created. To ensure the slight influence of mesh number on the results, the pump head was employed to test the mesh independence, as shown in Table 1. It is apparent from this table that there is a slight change in the pump head when the mesh number is bigger than 1,255,663. Thus, the mesh number was set as 1,255,663 for subsequent simulations.

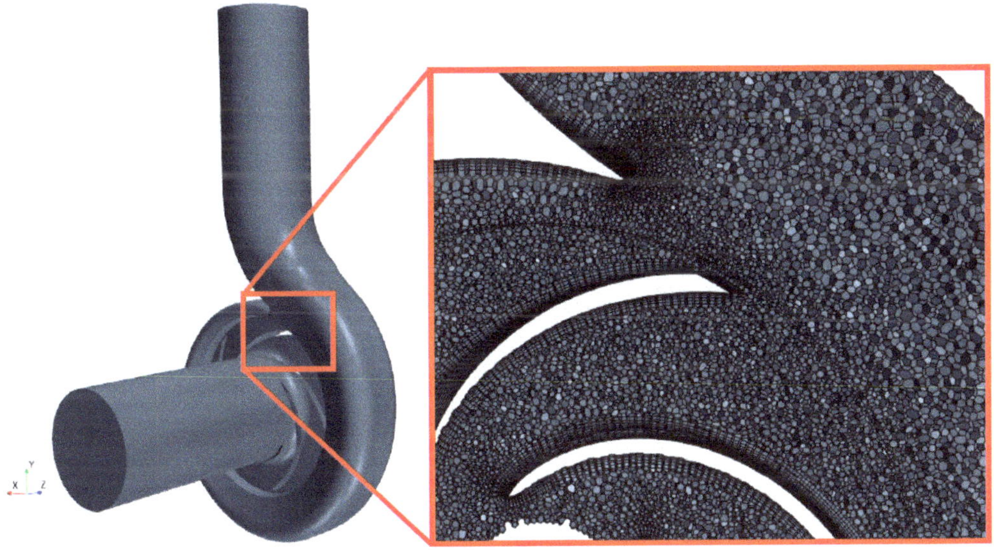

(**a**) (**b**)

Figure 1. (**a**) Geometric model. (**b**) Polyhedron meshes and prism layers.

Table 1. Mesh independence analysis.

Mesh Number	Pump Head (m)	Deviation (%)
650,584	8.47	
833,197	8.79	3.83
1,042,510	8.87	0.92
1,255,663	8.89	0.23
1,349,142	8.90	0.15

2.2. Governing Equations of Solid-Liquid

In the CFD-DEM modeling of this work, the liquid phase is assumed to be a macro-scaled continuum, and the governing equations of the liquid phase according to the mass and momentum conservation are shown by the following equation.

Mass conservation equation:

$$\frac{\partial}{\partial t}\left(\alpha_f \rho_f\right) + \frac{\partial}{\partial x_j}\left(\alpha_f \rho_f u_j\right) = 0, \quad (1)$$

Momentum conservation equation:

$$\frac{\partial}{\partial t}\left(\alpha_f \rho_f u_i\right) + \frac{\partial}{\partial x_j}\left(\alpha_f \rho_f u_i u_j\right) = -\frac{\partial p}{\partial x_i} + \frac{\partial}{\partial x_j}\left[\alpha_f \mu_{eff}\left(\frac{\partial u_i}{\partial x_j} + \frac{\partial u_j}{\partial x_i}\right)\right] + \alpha_f \rho_f \mathbf{g} + \mathbf{F}_{drag} + \mathbf{F}_{sl} + \mathbf{F}_{pg}, \quad (2)$$

where scalars t, ρ_f, x, u, p, and μ_{eff} are the time, fluid density, coordinates, fluid velocity, fluid pressure, and fluid effective viscosity, respectively. The vectors $\mathbf{g}$, $\mathbf{F}_{drag}$ $\mathbf{F}_{sl}$, and $\mathbf{F}_{pg}$ denote the gravity and particles-liquid interaction forces, namely the drag force, Saffman's lift force, and pressure gradient force. Furthermore, α_f is the fluid volume fraction in each cell, which is also called porosity or void fraction. All the particle centers are assumed to be located in a selected computational cell, and α_f can be estimated through the equation:

$$\alpha_f = 1 - \sum_{i=1}^{n} V_{p,i} / V_{cell}, \tag{3}$$

where n is the number of particles within the selected cell and $V_{p,i}$ denotes the volume of particle i within the computational cell. V_{cell} is the cell volume.

The translational motion of the solid particles is governed by Newton's second law, while the rotational motion of the particles is controlled by Euler's second law in DEM. The main governing equations for particles in the pump are shown by the following.

$$m\frac{d\mathbf{v}}{dt} = m\mathbf{g} + \sum \mathbf{F}_c + \mathbf{F}_{drag} + \mathbf{F}_{sl} + \mathbf{F}_{pg}, \tag{4}$$

$$\mathbf{I}\frac{d\boldsymbol{\omega}}{dt} = \sum \mathbf{T}_c + \mathbf{T}_f, \tag{5}$$

where the scalar m is the particle mass, while the vector $\mathbf{I}$ denotes the moment of inertia of the particle. The $d\mathbf{v}/dt$ and $d\boldsymbol{\omega}/dt$ are translational and angular acceleration of the particle. The vector $\mathbf{F}_c$ denotes the collisional forces of a particle with other particles or wall boundaries. The vector $\mathbf{T}_c$ is the sum of contact torques produced by particle-particle and particle-wall collision, and $\mathbf{T}_f$ denotes the particle torque produced by the surrounding liquid.

2.3. Particle and Wear Model

One sphere and four regular polyhedrons (see Figure 2a–d, respectively), namely the tetrahedron, hexahedron, octahedron, and dodecahedron, were mainly modeled to represent different-shaped particles in this study. The polyhedron particles were approximated by fixing a number of spheres together, which can overlap with each other. In addition, these spheres could hold various sizes and do not separate during contacts. The properties of composited polyhedron particles are listed in Table 2, where ξ denotes the particle sphericity, D_e denotes the equivalent spherical diameter, and ρ_p denotes the particle density. Sphericity is the most commonly used measure accounting for the shape of a non-spherical particle. It can be defined as:

$$\zeta = \frac{Ss}{Sp}, \tag{6}$$

where S_s and S_p, respectively, denote the superficial areas of a sphere and a non-spherical particle. Furthermore, the volume concentration of these particles above is set as 2%, and particles are produced in the vicinity of the pump inlet with arbitrary positions and orientations. The Hertz-Mindlin contact model [13] and a soft-sphere model [14] are applied to model the collisional forces of a particle with other particles or wall boundaries in DEM. Table 3 summarizes the coefficients of interactions included in the applied models. Moreover, the abrasive wear in this work is modeled using the Archard wear model [15], while the erosive wear is modeled using the Oka wear model [16,17], which has been widely adopted.

Table 2. Particle configurations.

	Tetrahedron	Hexahedron	Octahedron	Dodecahedron	Sphere
ξ	0.709	0.802	0.829	0.923	1
D_e (mm)	1	1	1	1	1
ρ_p (kg/m^3)	2600	2600	2600	2600	2600

Table 3. Collison configurations.

Collision Coefficient	Particle-Particle	Particle-Wall
Restitution	0.5	0.7
Static friction	0.3	0.15
Rolling friction	0.01	0.01

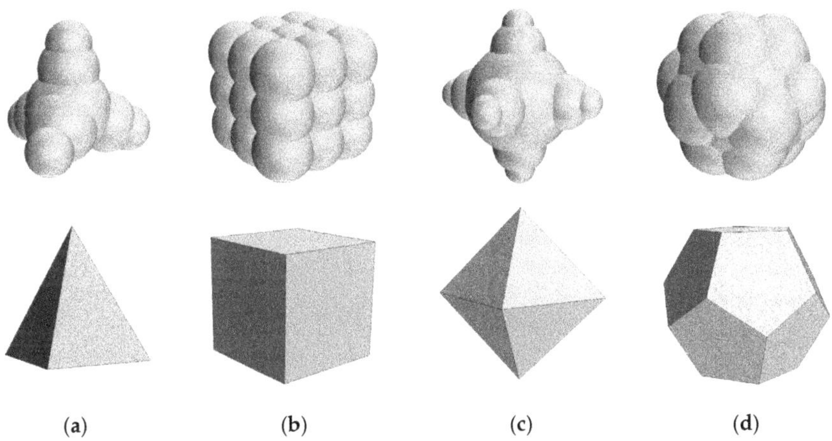

Figure 2. Particle DEM and geometrical models: (**a**) tetrahedron, (**b**) hexahedron, (**c**) octahedron, (**d**) dodecahedron.

2.4. Liquid Phase Setup

In this work, one major consideration in the CFD-DEM coupling method is the calculation time-cost to obtain an unsteady flow field within the entire pump. In addition, the realizable k-ε model has strengths in rotating and boundary-free shear flows with considerable computational efficiency and stability [18]. Therefore, the transient Reynolds averaged Navier-Stokes (RANS) equation with the realized two-layer k-ε turbulence model was solved, according to the SIMPLEC algorithm in STAR-CCM+ software. The liquid was defined as incompressible water, and the dynamic viscosity and density are 8.887×10^{-3} Pa·s and 998 kg/m^3, respectively. Moreover, the pump walls were determined as a no-slip boundary condition, and the two-Layer all y+ wall treatment was adopted as a wall function. The pump inlet was defined as a velocity inlet with a constant profile of 1.829 m/s, and the pump outlet was defined as a pressure outlet. The residual value of each variable for convergence was determined as 10^{-4}. In addition, the pump head fluctuations as a function of physical time were monitored in Figure 3. It is apparent that pump head fluctuations achieve stability with a regular vibration at about t = 0.24 s, which indicates a spatial convergence.

3. Results

3.1. Validation

As shown in Figure 4, in order to validate the CFD-DEM coupling, the numerical head and efficiency of the centrifugal pump as a function of liquid flow rate were compared with the experimental data from one of the pump companies in China [19]. The computational pump model kept the same parameters as that tested in the experiment. Overall, the numerical curves follow the tendency of experimental results with acceptable accuracy, which indicates the feasibility of the CFD-DEM coupling method. Moreover, the CFD-DEM coupling data are slightly greater than the experimental ones for all flow conditions. This discrepancy could be due to the neglect of mechanical and volume losses caused by bearings and seals in the simulation.

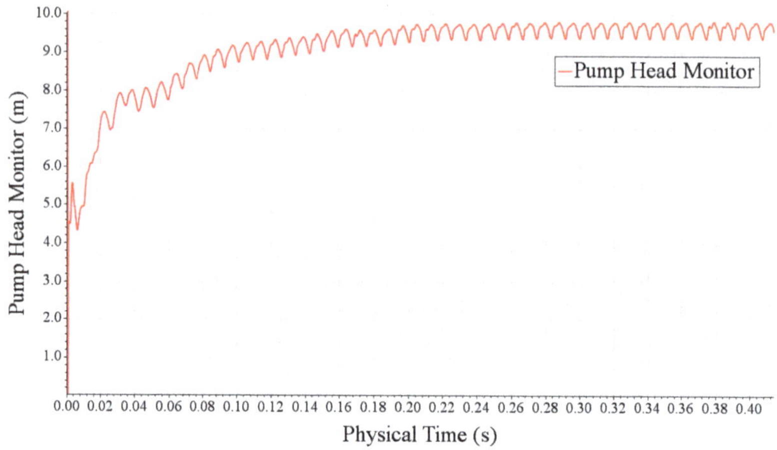

Figure 3. Variations in pump head for the sphere particle condition.

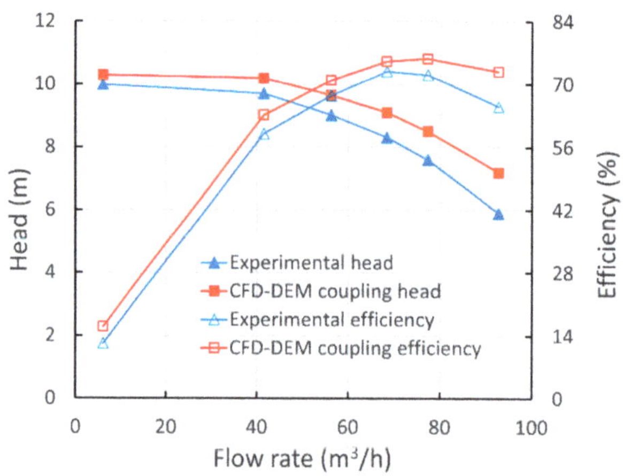

Figure 4. Comparison of the pump head and efficiency between the simulation and experimental data.

3.2. Particle Distribution

Figure 5 shows the trajectory and distribution of four regular polyhedron and spherical particles with the same equivalent diameter under the design flow rate. For a better view of particle distribution inside the impeller and volute, the inlet pipe is hidden in the current vision. Generally, most particles have a tendency to sustain a stable spiral trajectory following the volute. Moreover, particle trajectories basically correspond to the profile of the impeller blade. In the volute, most particles move along the wall surface and flow downstream to the pump outlet. In addition, particle concentrations are non-uniform within the impeller passage, with concentrations higher near the impeller inlet section and the blade pressure side.

However, there are a number of interesting differences in the distribution among different-shaped particles. Particles with low sphericity (see Figure 5a,b) are evenly distributed in the vicinity of each blade pressure side and volute wall. By contrast, with higher sphericity, particles tend to cluster in specific areas (see Figure 5c–e). One prominent feature is that the white spaces shown in the image within the volute are becoming apparent. Another feature is that more particles move closer to the blade pressure side (compare Figure 5a,e). Since the major changing factor is the shape of particles, an implication of these results is the possibility that the various drag fore and collision behavior of different-shaped particles may affect their trajectory and distribution in a centrifugal pump. Moreover, these results are associated with some wear results in the following discussion.

3.3. Wear of Flow Parts

For investigating the influence of particle shape on the wear characteristics in the centrifugal pump, the overall impact wear and abrasive wear rate of each flow part are monitored through time, as shown in Figures 6 and 7, respectively. Only the wear data on the hub, shroud, leading edge of the blade, pressure side of the blade, tongue, and volute have been detected. This means no wear erosion or the wear rate less than computational accuracy exists on the suction side of the blade and trailing edge of the blade. The most likely cause of no wear data there could be a relatively small particle flow rate.

It can be seen from the data in Figure 6 that the uncertainty of scattered data points from different flow parts varies significantly. In general, the overall impact wear rate of the hub has experienced minimal fluctuations for all particle conditions. A possible explanation for this might be that volute is a stationary part and has the largest surface area. However, the time-averaged data curves of all flow parts have achieved acceptable stability within computation time, indicating the convergence and subsequent wear analysis. Moreover, it can be found that the order of impact wear rate on the volute wall, leading edge, and hub remains the same for all particle conditions. On the other hand, the order of impact wear rate on the shroud, pressure side, and tongue varies with the particle shape. Furthermore, the same chronological order of curve starting points indicates the trajectory of particles colliding with flow parts. These results of impact wear rate are similar to those of abrasive wear rate shown in Figure 7.

Based on the above overall wear rate results, the relative impact, and abrasive wear amounts, a ratio of each part's wear rate to the overall pump impact wear rate, are presented for different practice conditions in Figures 8 and 9. What stands out in Figure 8 is that the impact wear amount of volute accounts for more than 65% of the overall pump impact wear amount per unit of time, which is generally consistent with the numerical results by Huang et al. [6]. Likewise, the abrasive wear amount of volute accounts for more than 71%. This finding suggests that no matter what the particles and wear model is, the volute suffers the principal wear erosion.

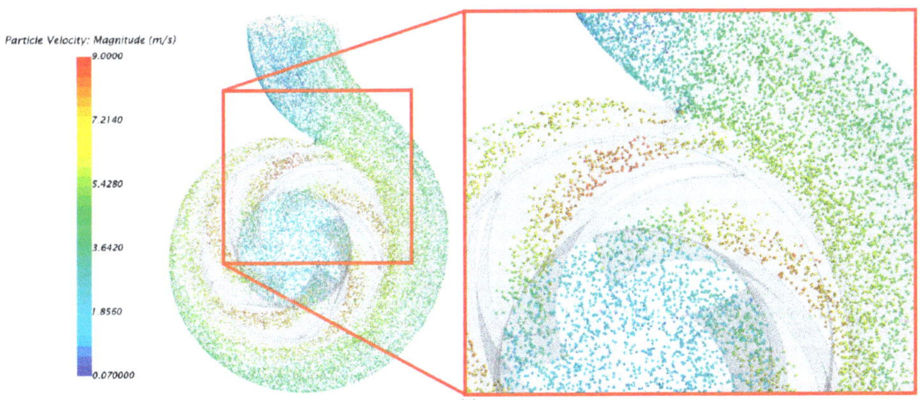

(**a**)

Figure 5. *Cont.*

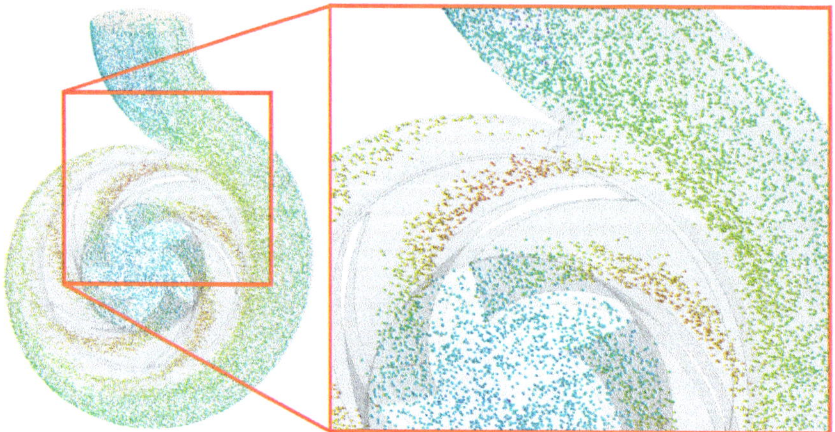

(b)

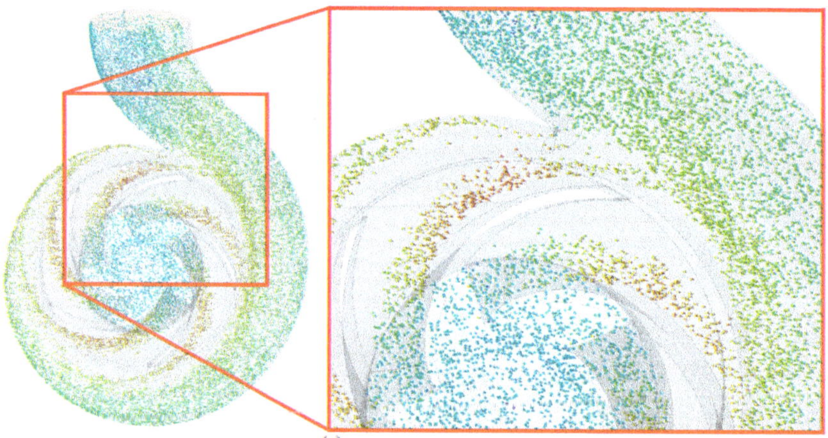

(c)

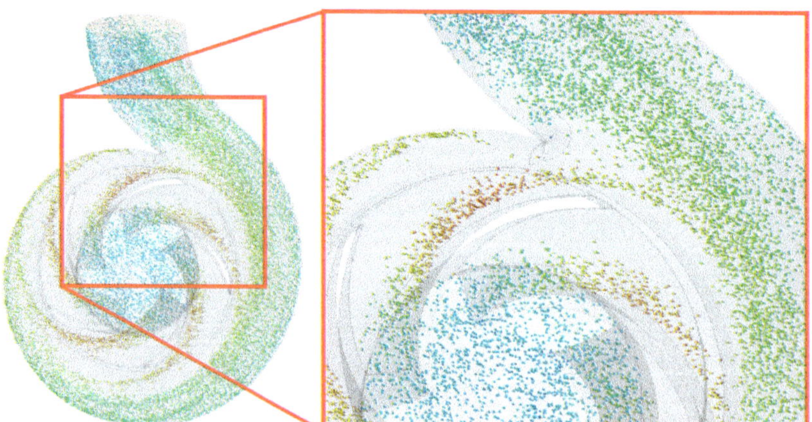

(d)

Figure 5. *Cont.*

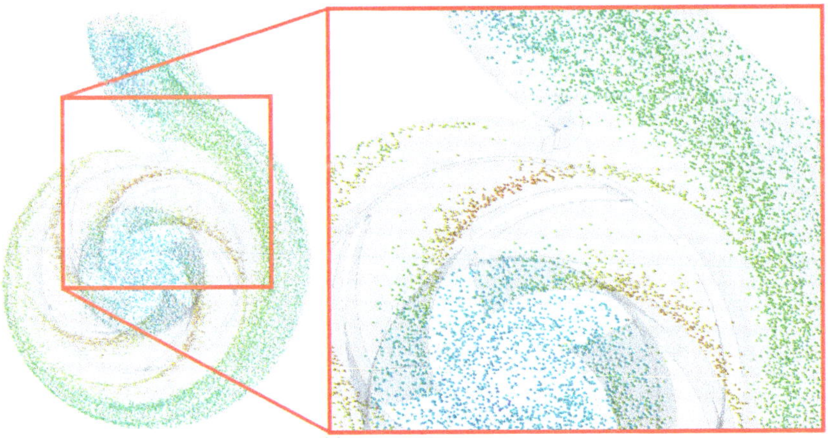

(e)

Figure 5. Trajectory and distribution of different particles: (**a**) tetrahedron, (**b**) hexahedron, (**c**) octahedron, (**d**) dodecahedron, and (**e**) sphere.

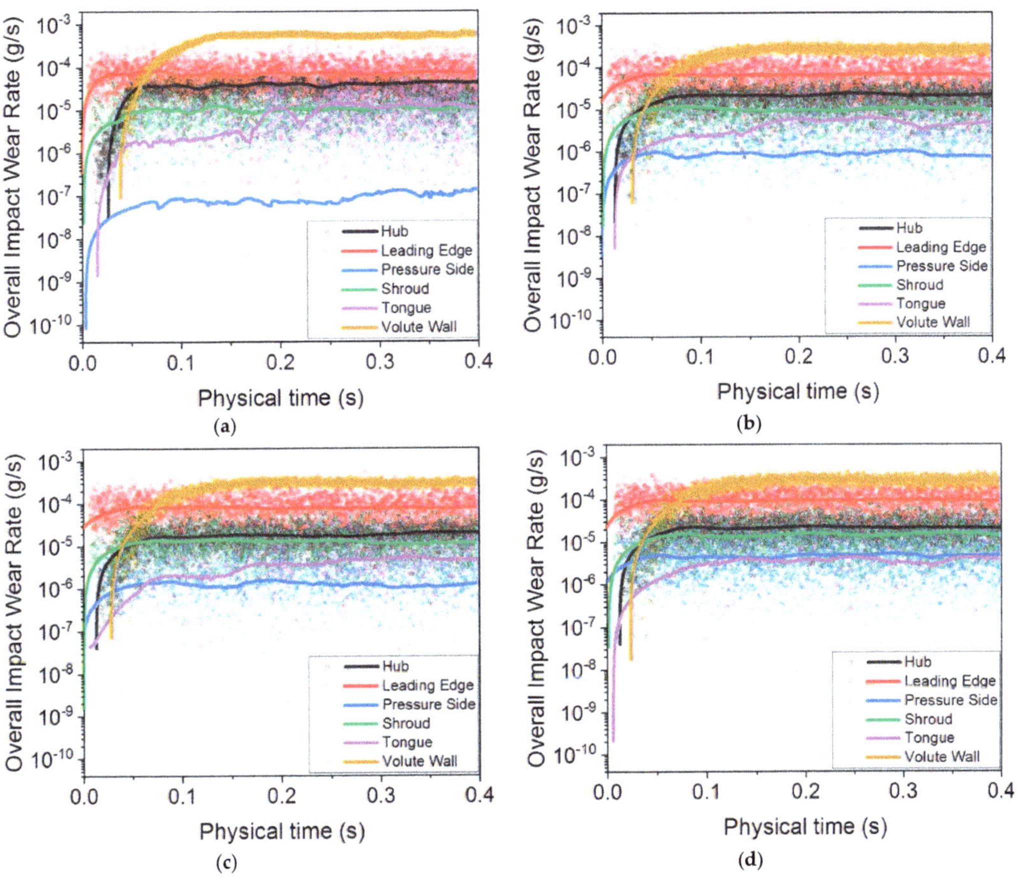

Figure 6. *Cont.*

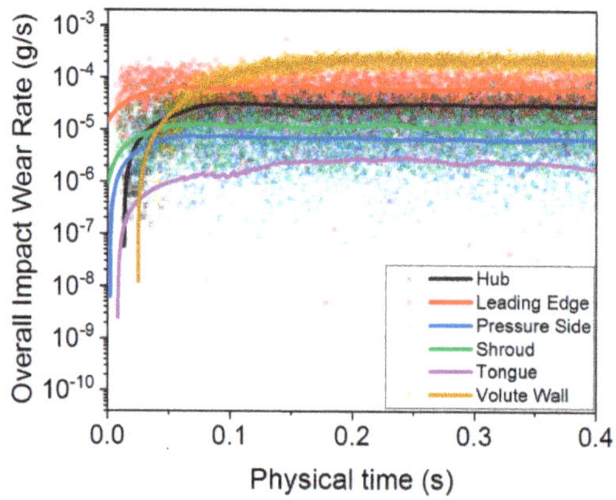

(e)

Figure 6. Overall impact wear rate of flow parts as a function of time: (**a**) tetrahedron, (**b**) hexahedron, (**c**) octahedron, (**d**) dodecahedron, and (**e**) sphere.

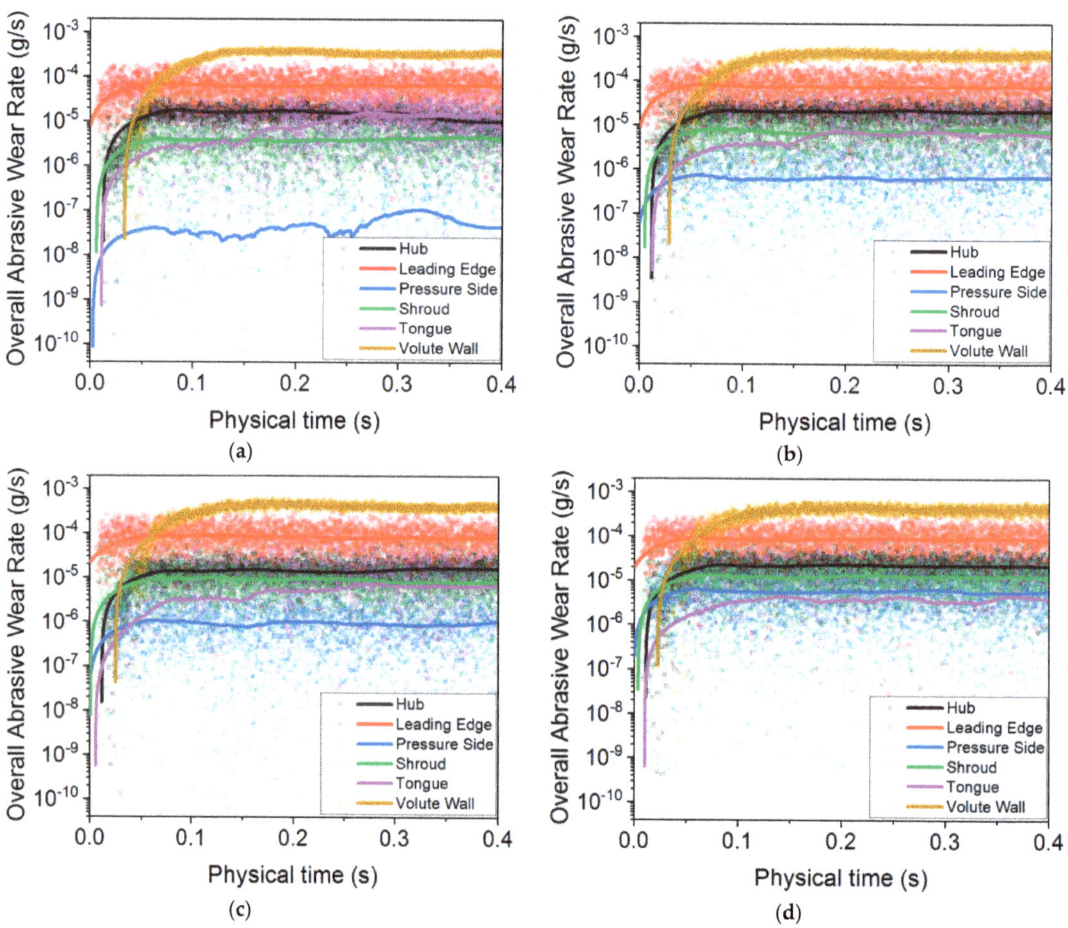

Figure 7. *Cont.*

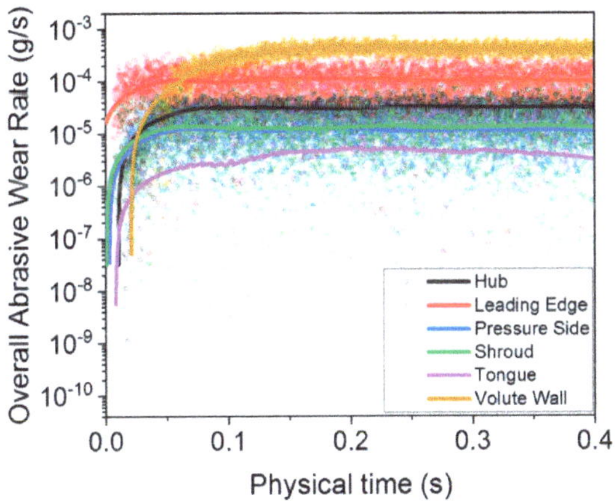

(e)

Figure 7. Overall abrasive wear rate of flow parts as a function of time: (**a**) tetrahedron, (**b**) hexahedron, (**c**) octahedron, (**d**) dodecahedron, and (**e**) sphere.

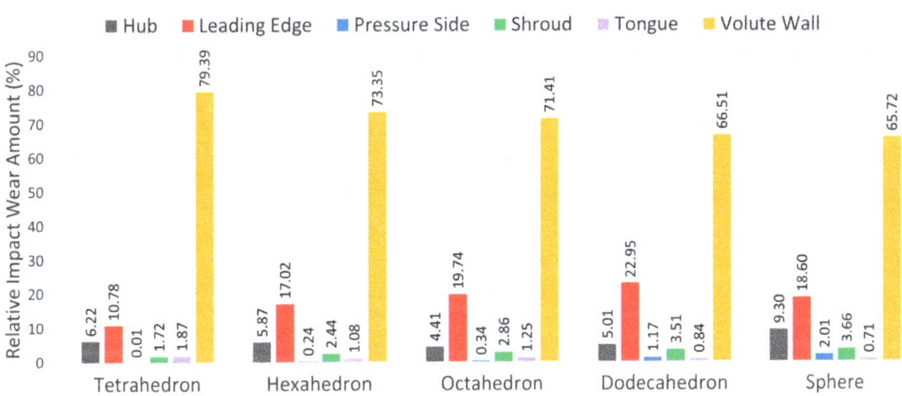

Figure 8. Relative impact wear amount of flow parts.

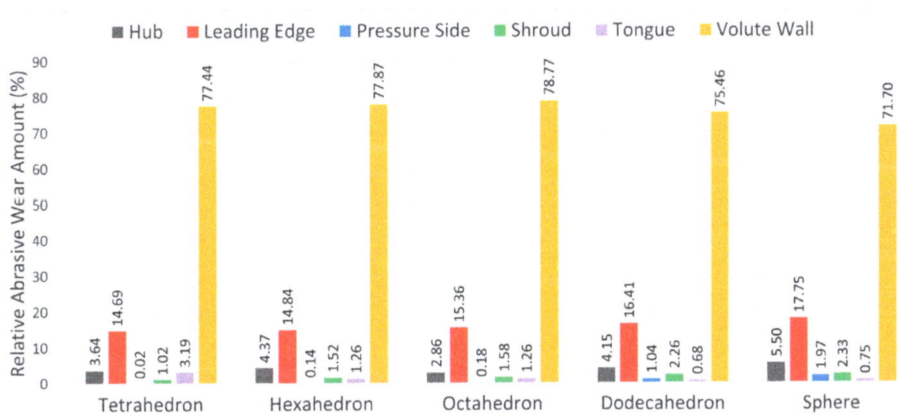

Figure 9. Relative abrasive wear amount of flow parts.

Both in Figures 8 and 9, it can be concluded that the wear in the impeller occurs mostly on the blade's leading edge and hub, which are consistent with the results of numerical simulation and paint wear test conducted by Ahmad et al. [3]. Moreover, the wear amount of the blade's leading edge is significantly larger than that of the blade pressure side. This result could be attributed to the narrow cross-sectional area of the impeller passage near the blade leading edge, which would cause the relatively high-particle number density and further lead to another collision here. Another possible explanation for this is that most particles tend to bounce toward the impeller passage after colliding with the leading edge and do not collide with the blade pressure side anymore. Similarly, the wear amount of the hub is considerably greater than that of the shroud. This result may be due to the larger inertia of particles, causing particle trajectory biased toward the hub instead of following the flow streamline.

Figures 10 and 11 quantitatively compare the overall impact and abrasive wear rate between different flow parts. In Figure 10, there is an evident climb in the impact wear rate of the pressure side as the particle sphericity increases (from the tetrahedron particle to the sphere particle). In the same way, the abrasive wear rate of the pressure side has experienced a dramatic rise with particle sphericity increasing in Figure 11. These similar trends are likely to be related to the results in Figure 5. An increasing number of particles move closer to the blade pressure side as particle sphericity rises, which results in a growing particle collision with the pressure side.

3.4. Effect of Particle Shape

The relative wear rate is defined as the pump's overall wear rate for polyhedron particles to the pump's overall impact wear rate for the spherical particle. Based on the numerical results of four regular polyhedron and another two polyhedron particles, the regression equations of relative impact wear rate and abrasive wear rate for the particle sphericity are modeled in Figure 12. The fitting parameters of the parabolic function are listed in the graphs. As can be seen from the first curve (see Figure 12a), the impact wear rate decreases first and then increases. There is a clear lowest impact wear rate when the sphericity is approximately equal to 0.86. This result may be because the impact wear is mainly affected by the impact velocity, impact angle, and other factors, not just particle sphericity. By contrast, the abrasive wear is principally subject to the particle rolling motion along the surface wall. Therefore, the relative abrasive wear rate grows steadily with increasing sphericity, as shown in Figure 12b. These findings could benefit the pump design by predicting the wear under multi-shaped particles.

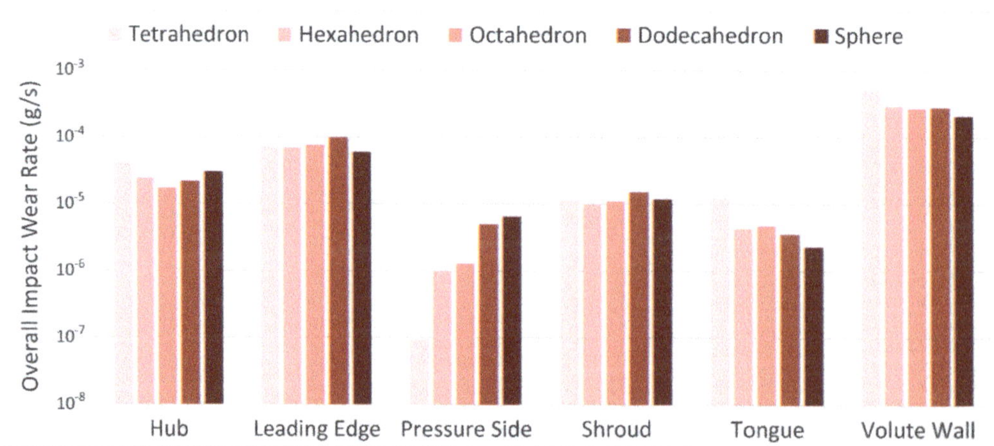

Figure 10. Overall impact wear rate of flow parts.

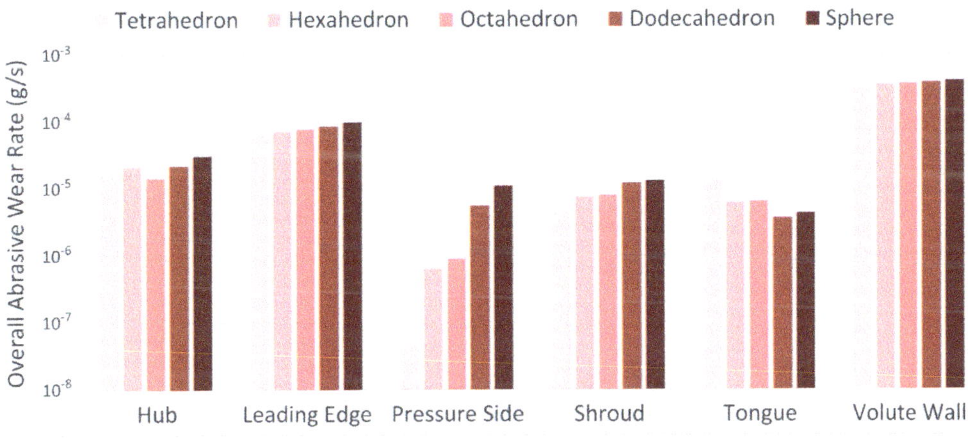

Figure 11. Overall abrasive wear rate of flow parts.

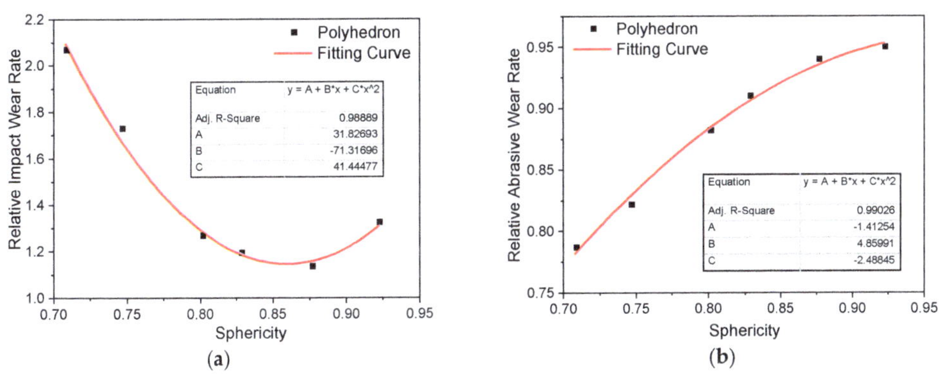

Figure 12. Regression equation for particle sphericity: (**a**) relative impact wear rate and (**b**) relative abrasive wear rate.

4. Conclusions

In this work, the CFD-DEM coupling method was employed to study the abrasive wear and impact wear of centrifugal pump flow parts for different-shaped particle conditions. The distribution and motion features of five different-shaped particles were compared. Two different wear models were adopted to quantitatively and qualitatively analyze the wear characteristics. The effect of particle shape on two types of wear rate was summarised in the regression equations. From the results, the following conclusions could be made as follows.

(1) All different-shaped particles have a tendency to sustain a stable spiral trajectory following the volute. Those particle trajectories basically correspond with the profile of the impeller blade. Most particles move along the wall surface and flow downstream to the pump outlet within the volute.

(2) Particles with low sphericity are evenly distributed in the vicinity of each blade pressure side and volute wall. However, particles tend to cluster in specific areas within the volute as sphericity rises. On the other hand, more particles move closer to the blade pressure side in the impeller passage with increasing sphericity.

(3) No matter what shapes the particles and wear model are, the volute suffers the principal wear erosion. Both the impact and abrasive wear in the impeller mostly occur on the blade's leading edge. The wear amount of the hub is considerably greater than that of

the shroud, likely due to the larger inertia of particles. Therefore, the flow parts mentioned above could be manufactured with more wear-resistance material than other parts.

(4) As the particle sphericity rises, there is a significant increase in the impact and abrasive wear on the pressure side due to particle trajectories biased toward the blade pressure side.

(5) The pump's overall impact wear rate decreases first and then increases with particle sphericity rising. By contrast, the pump's overall abrasive wear rate grows steadily with increasing sphericity.

(6) This study was limited to the effect of multi-shaped particles on abrasive and impact wear. Therefore, it would be desirable to study the influence of the particle concentration, particle size, inlet velocity, and properties of conveying fluid on the resulting wear in further study. In addition, although this paper explored the relationship between wear and particle shape, the geometries of particles were limited to correct unified shapes. Given this, some mixed and asymmetrical shapes would be considered in further studies.

Author Contributions: Conceptualization, C.T. and Y.-J.K. Methodology, C.T. and Y.-J.K. Software, C.T. and P.-Z.L. Validation, C.T. and Y.-C.Y. Formal analysis, C.T. Investigation, C.T. Resources, C.T. and Y.-C.Y. Data curation, C.T. Writing—original draft preparation, C.T. and Y.-J.K. Writing—review and editing, C.T. and Y.-J.K. Supervision, Y.-J.K. All the authors have read and agreed to the published version of the manuscript. All authors have read and agreed to the published version of the manuscript.

Funding: This work was supported by the Korea Agency for Infrastructure Technology Advancement (KAIA) grant funded by the Ministry of Land, Infrastructure, and Transport (Grant 21CTAP-C157760-02).

Conflicts of Interest: The authors declare no conflict of interest.

References

1. Finnie, I. Erosion of surfaces by solid particles. *Wear* **1960**, *3*, 87–103. [CrossRef]
2. Tao, Y.; Yuan, S.; Zhang, J.; Zhang, F.; Tao, J. Numerical simulation and test on impeller wear of slurry pump. *Trans. Chin. Soc. Agric. Eng.* **2014**, *30*, 63–69.
3. Ahmad, K.; Baker, R.; Goulas, A. Computation and experimental results of wear in a slurry pump impeller. *Proc. Inst. Mech. Eng. Part C J. Mech. Eng. Sci.* **1986**, *200*, 439–445. [CrossRef]
4. Lei, H.M.; Xiao, Y.X.; Chen, F.N.; Ahn, S.H.; Wang, Z.W.; Gui, Z.H.; Luo, Y.Y.; Zhao, X.R. Numerical simulation of solid-liquid two-phase flow in a centrifugal pump with different wear blades degree. In *Proceedings of the IOP Conference Series: Earth and Environmental Science*; IOP Publishing: England, UK, 2018; Volume 163, p. 012027.
5. Noon, A.A.; Kim, M.-H. Erosion wear on centrifugal pump casing due to slurry flow. *Wear* **2016**, *364*, 103–111. [CrossRef]
6. Huang, S.; Huang, J.; Guo, J.; Mo, Y. Study on Wear Properties of the Flow Parts in a Centrifugal Pump Based on EDEM–Fluent Coupling. *Processes* **2019**, *7*, 431. [CrossRef]
7. Zeng, D.; Zhang, E.; Ding, Y.; Yi, Y.; Xian, Q.; Yao, G.; Zhu, H.; Shi, T. Investigation of erosion behaviors of sulfur-particle-laden gas flow in an elbow via a CFD-DEM coupling method. *Powder Technol.* **2018**, *329*, 115–128. [CrossRef]
8. Tsuji, Y.; Kawaguchi, T.; Tanaka, T. Discrete particle simulation of two-dimensional fluidized bed. *Powder Technol.* **1993**, *77*, 79–87. [CrossRef]
9. Kafui, K.; Thornton, C.; Adams, M. Discrete particle-continuum fluid modelling of gas–solid fluidised beds. *Chem. Eng. Sci.* **2002**, *57*, 2395–2410. [CrossRef]
10. Cleary, P.W.; Sawley, M.L. DEM modelling of industrial granular flows: 3D case studies and the effect of particle shape on hopper discharge. *Appl. Math. Model.* **2002**, *26*, 89–111. [CrossRef]
11. Santamarina, J.; Cho, G.-C. Soil behaviour: The role of particle shape. In *Advances in Geotechnical Engineering: The Skempton Conference. In Proceedings of the Three Day Conference on Advances in Geotechnical Engineering, Organised by the Institution of Civil Engineers and Held at the Royal Geographical Society, London, UK, 29–31 March 2004*; Advances in Geotechnical Engineering: London, UK; pp. 604–617.
12. Pena, A.; Garcia-Rojo, R.; Herrmann, H.J. Influence of particle shape on sheared dense granular media. *Granul. Matter* **2007**, *9*, 279–291. [CrossRef]
13. Mindlin, R.D. Compliance of elastic bodies in contact. *J. Appl. Mech. ASME* **1949**, *16*, 259–268. [CrossRef]
14. Hertz, H. Über die Berührung fester elastischer Körper. *J. Für Die Reine Und Angew. Math.* **1882**, *92*, 22.
15. Archard, J. Contact and rubbing of flat surfaces. *J. Appl. Phys.* **1953**, *24*, 981–988. [CrossRef]
16. Oka, Y.I.; Okamura, K.; Yoshida, T. Practical estimation of erosion damage caused by solid particle impact: Part 1: Effects of impact parameters on a predictive equation. *Wear* **2005**, *259*, 95–101. [CrossRef]

17. Oka, Y.; Yoshida, T. Practical estimation of erosion damage caused by solid particle impact: Part 2: Mechanical properties of materials directly associated with erosion damage. *Wear* **2005**, *259*, 102–109. [CrossRef]
18. Shih, T.H.; Liou, W.W.; Shabbir, A.; Yang, Z.; Zhu, J. A new k-ϵ eddy viscosity model for high reynolds number turbulent flows. *Comput. Fluids* **1995**, *24*, 227–238. [CrossRef]
19. Yu, D.H.; Chen, Y.; Wang, C.; Bai, X.B.; Yang, Y.C. Simulation and experimental study on external performance of the volute centrifugal pumps. *China Rural Water Hydropower* **2020**, *1*, 181–187.

MDPI
St. Alban-Anlage 66
4052 Basel
Switzerland
Tel. +41 61 683 77 34
Fax +41 61 302 89 18
www.mdpi.com

Energies Editorial Office
E-mail: energies@mdpi.com
www.mdpi.com/journal/energies

www.ingramcontent.com/pod-product-compliance
Lightning Source LLC
Chambersburg PA
CBHW061515210926
51850CB00008B/1926
9783036583594